Applications of Mathematics

Stochastic Modelling
and Applied Probability

42

Springer-Science+Business Media, LLC

Applications of Mathematics

(continued after index)

Onésimo Hernández-Lerma
Jean Bernard Lasserre

Further Topics on Discrete-Time Markov Control Processes

 Springer

Onésimo Hernández-Lerma
CINVESTAV-IPN
Departamento de Matemáticas
Apartado Postal 14-740
07000 Mexico DF, Mexico
ohernard@math.cinvestav.mx

Jean Bernard Lasserre
LAAS-CNRS
7 Av. du Colonel Roche
31077 Toulouse Cedex, France
lasserre@laas.fr

Managing Editors

I. Karatzas
Departments of Mathematics and Statistics
Columbia University
New York, NY 10027, USA

M. Yor
CNRS, Laboratoire de Probabilités
Université Pierre et Marie Curie
4, Place Jussieu, Tour 56
F-75252 Paris Cedex 05, France

Mathematics Subject Classification (1991): 49L20, 90C39, 90C40, 93-02, 93E20

Library of Congress Cataloging-in-Publication Data
Hernández-Lerma, O. (Onésimo)
     Further topics on discrete-time Markov control processes/Onésimo
Hernández-Lerma, Jean B. Lasserre.
          p.   cm. — (Applications of mathematics; 42)
     Includes bibliographical references and index.
     ISBN 978-0-387-98694-4   ISBN 978-1-4612-0561-6 (eBook)
     DOI 10.1007/978-1-4612-0561-6
     1. Markov processes.   2. Discrete-time systems.   3. Control
theory.   I. Lasserre, Jean Bernard, 1953–   II. Title.   III. Series.
QA274.7.H473   1999
519.233—dc21   99-12351

Printed on acid-free paper.

Production managed by Frank McGuckin; manufacturing supervised by Jacqui Ashri.
Photocomposed copy prepared from the authors' LaTeX files.

9 8 7 6 5 4 3 2 1

ISBN 978-0-387-98694-4   SPIN 10707298

For Marina, Adrián, Gerardo, and Andrés
To Julia and Marine

Preface

This book presents the second part of a two-volume series devoted to a systematic exposition of some recent developments in the theory of discrete-time Markov control processes (MCPs). As in the first part, hereafter referred to as "Volume I" (see Hernández-Lerma and Lasserre [1]), interest is mainly confined to MCPs with *Borel* state and control spaces, and possibly *unbounded* costs. However, an important feature of the present volume is that it is essentially self-contained and can be read independently of Volume I. The reason for this independence is that even though both volumes deal with similar classes of MCPs, the assumptions on the control models are usually different. For instance, Volume I deals only with *nonnegative* cost-per-stage functions, whereas in the present volume we allow cost functions to take positive or negative values, as needed in some applications. Thus, many results in Volume I on, say, discounted or average cost problems are not applicable to the models considered here.

On the other hand, we now consider control models that typically require more restrictive classes of control-constraint sets and/or transition laws. This loss of generality is, of course, deliberate because it allows us to obtain more "precise" results. For example, in a very general context, in §4.2 of Volume I we showed the convergence of the value iteration (VI) procedure for discounted-cost MCPs, whereas now, in a somewhat more restricted setting, we actually get a lot more information on the VI procedure, such as the *rate of convergence* (§8.3), which in turn is used to study "rolling horizon" procedures, as well as the existence of "forecast horizons", and criteria for the elimination of nonoptimal control actions. Similarly, in Chapter 10 and Chapter 11, which deal with average cost problems, we obtain many interesting results that are virtually impossible to hold in a context as general as that of Volume I. In the Introduction of each chapter dealing with problems already studied in Volume I we clearly spell out the difference between the corresponding settings in each volume.

Volume I comprises Chapter 1 to Chapter 6, and the present volume contains Chapter 7 to Chapter 12. Chapter 7 introduces background material on weighted-norm spaces of functions and spaces of measures, and on noncontrolled Markov chains. In particular, it introduces the concept of w-geometric ergodicity of Markov chains, where w is a given "weight" (or bounding) function, and also the Poisson equation associated to a transition kernel and a given "charge". Chapter 8 studies α-*discounted cost* (abbreviated α-DC or simply DC) MCPs. The basic idea is to give conditions under which the dynamic programming (DP) operator is a *contraction* with respect to a suitable weighted norm. This contraction property is used to obtain the α-discount optimality (or DP-equation), and to make a detailed analysis of the VI procedure. In Chapter 9 we turn our attention to the *expected total cost* (ETC) criterion. Conditions are given for the ETC function to be well defined (as an extended-real-valued measurable function) and for the existence of ETC-optimal control policies, among other things. A quite complete analysis of the so-called "transient" case is also included (§9.6). Chapter 10 deals with *undiscounted* cost criteria, from average cost (AC) problems to overtaking optimality, passing through the existence of canonical policies, bias-optimal policies, Flynn's opportunity cost, and other "intermediate" criteria. AC problems are also dealt with in Chapters 11 and 12, but from very different viewpoints. Chapter 11 studies *sample path* optimality and *variance* minimization, essentially using probabilistic methods, while Chapter 12 concerns the *expected* AC problem using the linear programming approach. Chapter 12 includes in particular a procedure to approximate by *finite* linear programs the infinite-dimensional AC-related linear programs.

Acknowledgments. We wish to thank the Consejo Nacional de Ciencia y Tecnología (CONACYT, México) and the Centre National de la Recherche Scientifique (CNRS, France) for their generous support of our research work through the CONACYT-CNRS Scientific Cooperation Program. The work of the first author (OHL) has also been partially supported by CONACYT grant 3115P–E9608 and the Sistema Nacional de Investigadores (SNI). Thanks also due to Fernando Luque-Vásquez and Oscar Vega-Amaya for helpful comments on several chapters, and to Gerardo Hernández del Valle for his efficient typing of the manuscript.

March 1999   Onésimo Hernández-Lerma
                                                  Jean Bernard Lasserre

Contents

Contents of Volume I

Contents

7
Ergodicity and Poisson's Equation

7.1 Introduction

This chapter deals with *noncontrolled* Markov chains and presents important background material used in later chapters. The reader may omit it and refer to it as needed.

There are in particular two key concepts we wish to arrive at in this chapter, and which will be used to study several classes of Markov control problems. One is the concept of *w-geometric ergodicity* with respect to some *weight function w*, and the other is the *Poisson equation* (P.E.), which can be seen as a special case of the Average-Cost Optimality Equation. The former is introduced in §7.3.D and the latter in §7.5. First we introduce, in §7.2, some general notions on weighted norms and signed kernels, and then, in §§7.3.A–7.3.C, we review some standard results from Markov chain theory. The latter results are presented without proofs; they are introduced here mainly for ease of reference. Finally, §§7.4 and 7.5.C contain some examples on *w*-geometric ergodicity and the P.E., respectively.

Throughout the following X denotes a *Borel space* (that is, a Borel subset of a complete and separable metric space), unless explicitly stated otherwise. Its Borel σ-algebra is denoted by $\mathcal{B}(X)$.

7.2   Weighted norms and signed kernels

Let X be a Borel space, and let $\mathbb{B}(X)$ be the Banach space of real-valued bounded measurable functions u on X, with the **sup** (or **supremum**) **norm**

$$\|u\| := \sup_{x \in X} |u(x)|.$$

A. Weighted-norm spaces

We assume throughout that $w : X \to [1, \infty)$ denotes a given measurable function that will be referred to as a **weight function** (some authors call it a **bounding function**). If u is a real-valued function on X we define its w-**norm** as

$$\|u\|_w := \|u/w\| = \sup_{x \in X} |u(x)|/w(x). \tag{7.2.1}$$

Of course, if w is the constant function identical to 1, $w(\cdot) \equiv 1$, the w-norm and the sup norm coincide.

A function u is said to be **bounded** if $\|u\| < \infty$ and w-**bounded** if $\|u\|_w < \infty$. In general, the weight function w will be unbounded, although it is obviously w-bounded since $\|w\|_w = 1$. On the other hand, if u is a bounded function then it is w-bounded, as $w \geq 1$ yields

$$\|u\|_w \leq \|u\| < \infty \qquad \forall u \in \mathbb{B}(X). \tag{7.2.2}$$

Let $\mathbb{B}_w(X)$ be the normed linear space of w-bounded measurable functions u on X. This space is also *complete* because if $\{u_n\}$ is a Cauchy sequence in the w-norm, then $\{u_n/w\}$ is Cauchy in the sup norm; hence, as $\mathbb{B}(X)$ is a Banach space, one can deduce the existence of a function u in $\mathbb{B}_w(X)$ that is the w-limit of $\{u_n\}$. Combining this fact with (7.2.2) we obtain:

7.2.1 Proposition. $\mathbb{B}_w(X)$ *is a Banach space that contains* $\mathbb{B}(X)$.

Consider now the Banach space $\mathbb{M}(X)$ of finite signed measures μ on $\mathcal{B}(X)$ endowed with the **total variation norm**

$$\|\mu\|_{TV} := \sup_{\|u\| \leq 1} \left| \int_X u \, d\mu \right| = |\mu|(X), \tag{7.2.3}$$

where $|\mu| = \mu^+ + \mu^-$ denotes the *total variation* of μ, and μ^+, μ^- stand for the positive and negative parts of μ, respectively. By analogy, the w-**norm** of μ is defined by

$$\|\mu\|_w := \sup_{\|u\|_w \leq 1} \left| \int_X u \, d\mu \right| = \int_X w \, d|\mu|, \tag{7.2.4}$$

which reduces to (7.2.3) when $w(\cdot) \equiv 1$. Moreover, as $w \geq 1$, we obtain

$$\|\mu\|_w \geq \|\mu\|_{TV}.$$

The latter inequality can be used to show that the normed linear space $\mathbb{M}_w(X)$ of finite signed measures with a finite w-norm is a Banach space. Summarizing:

7.2.2 Proposition. $\mathbb{M}_w(X)$ *is a Banach space and it is contained in* $\mathbb{M}(X)$.

Remark. (a) In Chapter 6 we have already seen a particular class of weighted norms; see Definitions 6.3.2 and 6.3.4.

(b) $\mathbb{B}_w(X)$ and $\mathbb{M}_w(X)$ are in fact **ordered** Banach spaces. Namely, as usual, for functions in $\mathbb{B}_w(X)$ "$u \leq v$" means $u(x) \leq v(x)$ for all x in X, and for measures "$\mu \leq \nu$" means $\mu(B) \leq \nu(B)$ for all B in $\mathcal{B}(X)$.

B. Signed kernels

7.2.3 Definition. (a) Let X and Y be two Borel spaces. A **stochastic kernel on X given Y** is a function $Q = \{Q(B|y) : B \in \mathcal{B}(X), y \in Y\}$ such that

(i) $Q(\cdot|y)$ is a p.m. (probability measure) on $\mathcal{B}(X)$ for every fixed $y \in Y$, and

(ii) $Q(B|\cdot)$ is a measurable function on Y for every fixed $B \in \mathcal{B}(Y)$.

If the condition (i) is replaced by

(i)' $Q(\cdot|y)$ is a finite signed measure on $\mathcal{B}(X)$ for every fixed $y \in Y$, then Q is called a **signed kernel on X given Y**.

(b) Suppose that $X = Y$. If Q satisfies (i) and (ii) it is then called a **stochastic kernel** (or **transition probability function**) **on** X, whereas if it satisfies (i)' and (ii) it is said to be a **signed kernel on** X.

Signed kernels typically appear as a difference $P_1 - P_2$ of stochastic kernels.

In the remainder of this section we consider the case $X = Y$. Moreover, a measure μ in $\mathbb{M}_w(X)$ will be identified with the "constant" kernel $Q(\cdot|x) \equiv \mu(\cdot)$.

A signed kernel Q on X defines linear maps $u \mapsto Qu$ on $\mathbb{B}_w(X)$ and $\mu \mapsto \mu Q$ on $\mathbb{M}_w(X)$ as follows:

$$Qu(x) := \int_X u(y)Q(dy|x), \quad u \in \mathbb{B}_w(X), \ x \in X, \qquad (7.2.5)$$

and

$$\mu Q(B) := \int_X Q(B|x)\mu(dx), \quad \mu \in \mathbb{M}_w(X), \ B \in \mathcal{B}(X). \qquad (7.2.6)$$

7.2.4 Remark. If Q is the *transition probability function* of a time-homogeneous X-valued Markov chain $\{x_t, t = 0, 1, \ldots\}$, so that

$$Q(B|x) = \text{Prob}(x_{t+1} \in B|x_t = x) \quad \forall t = 0, 1, \ldots, B \in \mathcal{B}(X), \ x \in X,$$

then $Qu(x)$ in (7.2.5) is the conditional expectation of $u(x_{t+1})$ given $x_t = x$, that is,

$$Qu(x) = E\left[u(x_{t+1})|x_t = x\right], \quad t = 0, 1, \dots.$$

On the other hand, if μ is a p.m. that denotes the distribution of x_t, i.e.,

$$\mu(B) = \mathrm{Prob}(x_t \in B), \quad B \in \mathcal{B}(X),$$

then μQ in (7.2.6) denotes the distribution of x_{t+1}:

$$\mu Q(B) = \mathrm{Prob}(x_{t+1} \in B), \quad B \in \mathcal{B}(X). \quad \square$$

Let Q be a *signed kernel* on X. To ensure that (7.2.5) and (7.2.6) define operators with values on $\mathbb{B}_w(X)$ and $\mathbb{M}_w(X)$, respectively, we shall assume that Q has a finite w-**norm**, which is defined (as usual for "operator norms") by

$$\|Q\|_w := \sup\{\|Qu\|_w : \|u\|_w \le 1 \quad (\text{i.e., } |u| \le w) \}. \tag{7.2.7}$$

We can write the w-norm of Q in several equivalent forms. For instance, by (7.2.5) and (7.2.1),

$$\|Qu\|_w = \sup_X w(x)^{-1}|Qu(x)|,$$

which combined with (7.2.4) yields

$$\begin{aligned} \|Q\|_w &= \sup_X w(x)^{-1}\|Q(\cdot|x)\|_w \\ &= \sup_X w(x)^{-1} \int_X w(y)|Q(dy|x)|. \end{aligned} \tag{7.2.8}$$

On the other hand, replacing μ in (7.2.6) and (7.2.4) by the **Dirac measure** δ_x at the point $x \in X$ [that is, $\delta_x(B) := 1$ if $x \in B$ and $:= 0$ otherwise], we see that

$$\delta_x Q(\cdot) = Q(\cdot|x) \quad \text{and} \quad \|\delta_x\|_w = w(x). \tag{7.2.9}$$

Then a direct calculation shows that (7.2.7) can also be written in the following equivalent form using measures μ in $\mathbb{M}_w(X)$:

$$\|Q\|_w = \sup\{\|\mu Q\|_w : \|\mu\|_w \le 1\}. \tag{7.2.10}$$

Moreover, the usual arguments for "operator norms" yield:

7.2.5 Proposition. *Let Q be a signed kernel with a finite w-norm, i.e.,*

$$\|Q\|_w < \infty. \tag{7.2.11}$$

Then

$$\|Qu\|_w \le \|Q\|_w\|u\|_w < \infty \quad \forall u \in \mathbb{B}_w(X), \tag{7.2.12}$$

and, therefore, (7.2.5) defines a linear map from $\mathbb{B}_w(X)$ *into itself. Similarly, (7.2.6) defines a linear map from* $\mathbb{M}_w(X)$ *into itself, since*

$$\|\mu Q\|_w \le \|Q\|_w \|\mu\|_w < \infty \quad \forall \mu \in \mathbb{M}_w(X). \tag{7.2.13}$$

Finally, we consider the **product** (or **composition**) of two signed kernels Q and R:

$$(QR)(B|x) := \int_X R(B|y)Q(dy|x), \quad B \in \mathcal{B}(X), \ x \in X. \tag{7.2.14}$$

Further, if Q^0 denotes the **identity kernel** given by

$$Q^0(B|x) = \delta_x(B),$$

we define Q^n recursively as

$$Q^n := QQ^{n-1}, \quad n = 1, 2, \ldots. \tag{7.2.15}$$

Then, from Proposition 7.2.5 and standard results for bounded linear operators, we obtain:

7.2.6 Proposition. *If* Q *and* R *are signed kernels with finite* w-*norms, then*

$$\|QR\|_w \le \|Q\|_w \|R\|_w. \tag{7.2.16}$$

In particular,

$$\|Q^n\|_w \le \|Q\|_w^n, \quad n = 1, 2, \ldots.$$

C. Contraction maps

7.2.7 Definition. Let (S, d) be a metric space. A map $T : S \to S$ is called

(a) **nonexpansive** if $d(Ts_1, Ts_2) \le d(s_1, s_2)$ for all s_1, s_2 in S; and

(b) a **contraction** if there is a number $0 \le \gamma < 1$ such that

$$d(Ts_1, Ts_2) \le \gamma d(s_1, s_2) \text{ for all } s_1, \ s_2 \text{ in } S.$$

In this case, γ is called the **modulus** of T.

Further, a point s^* in S is said to be a **fixed point** of T if $Ts^* = s^*$.

Many applications of contraction maps are based on the following theorem of Banach.

7.2.8 Proposition. (Banach's Fixed Point Theorem.) *A contraction map* T *on a complete metric space* (S, d) *has a unique fixed point* s^*. *Moreover,*

$$d(T^n s, s^*) \le \gamma^n d(s, s^*) \quad \forall s \in S, \ n = 0, 1 \ldots. \tag{7.2.17}$$

where γ is the modulus of T, and $T^n := T(T^{n-1})$ for $n = 1, 2, \ldots$, with $T^0 :=$ identity.

Proof. See Ross [1] or Luenberger [1], for instance. $\square$

As an example, the maps defined by (7.2.5) and (7.2.6) are both nonexpansive if $\|Q\|_w = 1$, and they are contractions with modulus $\gamma := \|Q\|_w$ if

$$\|Q\|_w < 1. \tag{7.2.18}$$

This follows from (7.2.12) and (7.2.13). For instance, viewing Q as a map on $\mathbb{B}_u(X)$, (7.2.12) gives

$$\|Qu - Qv\|_w = \|Q(u - v)\|_w \le \|Q\|_w \|u - v\|_w,$$

and similarly for Q on $\mathbb{M}_w(X)$.

If T is a contraction map on $\mathbb{M}_w(X)$, with modulus γ, then Proposition 7.2.8 ensures the existence of a unique fixed point $\mu \in \mathbb{M}_w(X)$ of T and that

$$\|T^n \nu - \mu\|_w \le \gamma^n \|\nu - \mu\|_w \quad \forall \nu \in \mathbb{M}_w(X), \quad n = 0, 1, \ldots. \tag{7.2.19}$$

This property is closely related to the *w-geometric ergodicity* introduced in Definition 7.3.9 below. On the other hand, in later chapters we will see that *dynamic programming* (DP) operators on $\mathbb{B}_w(X)$, under suitable assumptions, turn out to be contractions. In the latter case, we will use the fact that typically a DP operator T is **monotone**, that is, $u \le v$ implies $Tu \le Tv$, and so we can apply the following criterion for T to be a contraction.

7.2.9 Proposition. *Let T be a monotone map from $\mathbb{B}_w(X)$ into itself. If there is a positive number $\gamma < 1$ such that*

$$T(u + rw) \le Tu + \gamma rw \quad \forall u \in \mathbb{B}_w(X), \quad r \in \mathbb{R}, \tag{7.2.20}$$

then T is a contraction with modulus γ.

Proof. For any two functions u and v in $\mathbb{B}_w(X)$ we have $u \le v + w \cdot \|u - v\|_w$. Thus, the monotonicity of T and (7.2.20) with $r = \|u - v\|_w$ yield

$$Tu \le T(v + rw) \le Tv + \gamma rw,$$

i.e.,

$$Tu - Tv \le \gamma w \|u - v\|_w.$$

If we now interchange u and v we get $Tu - Tv \ge -\gamma w \|u - v\|_w$, so that

$$|Tu - Tv| \le \gamma w \|u - v\|_w.$$

Hence, $\|Tu - Tv\|_w \le \gamma \|u - v\|_w$. $\square$

Notes on §7.2

1. Since the material for this chapter comes from several sources, to avoid confusions with related literature it is important to keep in mind the *equivalence* of the several forms (7.2.7), (7.2.8) and (7.2.10) for the w-norm of Q. For instance, Kartashov [1]–[5] uses (7.2.10), while, say, Meyn and Tweedie [1] use (7.2.7).

2. For Markov control processes with a nonnegative cost function c, in Chapter 6 we used a weight function of the form $w := 1 + c$.

7.3   Recurrence concepts

In this section, $\{x_t, \ t = 0, 1, \dots\}$ denotes a time-homogeneous, X-valued Markov chain with transition probability function (or stochastic kernel) $P(B|x)$, i.e.,

$$P(B|x) = \mathrm{Prob}(x_{t+1} \in B | x_t = x) \quad \forall t = 0, 1, \dots, B \in \mathcal{B}(X), \ x \in X.$$

See Remark 7.2.4 (with $Q = P$) for the meaning of $Pu(x)$ and $\mu P(B)$. All of the concepts and results presented in subsections A, B, C are well known in Markov chain theory and will be stated without proofs. General references are the books by Duflo [1], Meyn and Tweedie [1], and Nummelin [1].

A. Irreducibility and recurrence

For any Borel set $B \in \mathcal{B}(X)$ we define the **hitting time**

$$\tau_B := \inf\{t \geq 1 \ : \ x_t \in B\}$$

and the **occupation time**

$$\eta_B := \sum_{t=1}^{\infty} I_B(x_t),$$

where I_B denotes the indicator function of B. Moreover, given an initial state $x_0 = x$ we define

$$L(x, B) := P_x(\tau_B < \infty) = P_x(x_t \in B \text{ for some } t \geq 1),$$

and

$$U(x, B) := E_x(\eta_B) = \sum_{t=1}^{\infty} P^t(B|x),$$

where $P^t(B|x) = P_x(x_t \in B)$ denotes the t-step transition probability. [This can be defined recursively as in (7.2.15).]

7.3.1 Definition. The Markov chain $\{x_t\}$ (or the corresponding stochastic kernel P) is called

(a) λ-**irreducible** if there exists a σ-finite measure λ on $\mathcal{B}(X)$ such that $L(x, B) > 0$ for all $x \in X$ whenever $\lambda(B) > 0$, and in this case λ is called an **irreducibility measure**;

(b) **Harris recurrent** if there exists a σ-finite measure μ on $\mathcal{B}(X)$ such that $L(x, B) = 1$ for all $x \in X$ whenever $\mu(B) > 0$.

Each of the following two conditions is equivalent to λ-irreducibility:

(a_1) $U(x, B) > 0$ for all $x \in X$ whenever $\lambda(B) > 0$.

(a_2) If $\lambda(B) > 0$, then for each $x \in X$ there exists some $n = n(B, x) > 0$ such that $P^n(B|x) > 0$.

If P is λ-irreducible, then P is λ'-irreducible for any measure λ' equivalent to λ. Moreover, if P is λ-irreducible, then there exists a *maximal irreducibility measure* ψ, that is, an irreducibility measure ψ such that, for any measure λ', the chain is λ'-irreducible if and only if λ' is absolutely continuous with respect to ψ (in symbols: $\lambda' << \psi$). We shall assume that the measure λ in Definition 7.3.1(a) is a maximal irreducibility measure, and we shall denote by $\mathcal{B}(X)^+$ the family of sets $B \in \mathcal{B}(X)$ for which $\lambda(B) > 0$.

We next define another important notion of recurrence, weaker than Harris recurrence.

7.3.2 Definition.

(a) A set $B \in \mathcal{B}(X)$ is called **recurrent** if $U(x, B) = \infty$ for $x \in B$, and the Markov chain (or the stochastic kernel P) is said to be **recurrent** if it is λ-irreducible and every set in $\mathcal{B}(X)^+$ is recurrent.

(b) We call a set $B \in \mathcal{B}(X)$ **uniformly transient** if there exists a constant M such that $U(x, B) \leq M$ for all $x \in B$, and simply **transient** if it can be covered by countably many uniformly transient sets. Moreover, the chain itself is said to be **transient** if the state space X is a transient set.

There is a dichotomy between recurrence and transience in the following sense.

7.3.3 Theorem. *If $\{x_t\}$ is λ-irreducible, then it is either recurrent or transient.*

B. Invariant measures

Recurrence is closely related to the existence of invariant measures. A σ-finite measure μ on $\mathcal{B}(X)$ is said to be **invariant** for $\{x_t\}$ (or for P) if

$$\mu(B) = \int_X P(B|x)\mu(dx) \quad \forall B \in \mathcal{B}(X). \tag{7.3.1}$$

Extending the notation (7.2.6) to σ-finite measures, we may rewrite (7.3.1) as

$$\mu P = \mu. \tag{7.3.2}$$

Thus μ is an invariant measure if it is a "fixed point" of P.

The following theorem summarizes some relations between recurrence and invariant measures. We shall use the abbreviation **i.p.m.** for invariant probability measure, and (sometimes) we write $\mu(v)$ for an integral $\int v d\mu$. In chapters dealing with linear programming problems we usually write integrals $\int v d\mu$ as an "inner product" $\langle \mu, v \rangle$, so we have

$$\int v d\mu \equiv \mu(v) \equiv \langle \mu, v \rangle.$$

7.3.4 Theorem.

(a) *If P is recurrent (in particular, Harris-recurrent), then there exists a nontrivial invariant measure μ, which is unique up to a multiplicative constant.*

(b) *If there exists an i.p.m. μ for P and P is λ-irreducible, then P is recurrent; hence, by (a), μ is the unique i.p.m. for P.*

(c) *Suppose that P is Harris-recurrent and let μ be as in (a). Then there exists a triplet (n, ν, l) consisting of an integer $n \geq 1$, a p.m. ν, and a measurable function $0 \leq l \leq 1$ such that:*

(c$_1$) *$P^n(B|x) \geq l(x)\nu(B)$ $\forall B \in \mathcal{B}(X)$ and $x \in X$;*

(c$_2$) *$\nu(l) > 0$; and*

(c$_3$) *$0 < \mu(l) < \infty$.*

When $n = 1$ in Theorem 7.3.4(c), the conditions (c$_1$)–(c$_3$) are related to w-geometric ergodicity. (See Theorem 7.3.11.)

If the Markov chain is (Harris) recurrent and the invariant measure μ in Theorem 7.3.4(a) is *finite*, then the chain is called **positive (Harris) recurrent**; otherwise it is called **null**.

C. Conditions for irreducibility and recurrence

We next introduce some concepts that will be used to give conditions for irreducibility and recurrence of Markov chains.

Let $\mathbb{B}(X)$ be as in §7.2, the Banach space of bounded measurable functions on X endowed with the sup norm, and let $C_b(X)$ be the subspace of *continuous* bounded functions. We shall use the notation Pu as in (7.2.5) with $Q = P$.

7.3.5 Definition. The Markov chain $\{x_t\}$ (or its stochastic kernel P) is said to satisfy the **weak Feller** property if P leaves $C_b(X)$ invariant, i.e.,

$$Pu \in C_b(X) \quad \forall u \in C_b(X),$$

and the **strong Feller** property if P maps $\mathbb{B}(X)$ into $C_b(X)$, i.e.,

$$Pu \in C_b(X) \quad \forall u \in \mathbb{B}(X).$$

The weak Feller property has important implications but, unfortunately, many results for MCPs (Markov control processes) require the much more restrictive strong Feller condition. To mitigate this fact, we shall sometimes replace the latter condition by another which is easier to verify. First we need some notation.

Let $q = \{q(t), \ t = 0, 1, \ldots\}$ be a probability distribution on the set $\mathbb{N}_+$ of nonnegative integers and define the stochastic kernel

$$K_q(B|x) := \sum_{t=0}^{\infty} q(t)P^t(B|x), \quad x \in X, \ B \in \mathcal{B}(X). \tag{7.3.3}$$

q is referred to as a **sampling distribution**. Note that if β is a discrete random variable with values in $\mathbb{N}_+$ and β has distribution q, then $K_q(B|x) = P_x(x_\beta \in B)$. A usual choice for q is the geometric distribution

$$q(t) := (1 - \alpha)\alpha^t, \quad t = 0, 1, \ldots; \ 0 < \alpha < 1,$$

in which case K_q corresponds to the **resolvent** of P.

7.3.6 Definition. Let $T(B|x)$ be a substochastic kernel on X, that is, T is a nonnegative kernel such that $T(X|x) \leq 1$ for all $x \in X$, and let q be a sampling distribution.

(a) T is called a **continuous component** of K_q if

$$K_q(B|x) \geq T(B|x) \quad \forall x \in X, \ B \in \mathcal{B}(X),$$

and $T(B|\cdot)$ is a l.s.c. (lower semicontinuous) function on X for every $B \in \mathcal{B}(X)$.

(b) The Markov chain $\{x_t\}$ is said to be a T-**chain** if there exists a sampling distribution q such that K_q admits a continuous component T with $T(X|x) > 0$ for all x.

Of course, if $\{x_t\}$ satisfies the *strong* Feller property, then it is a T-chain; it suffices to take $q(1) = 1$ and $T(B|x) = P(B|x)$. Before considering other—less obvious—relations between the concepts introduced above, let us recall that the **support** of a measure μ on $\mathcal{B}(X)$ is the unique closed set $F \subset X$ such that $\mu(X - F) = 0$ and $\mu(F \cap G) > 0$ for every open set G

that intersects F. In particular, the *support of a p.m.* μ is the intersection of all closed sets C in X such that $\mu(C) = 1$.

7.3.7 Theorem. *If $\{x_t\}$ is a λ-irreducible Markov chain that satisfies the weak Feller property and the support of λ has nonempty interior, then $\{x_t\}$ is a λ-irreducible T-chain.*

For a λ-irreducible Markov chain we have:

(1) The chain has a **cycle of period** $d \geq 1$, that is, a collection $\{C_1, \ldots, C_d\}$ of disjoint Borel sets such that $P(C_{i+1}|x) = 1$ for $x \in C_i$, $i = 1, \ldots, d-1$, and $P(C_1|x) = 1$ for $x \in C_d$. When $d = 1$, the chain is called **aperiodic**.

(2) Every set in $\mathcal{B}(X)^+$ contains a **small set** (also known as a *C*-set), that is, a Borel set C such that for some integer n and some nontrivial measure μ

$$P^n(\cdot|x) \geq \mu(\cdot) \quad \forall x \in C. \tag{7.3.4}$$

A more general concept than that of a small set is the following. A set C in $\mathcal{B}(X)$ is called **petite** if there is a sampling distribution q and a nontrivial measure μ' such that

$$K_q(\cdot|x) \geq \mu'(\cdot) \quad \forall x \in C. \tag{7.3.5}$$

If the chain $\{x_t\}$ is λ-irreducible, then a Borel set C is petite if and only if for some $n \geq 1$ and a nontrivial measure ν,

$$\sum_{t=1}^{n} P^t(\cdot|x) \geq \nu(\cdot) \quad \forall x \in C. \tag{7.3.6}$$

Further, for an *aperiodic* λ-irreducible chain, the class of petite sets is the same as the class of small sets; however, the measures μ and μ' in (7.3.4) and (7.3.5) need not coincide.

Finally, we have the following important result, which will be used later on in conjuntion with Theorem 7.3.7.

7.3.8 Theorem. *If every compact set is petite, then $\{x_t\}$ is a T-chain. Conversely, if $\{x_t\}$ is a λ-irreducible T-chain, then every compact set is petite.*

D. w-Geometric ergodicity

7.3.9 Definition. Let $w \geq 1$ be a weight function such that $\|P\|_w < \infty$. Then P (or the Markov chain $\{x_t\}$) is called w-**geometrically ergodic** if there is a p.m. μ in $\mathbb{M}_w(X)$ and nonnegative constants R and ρ, with $\rho < 1$, that satisfy

$$\|P^t - \mu\|_w \leq R\rho^t \quad \forall t = 0, 1, \ldots; \tag{7.3.7}$$

more explicitly [by (7.2.7)], for every $u \in \mathbb{B}_w(X)$, $x \in X$, and $t = 0, 1, \ldots$,

$$\left| \int u(y) P^t(dy|x) - \mu(u) \right| \le \|u\|_w w(x) R \rho^t. \tag{7.3.8}$$

The so-called **limiting p.m.** μ is necessarily the unique i.p.m. in $\mathbb{M}_w(X)$ that satisfies (7.3.7).

To see that μ is indeed an i.p.m. for P, write P^t as $P^{t-1}P$ and use (7.3.7) and (7.2.16) to get

$$\begin{aligned} \|P^t - \mu P\|_w &= \|(P^{t-1} - \mu)P\|_w \\ &\le \|P^{t-1} - \mu\|_w \|P\|_w \to 0 \quad \text{as} \quad t \to \infty. \end{aligned}$$

Using this fact and (7.3.7) again we obtain

$$\|\mu - \mu P\|_w \le \|\mu - P^t\|_w + \|P^t - \mu P\|_w \to 0.$$

Thus $\mu = \mu P$.

We next present several results that guarantee (7.3.7).

7.3.10 Theorem. *Let $w \ge 1$ be a weight function. Suppose that the Markov chain is λ-irreducible and either*

(i) *there is a petite set $C \in \mathcal{B}(X)$ and constants $\beta < 1$ and $b < \infty$ such that*

$$Pw(x) \le \beta w(x) + b I_C(x) \quad \forall x \in X; \tag{7.3.9a}$$

or

(ii) *w is unbounded off petite sets (that is, the level set $\{x \in X | w(x) \le n\}$ is petite for every $n < \infty$) and, for some $\beta < 1$ and $b < \infty$,*

$$Pw(x) \le \beta w(x) + b \quad \forall x \in X. \tag{7.3.9b}$$

Then:

(a) *The chain is positive Harris recurrent, and*

(b) *$\mu(w) < \infty$, where μ denotes the unique i.p.m. for P.*

If, in addition, the chain is aperiodic then

(c) *The chain is w-geometrically ergodic, with limiting p.m. μ as in (b).*

Proof. See Meyn and Tweedie [1], §16.1. [Lemma 15.2.8 of the latter reference shows the *equivalence of the conditions* (i) *and* (ii).] $\square$

7.3.11 Theorem. *Let $w \ge 1$ be a weight function. Suppose that the Markov chain has a unique i.p.m. μ and, further, there exists a p.m. ν, a measurable function $0 \le l(\cdot) \le 1$, and a number $0 < \beta < 1$ such that*

(i) $P(B|x) \geq l(x)\nu(B) \quad \forall B \in \mathcal{B}(X)$ *and* $x \in X$;

(ii) $\nu(w) < \infty$ *and* $\nu(l) \cdot \mu(l) > 0$; *and*

(iii) $Pw(x) \leq \beta w(x) + \nu(w)l(x) \quad \forall x \in X.$ \hfill (7.3.10)

Then the chain is w-geometrically ergodic with limiting p.m. μ.

Proof. See Kartashov [3, Theorem 6]. $\square$

7.3.12 Remark. Observe that each of the inequalities (7.3.9), (7.3.10) trivially implies the condition $\|P\|_w < \infty$ required in Definition 7.3.9. For instance, since $w \geq 1$, multiplying both sides of (7.3.9b) by w^{-1} we get

$$(Pw)/w \leq \beta + b;$$

hence, by (7.2.8), $\|P\|_w \leq \beta + b$.

7.3.13 Remark. (See Meyn and Tweedie [1], Theorem 16.0.1.) If the Markov chain is λ-irreducible and aperiodic, then the following conditions are equivalent for any weight function $w \geq 1$:

(a) The chain is w-geometrically ergodic; that is, (7.3.7) holds.

(b) The inequality (7.3.9a) holds for some function w_0 which is equivalent to w in the sense that $c^{-1}w \leq w_0 \leq cw$ for some constant $c \geq 1$.

On the other hand, Kartashov [2], [5, p. 20], studies (7.3.7) using the following *equivalent* conditions (c) and (d):

(c) There is an integer $n \geq 1$ and a number $0 < \rho < 1$ such that

$$\|\theta P^n\|_w \leq \rho\|\theta\|_w \quad \forall \theta \in \mathbb{M}_w^0(X), \hfill (7.3.11)$$

where $\mathbb{M}_w^0(X) \subset \mathbb{M}_w(X)$ is the subspace of signed measures θ in $\mathbb{M}_w(X)$ with $\theta(X) = 0$.

(d) There is an integer $n \geq 1$ and a number $0 < \rho < 1$ such that

$$\int_X w(y)|P^n(dy|x) - P^n(dy|x')| \leq \rho[w(x) + w(x')] \hfill (7.3.12)$$

for all x, x' in X.

He shows that if P has a finite w-norm, then (c) [or (d)] is equivalent to (a).

Statements (c) and (d) are nice because they are easy to interpret. For instance [by Definition 7.2.7(b)], (7.3.11) means that P^n is a *contraction map* (or, equivalently, P is a *n-contraction*) on the subspace $\mathbb{M}_w^0(X)$ of $\mathbb{M}_w(X)$. Furthermore, (c) or (d) can be directly "translated" to obtain geometric ergodicity in the *total variation norm*, as in (7.3.16) below. For example, if we take $w(\cdot) \equiv 1$ in (7.3.12) we get

$$\|P^n(\cdot|x) - P^n(\cdot|x')\|_{TV} \le 2\rho \quad \forall x, x' \in X, \qquad (7.3.13)$$

which is a well-known necessary and sufficient condition for (7.3.16) (see the references in Note 1). $\square$

The equivalence of (a) and (c) [or (d)] in Remark 7.3.13 requires to know *a priori* that P has a finite w-norm. However, specializing (7.3.11), (7.3.12) to $n = 1$ and adding a mild assumption we get the following.

7.3.14 Theorem. *Suppose there exists a weight function $w(\cdot) \ge 1$, a positive number $\rho < 1$, and a state $x^* \in X$ that satisfy*

(a) $u_* := Pw(x^*) = \int_X w(y)P(dy|x^*) < \infty$, *and*

(b_1) $\int_X w(y)|P(dy|x) - P(dy|x')| \le \rho[w(x) + w(x')] \quad \forall x, x'$ *in* X,

or, equivalently [*with* $\mathbb{M}_w^0(X)$ *as in Remark 7.3.13(c)*],

(b_2) $|\theta P\|_w \le \rho\|\theta\|_w \quad \forall \theta \in \mathbb{M}_w^0(X)$.

Then

(i) $Pw(x) \le \rho w(x) + b$ *for all* $x \in X$, *with* $b := \rho w(x^*) + w_*$;

(ii) P *has a finite w-norm; and*

(iii) P *has a unique i.p.m. μ with a finite w-norm $\|\mu\|_w \le b/(1 - \rho)$, and, moreover, (7.3.7) holds with $R := 1 + \|\mu\|_w \le 1 + b/(1 + \rho)$.*

Proof. (i) Let x^* and w_* be as in (a). Then, by (b_1),

$$\int w(y)P(dy|x) \le \int w(y)|P(dy|x) - P(dy|x^*)| + \int w(y)P(dy|x^*)$$
$$\le \rho[w(x) + w(x^*)] + w_*,$$

and (i) follows.

(ii) As noted in Remark 7.3.12, the inequality in (i) gives that P has a finite w-norm. Furthermore, iteration of the inequality $Pw \le \rho w + b$ in (i) yields, for all $m = 1, 2, \ldots$,

$$\begin{aligned} P^m w &\le \rho^m w + b(1 + \rho + \cdots + \rho^{m-1}) \\ &\le w + b/(1 - \rho) \\ &\le [1 + b/(1 - \rho)]w; \end{aligned}$$

thus

$$\|P^m\|_w \le R' \quad \forall m = 1, 2, \ldots, \text{ with } R' := 1 + b/(1 - \rho). \qquad (7.3.14)$$

(iii) To prove part (iii), we first show that for every $x \in X$ the sequence $\{P^t(\cdot|x)\}$ is a Cauchy sequence in (the Banach space) $\mathbb{M}_w(X)$. To prove this, fix an arbitrary state x and note that, for any given integer $m \geq 1$, the finite signed measure

$$\theta(\cdot) := \delta_x(\cdot) - P^m(\cdot|x) = \delta_x(\cdot) - \delta_x P^m \quad \text{[see (7.2.9)]}$$

is such that $\theta(X) = 0$ and so it belongs to $\mathbb{M}_w^0(X)$ since, by (7.3.14) and (7.2.9),

$$\|\theta\|_w \leq \|\delta_x\|_w + \|P^m(\cdot|x)\|_w \leq (1 + R')w(x) < \infty.$$

Furthermore, for every $n = 1, 2, \ldots$,

$$P^n(\cdot|x) - P^{n+m}(\cdot|x) = \delta_x P^n - \delta_x P^m P^n = \theta P^n,$$

whereas iteration of (b$_2$) gives

$$\|\theta P^n\|_w \leq \rho^n \|\theta\|_w, \tag{7.3.15}$$

so that $\|\theta P^n\|_w \leq \rho^n(1 + R')w(x)$. Combining these facts we see that

$$\|P^n(\cdot|x) - P^{n+m}(\cdot|x)\|_w \leq \rho^n(1 + R')w(x) \to 0 \quad \text{as} \quad n \to \infty,$$

which, as m was arbitrary, clearly shows that $\{P^t(\cdot|x)\}$ is a Cauchy sequence in $\mathbb{M}_w(X)$.

Therefore, as $\mathbb{M}_w(X)$ is a Banach space, $P^t(\cdot|x)$ converges in the w-norm to some measure μ^x in $\mathbb{M}_w(X)$. In fact, $\mu^x \equiv \mu$ is *independent of* x, because if we apply (7.3.15) to the signed measure $\theta'(\cdot) := \delta_x(\cdot) - \delta_{x'}(\cdot)$, we obtain

$$\|P^n(\cdot|x) - P^n(\cdot|x')\|_w = \|(\delta_x - \delta_{x'})P^n\|_w \leq \rho^n[w(x) + w(x')] \to 0$$

for all x, x' in X. Hence, as in the paragraph after Definition 7.3.9, we conclude that μ *is an i.p.m for* P. Moreover, integration of both sides of the inequality in (i) with respect to μ gives that $\|\mu\|_w = \int w d\mu$ satisfies $\|\mu\|_w \leq \rho\|\mu\|_w + b$, which yields the inequality

$$\|\mu\|_w \leq b/(1 - \rho).$$

Finally, to prove the last statement in (iii), we can use the invariance of μ to write $\mu = \mu P^n$ for all $n = 0, 1, \ldots$, so that

$$P^n(\cdot|x) - \mu = \delta_x P^n - \mu P^n = (\delta_x - \mu)P^n.$$

Thus, applying (7.3.15) to the signed measure $\delta_x(\cdot) - \mu(\cdot) \in \mathbb{M}_w^0(X)$,

$$\|P^n(\cdot|x) - \mu\|_w \leq \rho^n\|\delta_x - \mu\|_w \leq \rho^n[w(x) + \|\mu\|_w],$$

which implies the desired result. $\square$

Theorem 7.3.14 is essentially contained in the paper by Gordienko, Montes-de-Oca and Minjárez-Sosa [1], except that, in addition to the hypotheses (a) and (b), they assume that P has a unique i.p.m., and, further, the proof of some of the conclusions are referred to Kartashov [2]. We decided to include here a proof, independent of Kartashov's, because similar arguments are repeatedly used in the remainder of this book.

Notes on §7.3

1. For each of the Theorems 7.3.10 and 7.3.11 it is possible to estimate the values of R and ρ in (7.3.7): see Meyn and Tweedie [2] and Kartashov [3], respectively. For w-geometric ergodicity of *denumerable* Markov chains, see Dekker, Hordijk and Spieksma [1] or Spieksma and Tweedie [1].

On the other hand, an important special case is when $w(\cdot)$ is a *bounded* function. In such a situation, the w-geometric ergodicity (7.3.7) reduces to the standard (uniform) **geometric ergodicity** in the **total variation norm** (7.2.3); that is [instead of (7.3.7)],

$$\sup_X \|P^t(\cdot|x) - \mu(\cdot)\|_{TV} \leq R\rho^t \quad \forall t = 0, 1, \ldots. \tag{7.3.16}$$

Necessary and/or sufficient conditions for geometric ergodicity are well known; see, for instance, Hernández-Lerma [1], Hernández-Lerma, Montes-de-Oca and Cavazos-Cadena [1], and Meyn and Tweedie [1]. In particular, Theorem 16.0.2 of the latter reference shows that the following conditions—among others—are *equivalent*:

(a) The chain is (uniformly) geometrically ergodic;

(b) The chain is aperiodic and there is a *bounded* solution $w \geq 1$ to the inequality (7.3.9a);

(c) The chain is aperiodic and **Doeblin's condition** holds; that is, there is a probability measure γ on $\mathcal{B}(X)$, positive numbers $\varepsilon < 1$ and δ, and a positive integer m such that

$$\inf_X P^m(B|x) > \delta \quad \text{whenever} \quad \gamma(B) > \varepsilon.$$

2. The reader should be warned that the term "w-geometric ergodicity" (Definition 7.3.9) is not quite standard. For example, Meyn and Tweedie [1] call it w-*uniform ergodicity* (and use "V" instead of "w"), but in Kartashov [1]–[5] the latter term means that P has a unique i.p.m. μ and that (7.3.7) holds for the **Cesàro sums**

$$t^{-1} \sum_{k=0}^{t-1} P^k$$

in lieu of P^t. Kartashov does not use any specific term for (7.3.7).

3. Speaking of terminology, we have in mind weight functions w that satisfy

$$w(x) \to \infty \quad \text{as} \quad |x| \to \infty, \tag{7.3.17}$$

which in §5.7 were called *strictly unbounded* or *moment* functions. However, a nonnegative measurable function that satisfies (7.3.17) is also known as a *Lyapunov* (Duflo [1], Lasota and Mackey [1]) or *norm-like* (Meyn and Tweedie [1]) function. In fact, inequalities of the form (7.3.9) and (7.3.10) are called Lyapunov or Foster-Lyapunov criteria for "stability" (in this case w-geometric ergodicity) of P, hence the name "Lyapunov function" for w. There are Lyapunov-like criteria for other forms of "stability", such as the existence of i.p.m.'s. Another way to approach the latter problem is to view (7.3.2) as a "fixed point" equation or as a linear equation on a space of measures. Then one can use generalized (infinite-dimensional) versions of *Farkas Theorem* to obtain conditions for the existence of solutions to (7.3.2); see Hernández-Lerma and Lasserre [2]–[5], and Lasserre [1], [2].

7.4 Examples on w-geometric ergodicity

In this section we illustrate Theorems 7.3.10 and 7.3.11 on w-geometric ergodicity. When considering a X-valued Markov chain of the form

$$x_{t+1} = F(x_t, z_t), \quad t = 0, 1, \ldots, \tag{7.4.1}$$

we always suppose the following:

7.4.1 Assumption.

(a) The disturbance sequence $\{z_t\}$ consists of i.i.d. (independent and identically distributed) random variables with values in a Borel space Z, and $\{z_t\}$ is independent of the initial state x_0. The common distribution of the z_t is denoted by G

(b) $F : X \times Z \to X$ is a given measurable function.

By Assumption 7.4.1(a), the variables x_t and z_t are independent for all $t = 0, 1, \ldots$. Hence, as in Remark 7.2.4 (with $Q = P$), for a system of the form (7.4.1) we have

$$\begin{aligned} Pu(x) &= E[u(x_{t+1})|x_t = x] \\ &= Eu[F(x, z_0)] = \int_Z u[F(x, z)]G(dz). \end{aligned} \tag{7.4.2}$$

In particular, taking $u = I_B$, the indicator function of a Borel set $B \in \mathcal{B}(X)$, we obtain the transition probability function of $\{x_t\}$:

$$P(B|x) = PI_B(x) = EI_B[F(x, z_0)] = \int_Z I_B[F(x, z)]G(dz). \tag{7.4.3}$$

7.4.2 Example: Random walks. (Kartashov [4].) Consider the random walk in the interval $X = [0, \infty)$, namely,

$$x_{t+1} = (x_t + z_t)^+, \quad t = 0, 1, \ldots, \tag{7.4.4}$$

where $r^+ := \max(r, 0)$. We wish to verify that $\{x_t\}$ satisfies the hypotheses of Theorem 7.3.11 so that it is w-geometrically ergodic for some weight function w. We shall suppose that $E|z_0| < \infty$ and

$$E(z_0) < 0. \tag{7.4.5}$$

Under the condition $E|z_0| < \infty$, it is well known that (7.4.5) implies that $\{x_t\}$ is positive recurrent (see Meyn and Tweedie [1] or Nummelin [1]), and so $\{x_t\}$ has a unique i.p.m., which we shall denote by μ. Let us also suppose that the *moment generating function* of z_0, namely,

$$m(s) := E \exp(sz_0) = \int_Z \exp(sz)G(dz),$$

is finite for all s in some interval $[0, \bar{s}]$, with $\bar{s} > 0$. Then, as $m(0) = 1$ and $m'(0) = E(z_0) < 0$, there is a number $\hat{s}$ such that

$$\beta := m(\hat{s}) = E \exp(\hat{s}z_0) < 1. \tag{7.4.6}$$

Now, for $x \geq 0$, let

$$w(x) := e^{\widehat{s}x}, \quad l(x) := \text{Prob}(z_0 + x \leq 0), \quad \text{and} \quad \nu(\cdot) := \delta_0(\cdot), \tag{7.4.7}$$

where δ_0 is the Dirac measure at 0. To verify the *condition (i)* in Theorem 7.3.11 consider the kernel

$$Q(B|x) := P(B|x) - l(x)\nu(B).$$

If 0 is *not* in B, then of course $Q(B|x) = P(B|x) \geq 0$. On the other hand, if $0 \in B$ then

$$Q(B|x) = \text{Prob}[(x + z_0)^+ \in B, \ z_0 > -x] \geq 0.$$

Hence Q is a nonnegative kernel and the condition (i) follows.

To obtain the condition (ii) in Theorem 7.3.11 note that $\nu(w) = w(0) = 1$, and, by (7.4.5), $\nu(l) = l(0) > 0$. Further, as

$$l(x) = \text{Prob}(x_{t+1} = 0 | x_t = x) = P(\{0\}|x),$$

the invariance of μ [see (7.3.1)] yields

$$\mu(l) = \int l \, d\mu = \int_X P(\{0\}|x)\mu(dx) = \mu(\{0\}) > 0.$$

Finally, to obtain (7.3.10) observe first that

$$w[(x + z_0)^+] \quad = \quad w(x)\exp(\hat{s}z_0) \quad \text{if} \quad x + z_0 > 0,$$
$$= \quad 1 \quad \text{if} \quad x + z_0 \le 0.$$

Hence, with β as in (7.4.6), we get

$$Pw(x) \quad = \quad E[w(x_{t+1})|x_t = x]$$
$$= \quad \beta w(x) + \text{Prob}(z_0 + x \le 0)$$
$$= \quad \beta w(x) + \nu(w)l(x) \quad [\text{since } \nu(w) = 1],$$

which yields (7.3.10). Summarizing, the hypotheses of Theorem 7.3.11 are satisfied and, consequently, *the random walk (7.4.4) is w-geometrically ergodic.*

This result for (7.4.4) can be extended in an obvious manner to chains of the form

$$x_{t+1} = (x_t + \eta_t - \xi_t)^+, \quad t = 0, 1, \ldots, \tag{7.4.8}$$

where $\{\eta_t\}$ and $\{\xi_t\}$ are independent sequences of nonnegative i.i.d. random variables with finite means, and also independent of x_0. Defining $z_t := \eta_t - \xi_t$, the condition corresponding to (7.4.5) is

$$E(\eta_0) < E(\xi_0) \tag{7.4.9}$$

and we again require the moment generating function of $z_0 = \eta_0 - \xi_0$ to be finite in some interval $[0, \bar{s}]$, with $\bar{s} > 0$. Then the chain in (7.4.8) is w-geometrically ergodic with respect to the weight function w in (7.4.7).

The stochastic model (7.4.8) appears in many important applications. For instance, in a *random-release dam model*, x_t denotes the content of the dam at time t, η_t is the input, and ξ_t is the random release. On the other hand, in a *single-server queueing system*, $\{\xi_t\}$ is the sequence of interarrival times of customers, the η_t are the service times, and x_t is the waiting time of the tth customer. Finally, (7.4.8) can also be interpreted as representing an *inventory-production system* in which x_t is the stock level at time t, η_t is the quantity produced, and ξ_t is the demand during the tth period. $\square$

7.4.3 Example: A queueing system. The approach in Example 7.4.2, using the moment generating function $m(s) := E[\exp(sz_0)]$ to find a suitable weight function, often works in other cases. For instance, consider a discrete-time queueing system in which new customers are *not* admitted if the system is not empty, that is, when $x_t > 0$, where x_t is the number of customers at the beginning of time slot t. In this case, when $x_t > 0$ there is exactly one service completion (and the served customer leaves the system) in every time slot, and once the system is emptied ($x_t = 0$) new customers are accepted. Thus, denoting by z_t the number of arrivals during time slot t, the process is described by

$$x_{t+1} \quad = \quad x_t - 1 \quad \text{if} \quad x_t > 0$$
$$= \quad z_t \quad \text{if} \quad x_t = 0.$$

Equivalently, if I_0 stands for the indicator function of the set $\{0\}$, we can write x_{t+1} as

$$x_{t+1} = x_t - 1 + (z_t + 1)I_0(x_t), \quad t = 0, 1, \ldots,$$

which is an equation of the form (7.4.1). Further, in Assumption 7.4.1 we have $X = Z = \{0, 1 \ldots\}$ and G is the "arrival distribution":

$$G(B) = \sum_{i \in B} q(i) \quad \forall B \subset \{0, 1, \ldots\},$$

with $q(i) \geq 0$ and $\sum_i q(i) = 1$. The transition probability function $P(\{j\}|i) =: p(j|i)$ is given by

$$p(i - 1|i) = 1 \quad \forall i \geq 1 \qquad \text{and} \qquad p(j|0) = q(j) \quad \forall j \geq 0.$$

To guarantee the existence of a unique i.p.m., as required in Theorem 7.3.11, it suffices to assume that the arrival distribution has a finite mean $\bar{q} := E(z_0)$, i.e.,

(a) $\bar{q} = \sum_i i q(i) < \infty.$

Then the invariance equation (7.3.1) becomes

$$\mu(j) = \sum_{i=0}^{\infty} p(j|i)\mu(i), \quad \forall j \geq 0,$$

with $\mu(j) \geq 0$ and $\sum_j \mu(j) = 1$, which reduces to

$$\mu(j) = \mu(0)q(j) + \mu(j + 1) \quad \forall j \geq 0.$$

Hence the unique i.p.m. is given by

$$\mu(j) = \mu(0) \sum_{k=j}^{\infty} q(k) \quad \forall j \geq 0, \quad \text{with} \quad \mu(0) = 1/(1 + \bar{q}).$$

On the other hand, to ensure that the conditions (i)–(iii) in Theorem 7.3.11 hold, we shall suppose that z_0 has a finite moment generating function $m(s)$ in some interval $[0, \bar{s}]$, that is,

(b) $m(s) := E[\exp(sz_0)] < \infty \qquad \forall s$ in $[0, \bar{s}]$, for some $\bar{s} > 0.$

Now fix a number $\hat{s}$ in $(0, \bar{s})$, and define the number $\beta := \exp(-\hat{s})$ and the functions

$$l(i) := I_0(i) \quad \text{and} \quad w(i) := \exp(\hat{s}i), \quad i \geq 0,$$

where I_0 is the indicator function of $\{0\}$. Then taking the p.m. ν as the arrival distribution, that is, $\nu(i) = q(i)$ for all i, easy calculations show that the hypotheses (i)–(iii) of Theorem 7.3.11 are satisfied; hence, $\{x_t\}$ is w-geometrically ergodic. $\square$

7.4.4 Example: Linear systems. In this example the state and disturbance spaces are $X = Z = \mathbb{R}^d$ and (7.4.1) is replaced by

$$x_{t+1} = Fx_t + z_t, \qquad (7.4.10)$$

where F is a given $d \times d$ matrix. In addition to Assumption 7.4.1(a) we will suppose the following:

(a) all the eigenvalues of F have magnitude less than 1 (equivalently, the *spectral radius* of F is less than 1);

(b) the disturbance distribution G is nonsingular with respect to Lebesgue measure, with a nontrivial density, zero mean and finite second moment.

Under these hypotheses, the linear system (7.4.10) has many interesting properties. In particular, it turns out to be w_i-geometrically ergodic ($i = 1, 2, 3$) with respect to each of the functions

$$w_1(x) = |x| + 1, \quad w_2(x) = |x|^2 + 1, \quad \text{and} \quad w_3(x) = x'Lx + 1, \quad (7.4.11)$$

where "prime" denotes "transpose", and L in w_3 is a symmetric positive-definite matrix that satisfies the matrix equation

$$F'LF = L - I.$$

The existence of such a matrix L is ensured by the hypothesis (a); see, for instance, Chen [1], Theorem 8–22. We shall omit the calculations showing the w_i-geometric ergodicity and refer to the proofs of Proposition 12.5.1 and Theorem 17.6.2 of Meyn and Tweedie [1] for further details. Moreover, it is worth noting that these results are valid when (7.4.10) is replaced by the more general model

$$x_{t+1} = Fx_t + Hz_t, \qquad (7.4.12)$$

where the z_t are p-dimensional random vector and H is a $d \times p$ matrix, provided that the pair of matrices (F, H) is "controllable" in the sense that the matrix

$$[H|FH|\cdots|F^{d-1}H]$$

has rank d. $\square$

We now consider a general Markov chain (7.4.1) with state space $X = \mathbb{R}^d$; the disturbance space Z is an arbitrary Borel space. We first present conditions under which the inequalities (7.3.9) are satisfied, so that the chain will be w-geometrically ergodic if the additional hypotheses of Theorem 7.3.10 are satisfied. Conditions for the latter to be true are given in Example 7.4.6 for additive-noise systems.

7.4.5 Proposition. *Let $\{x_t\}$ be as in (7.4.1), with $X = \mathbb{R}^d$, and Z an arbitrary Borel space. Suppose that there exist positive constants β, γ and M such that*

(a) $\beta < 1$ and $\beta + \gamma \geq 1$;

(b) $E|F(x, z_0)| \leq \beta|x| - \gamma$ $\forall |x| > M$; and

(c) $\widehat{F} := \sup_{|x| \leq M} E|F(x, z_0)| < \infty.$

Then

$$Pw(x) \leq \beta w(x) + bI_C(x) \quad \forall x \in X, \tag{7.4.13}$$

where

$$w(x) := |x| + 1, \quad b := \gamma + \widehat{F}, \quad and \quad C := \{|x| \leq M\}. \tag{7.4.14}$$

Proof. The proof follows from direct calculations. Let $D := X - C$ be the complement of C. Then, since $I_C + I_D = 1$,

$$
\begin{aligned}
E|F(x, z_0)| &= E(|F(x, z_0)|[I_D(x) + I_C(x)]) \\
&\leq (\beta|x| - \gamma)I_D(x) + \widehat{F}I_C(x) \\
&= \beta|x| - \gamma + (\widehat{F} - \beta|x| + \gamma)I_C(x) \\
&\leq \beta w(x) - (\beta + \gamma) + bI_C(x).
\end{aligned}
$$

Hence,

$$Pw(x) = E(|F(x, z_0)| + 1) \leq \beta w(x) - (\beta + \gamma) + bI_C(x) + 1$$

and (7.4.13) follows since, by (a), $-(\beta + \gamma) + 1 \leq 0$. $\square$

As an application of Proposition 7.4.5, we shall combine it with Theorem 7.3.10 to give a set of conditions under which the *additive-noise* (or autoregressive) process

$$x_{t+1} = F(x_t) + z_t, \quad t = 0, 1, \ldots \tag{7.4.15}$$

is w-geometrically ergodic.

7.4.6 Example: Additive-noise nonlinear systems. Let $\{x_t\}$ be the Markov chain given by (7.4.15), with values in $X = \mathbb{R}^d$ and disturbances satisfying Assumption 7.4.1. In addition, suppose that

(a) $F : \mathbb{R}^d \to \mathbb{R}^d$ is a continuous function;

(b) The disturbance distribution G is absolutely continuous with respect to *Lebesgue measure* $\lambda(dz) = dz$, and its density g [i.e., $G(dz) = g(z)dz$] is positive λ-a.e., and has a finite mean value;

(c) There exist positive constants β, γ and M such that $\beta < 1$, $\beta + \gamma \geq 1$, and

$$E|F(x) + z_0| \leq \beta|x| - \gamma \quad \forall |x| > M. \tag{7.4.16}$$

Then $\{x_t\}$ is w-geometrically ergodic, where $w(x) := |x| + 1$.

Indeed, the conditions (a) and (c) yield that the hypotheses of Proposition 7.4.5 are satisfied, and, therefore, we obtain the inequality (7.4.13), which is the same as (7.3.9a). Hence, to obtain the desired conclusion it suffices to verify that the chain satisfies the other hypotheses of Theorem 7.3.10. To do this, first note that (7.4.15) and (7.4.2) yield

$$Pu(x) = Eu[F(x) + z_0] = \int u[F(x) + z]g(z)dz. \tag{7.4.17}$$

Thus, in particular, if u is a continuous and bounded function on $X = \mathbb{R}^d$, then so is Pu; in other words, $\{x_t\}$ *satisfies the weak Feller property*. On the other hand, taking $u = I_B$ for any Borel set B, a change of integration variable in (7.4.17) gives [cf. (7.4.3)]

$$P(B|x) = \int_B g(z - F(x))dz, \tag{7.4.18}$$

which, by hypothesis (b), shows that $\{x_t\}$ *is λ-irreducible*. Therefore, Theorem 7.3.7 yields that $\{x_t\}$ is a λ-irreducible T-chain, which in turn, by Theorem 7.3.8, shows that every compact set is petite. In particular, $C := \{|x| \leq M\}$ *is a petite set*. Finally, from (b) and (7.4.18) again, we see that *the chain is aperiodic*, so that all the assumptions of Theorem 7.3.10 are valid in the present case. $\square$

7.4.7 Example: Iterated function systems. An *iterated function system* (IFS) is a Markov chain of the form (7.4.1) with state space a closed subset X of $\mathbb{R}^d$ and a *finite* disturbance set Z, say $Z = \{1, \ldots, N\}$. Let $\{q_1, \ldots, q_n\}$ be the common distribution of the i.i.d. disturbances, that is,

$$q_i = \text{Prob}(z_0 = i), \quad i = 1, \ldots, N.$$

We assume that the probabilities q_i are positive. Then, writing $F(x, i)$ as $F_i(x)$, for $x \in X$ and $i = 1, \ldots, N$, the expressions (7.4.2) and (7.4.3) become

$$Pu(x) = \sum_{i=1}^{N} u[F_i(x)]q_i \tag{7.4.19}$$

and

$$P(B|x) = \sum_{i=1}^{N} I_B[F_i(x)]q_i, \tag{7.4.20}$$

respectively. It turns out that the hypothesis (H) below implies that the IFS $\{x_t\}$ is w-geometrically ergodic with respect to the weight function

$$w(x) := |x| + 1, \quad x \in X. \tag{7.4.21}$$

(H) The zero vector 0 is in X and the functions F_i satisfy the Lipschitz conditions

$$|F_i(x) - F_i(y)| \le L_i |x - y| \quad \forall x, y \in X, \quad i = 1, \dots, N,$$

where $L_1, \dots, L_N$ are nonnegative constants such that

$$\beta := \sum_{i=1}^{N} L_i q_i < 1. \tag{7.4.22}$$

By Proposition 12.8.1 of Lasota and Mackey [1], (H) implies that the IFS is *weakly asymptotically stable*, which means that the transition probability function P has a *unique i.p.m.* μ, and, in addition, $(\nu P^t)(u) \to \mu(u)$ for any initial distribution ν and any continuous bounded function u on X. (In other words, for any p.m. ν, νP^t converges *weakly* to μ.) On the other hand, the uniqueness of μ and the assumption that $q_i > 0$ yield that the IFS $\{z_t\}$ is μ-irreducible and aperiodic. Hence, by Theorem 7.3.10, to prove that $\{x_t\}$ is w-geometrically ergodic it suffices to show that the inequality (7.3.9b) holds. To see this observe that, by (H),

$$\begin{aligned} |F_i(x)| &\le |F_i(x) - F_i(0)| + |F_i(0)| \\ &\le L_i |x| + \overline{F} \quad \forall x \in X \text{ and } i = 1, \dots, N, \end{aligned}$$

where $\overline{F} := \max\{|F_i(0)|, \ i = 1, \dots, N\}$. Thus, from (7.4.21) and (7.4.19), we obtain

$$Pw(x) = \sum_{i=1}^{N} (|F_i(x)| + 1) q_i \le \beta w(x) + (\overline{F} + 1 - \beta),$$

which is the same as (7.3.9b) with $b := \overline{F} + 1 - \beta$. $\square$

7.5   Poisson's equation

Let $\{x_t\}$ be a X-valued Markov chain with transition probability function $P(B|x)$, and let $(\mathcal{X}, \|\cdot\|)$ be a normed vector space of real-valued measurable functions on X. In most applications, $\mathcal{X}$ is one of the spaces $\mathbb{B}(X)$, $\mathbb{B}_w(X)$ or $L_p(X, \mathcal{B}(X), \mu)$ for some measure μ. We assume that the usual linear map $u \mapsto Pu$ with

$$Pu(x) := \int_X u(y) P(dy|x), \quad x \in X,$$

maps $\mathcal{X}$ into itself.

7.5.1 Definition. Let c, g and h be functions in $\mathcal{X}$. Then the system of equations

$$\textbf{(a)} \ \ g = Pg, \quad \text{and} \quad \textbf{(b)} \ \ g + h = c + Ph \qquad (7.5.1)$$

is called the **Poisson equation** (P.E.) for P with **charge** c, and if there is a pair (g, h) of functions that satisfies (7.5.1) then (g, h) is said to be a **solution** to the P.E. for P with charge c. Further, if P admits a *unique i.p.m.*, then (7.5.1) is referred to as the **unichain P.E.**; otherwise, is called the **multichain P.E.** A function that satisfies (7.5.1)(a) is said to be **invariant** (or **harmonic**) with respect to P.

Remark. Since $Pk = k$ for any constant k, we can see that if (g, h) is a solution of (7.5.1), then so is $(g, h + k)$ for any constant k. Several results in this section deal with the question of "uniqueness" of a solution to the P.E. (See Corollary 7.5.6, and Theorems 7.5.7 and 7.5.10.) □

In this section we first present an elementary example showing, in particular, that the choice of the underlying space $\mathcal{X}$ is an important issue. The remaining material is divided into three parts. In part A we present some basic results on the multichain P.E., and in part B we consider the special unichain case in which P is w-geometrically ergodic with respect to a weight function w. Finally, in part C we present some examples.

7.5.2 Example. Let $\{x_t\}$ be the Markov chain in Example 7.4.3; that is, the state space is $X = \{0, 1, \ldots\}$ and the transition probabilities are

$$p(i - 1|i) = 1 \quad \forall i \geq 1, \quad \text{and} \quad p(j|0) = q(j) \quad \forall j \geq 0,$$

where $\{q(i), \ i \in X\}$ is a given probability distribution on X with finite mean value, i.e.,

$$\bar{q} := \sum_{i=0}^{\infty} iq(i) < \infty.$$

Let $\mathcal{X} = \mathbb{B}(X)$ be the Banach space of bounded functions on X with the sup norm, and consider the P.E. (7.5.1) with charge c in $\mathbb{B}(X)$ given by

$$c(0) := -\bar{q}, \quad \text{and} \quad c(i) := 1 \quad \forall i \geq 1.$$

Then it is easily verified that the pair of functions (g, h) with

$$g(\cdot) \equiv 0, \quad \text{and} \quad h(i) = i - \bar{q} \quad \forall i \in X$$

satisfies (7.5.1). However, since h is *unbounded*, it does not belong to the space $\mathcal{X} = \mathbb{B}(X)$ and, therefore, the pair (g, h) is *not* a "solution" of the P.E. in the sense of Definition 7.5.1. The latter situation can be remedied by introducing a suitable space larger than $\mathbb{B}(X)$. For instance, consider the weight function $w(i) := i + 1$. Then using (7.2.8) it is easily verified that the transition law is bounded in w-norm (in fact, $\|p\|_w \leq \bar{q} + 1$) and so, by Proposition 7.2.5, p maps $\mathbb{B}_w(X)$ into itself. Finally, replacing $\mathbb{B}(X)$

by $\mathbb{B}_w(X)$, we see that (g, h) is indeed a solution of (7.5.1) in the space $\mathbb{B}_w(X) \supset \mathbb{B}(X)$. $\square$

A. The multichain case

7.5.3 Remark. It is important to keep in mind the probabilistic meaning of equations such as (7.5.1), but sometimes it will be more expedient to do things "operationally". For instance, iterating the invariance equation (7.5.1)(a) we obtain

$$g = P^t g \quad \forall t = 0, 1, \ldots, \tag{7.5.2}$$

which in turn yields

$$ng = \sum_{t=0}^{n-1} P^t g \quad \forall n = 1, 2, \ldots. \tag{7.5.3}$$

On the other hand, from a "probabilistic" viewpoint, (7.5.1)(a) implies that $\{g(x_t), \ t = 0, 1, \ldots\}$ is a *martingale*. Indeed, for any t, $g(x_t)$ is measurable with respect to $\sigma(x_0, \ldots, x_t)$, the σ-algebra generated by $\{x_0, \ldots, x_t\}$, $E_x|g(x_t)| \le \|g\| < \infty$ for each x, whereas the Markov property yields

$$
\begin{aligned}
E[g(x_{t+1})|x_0, \ldots, x_t] &= E[g(x_{t+1})|x_t] \\
&= \int g(y) P(dy|x_t) \\
&= Pg(x_t) = g(x_t) \quad \text{by (7.5.1)(a).}
\end{aligned}
$$

This, of course, also yields (7.5.2), which can be written as

$$g(x) = P^t g(x) = \int g(y) P^t(dy|x) = E_x[g(x_t)] \quad \forall t = 0, 1, \ldots, x \in X.$$

Moreover, rewriting (7.5.1)(b) as $h = c - g + Ph$ and iterating we get

$$h = \sum_{t=0}^{n-1} P^t(c - g) + P^n h \quad \forall n = 1, 2, \ldots, \tag{7.5.4}$$

which "probabilistically" can be rewritten as

$$
\begin{aligned}
h(x) &= \sum_{t=0}^{n-1} [E_x c(x_t) - E_x g(x_t)] + E_x h(x_n) \\
&= E_x \left[\sum_{t=0}^{n-1} [c(x_t) - g(x_t)] + h(x_n) \right] \tag{7.5.5} \\
&= E_x(M_n),
\end{aligned}
$$

where $\{M_n\}$ is defined by

$$M_0 := h(x_0), \quad \text{and} \quad M_n := \sum_{t=0}^{n-1}[c(x_t) - g(x_t)] + h(x_n) \qquad (7.5.6)$$

for $n \geq 1$. $\square$

7.5.4 Definition. Let c, g and h be functions in $\mathcal{X}$. The pair (g, h) is called a **c-canonical pair** if

$$ng + h = \sum_{t=0}^{n-1} P^t c + P^n h \quad \forall n = 1, 2, \ldots. \qquad (7.5.7)$$

Canonical pairs, the sequence $\{M_n\}$ in (7.5.6), and the P.E. (7.5.1) are related as follows.

7.5.5 Theorem. *The following conditions are equivalent:*

(a) (g, h) *is a solution to the P.E. with charge c.*

(b) (g, h) *is a c-canonical pair.*

(c) $\{M_n\}$ *is a martingale and g is invariant.*

Proof. (a) $\Leftrightarrow$ (b). The implication (a) $\Rightarrow$ (b) follows from (7.5.4) and (7.5.3). Conversely, (7.5.1)(b) follows from (7.5.7) with $n = 1$. To obtain the invariance equation (7.5.1)(a), apply P to both sides of (7.5.1)(b) to get

$$P^2 h = Pg + Ph - Pc,$$

and, on the other hand, observe that for $n = 2$ (7.5.7) becomes

$$P^2 h = 2g + h - c - Pc.$$

The last two equations yield

$$Pg - g = g + h - (c + Ph) = 0, \quad \text{i.e.,} \quad g = Pg.$$

(a) $\Leftrightarrow$ (c). If (a) holds, then the invariance conditon on g is obvious, whereas, by its very definition, M_n is measurable with respect to $\sigma\{x_0, \ldots, x_n\}$ for every n, and $E_x|M_n| \leq n(\|c\| + \|g\|) + \|h\| < \infty$ for each $x \in X$. Moreover, by the Markov property,

$$\begin{aligned} E[M_{n+1} - M_n | x_0, \ldots, x_n] &= E[h(x_{n+1})|x_n] + c(x_n) - g(x_n) - h(x_n) \\ &= Ph(x_n) + c(x_n) - g(x_n) - h(x_n) \\ &= 0 \quad \forall n \geq 0, \quad \text{by (7.5.1)(b),} \end{aligned}$$

i.e., $\{M_n\}$ is a martingale. The converse follows from (7.5.5) [or (7.5.4)] with $n = 1$, and the invariance of g. $\square$

Although Theorem 7.5.5 is quite straightforward, it has important consequences. In particular, it allows us to derive additional necessary and/or
sufficient conditions for the existence of solutions to the P.E. In part (c)
of the following corollary we require the stochastic kernel P to be **power-
bounded**, which means that

$$\|P^n\| \le K \quad \text{for all} \quad n = 0, 1, \ldots, \quad \text{and some constant } K. \tag{7.5.8}$$

[Recall that the **norm** of an operator T on $(\mathcal{X}, \|\cdot\|)$ is defined as

$$\|T\| := \sup\{\|Tu\| : u \in \mathcal{X}, \|u\| \le 1\};$$

cf. (7.2.7) or (7.2.10).] The condition (7.5.8) holds, for instance, if $\|P\| \le 1$,
in which case P is **nonexpansive** (Definition 7.2.7) with respect to the
norm $\|\cdot\|$. For such a P, we have $\|P^n\| \le \|P\|^n \le 1$ [cf. (7.2.16)] and
(7.5.8) follows. On the other hand, note that obviously a power-bounded
operator is **bounded**, i.e., $\|P\| \le K$ for some constant K. An important
consequence of power-boundedness is given in part (c) of the following
corollary.

7.5.6 Corollary. *Let (g, h) be a solution of the P.E. with charge c. Then:*

(a) $g = \lim\limits_{n \to \infty} \frac{1}{n} \sum\limits_{t=0}^{n-1} P^t g$ *pointwise and in norm (that is, in the norm $\|\cdot\|$*
of $\mathcal{X}$).

(b) *If*
$$P^n h/n \to 0 \quad \text{pointwise or in norm}, \tag{7.5.9}$$

then
$$g = \lim_{n \to \infty} \frac{1}{n} \sum_{t=0}^{n-1} P^t g = \lim_{n \to \infty} \frac{1}{n} \sum_{t=0}^{n-1} P^t c \tag{7.5.10}$$

pointwise or in norm, respectively.

(c) *If, further, P is power-bounded [with a constant K as in (7.5.8)] then,
for all $n \ge 1$,*

$$\left\| \sum_{t=0}^{n-1} P^t(c - g) \right\| = \left\| \sum_{t=0}^{n-1} P^t c - ng \right\| \le (1 + K)\|h\|;$$

hence
$$\left\| \frac{1}{n} \sum_{t=0}^{n-1} P^t c - g \right\| \le (1 + K)\|h\|/n.$$

(d) *Uniqueness of solutions: Let (g_1, h_1) and (g_2, h_2) be two solutions of
the P.E. with charge c such that h_1 and h_2 satisfy (7.5.9). Then $g_1 =
g_2$ and*

$$h_1 - h_2 = \lim_{n \to \infty} \frac{1}{n} \sum_{t=0}^{n-1} P^t(h_1 - h_2) \quad \text{pointwise and in norm.} \quad (7.5.11)$$

Proof. Part (a) follows from (7.5.3). Moreover, from (7.5.3) and (7.5.7) we get

$$\sum_{t=0}^{n-1} P^t(c - g) = h - P^n h, \quad (7.5.12)$$

which, using part (a), yields (b). In fact, (7.5.12) also gives (c) because

$$\| \sum_{t=0}^{n-1} P^t(c - g) \| \le (1 + K) \| h \|,$$

with K as in (7.5.8).

(d) The equality $g_1 = g_2$ results from (b), since

$$g_1 = \lim \frac{1}{n} \sum_{t=0}^{n-1} P^t c = g_2.$$

Let $g := g_1 = g_2$. Then writing (7.5.1)(b) for (g, h_1) and for (g, h_2) and subtracting we see that $u := h_1 - h_2$ is invariant, i.e., $u = Pu$. Therefore, (7.5.11) follows from the same argument used in (a). $\square$

The following theorem gives another characterization of a solution to the P.E.

7.5.7 Theorem. *Let c, g and h be functions in $(\mathcal{X}, \| \cdot \|)$ and suppose that*

(a) *P is bounded (i.e., $\|P\| \le K$ for some constant K), and*

(b) *(7.5.9) holds in norm for every u in $\mathcal{X}$, i.e., $\|P^n u\|/n \to 0$.*

Then the two following assertions are equivalent:

(i) *(g, h) is the unique solution of the P.E. with charge c for which*

$$\lim \frac{1}{n} \sum_{t=0}^{n-1} P^t h = 0 \quad \text{in norm.} \quad (7.5.13)$$

(ii) *g satisfies (7.5.10) in norm and*

$$h = \lim_{N \to \infty} \frac{1}{N} \sum_{n=1}^{N} \sum_{t=0}^{n-1} P^t(c - g) \quad \text{in norm.} \quad (7.5.14)$$

Proof. (i) $\Rightarrow$ (ii). Suppose that (i) holds. Then, by the hypothesis (b), h satisfies (7.5.9) and so the requirement on g follows from Corollary 7.5.6(b). On the other hand, by (7.5.7) and (7.5.3),

$$Nh = \sum_{n=1}^{N} \sum_{t=0}^{n-1} P^t(c - g) + \sum_{n=1}^{N} P^n h \quad \forall N = 1, 2, \ldots. \quad (7.5.15)$$

Hence, (7.5.13) gives (7.5.14).

(ii) $\Rightarrow$ (i). Suppose that (ii) holds. By the assumption (a), we can interchange P and limits in norm; that is,

$$P(\lim f_n) = \lim P f_n \quad \text{if} \quad f_n \quad \text{converges in norm.} \tag{7.5.16}$$

Applying this fact to the first equality in (7.5.10) we obtain $Pg = g$; that is, g satisfies (7.5.1)(a). Moreover, note that the hypothesis on g gives

$$\lim \frac{1}{n} \sum_{t=0}^{n-1} P^t(c - g) = 0 \quad \text{in norm.} \tag{7.5.17}$$

Now, to obtain (7.5.1)(b), observe first that

$$(I - P) \sum_{t=0}^{n-1} P^t = I - P^n \quad \forall n = 1, 2, \ldots,$$

so that

$$(I - P) \sum_{t=0}^{n-1} P^t(c - g) = (I - P^n)(c - g). \tag{7.5.18}$$

Therefore, applying $I - P$ to (7.5.14) and using (7.5.16) and (7.5.18) we get

$$
\begin{aligned}
(I - P)h &= \lim_N \frac{1}{N} \sum_{n=1}^{N} (I - P) \sum_{t=0}^{n-1} P^t(c - g) \\
&= (c - g) - \lim_N \frac{1}{N} \sum_{n=1}^{N} P^n(c - g) \\
&= c - g \quad \text{by (7.5.17)};
\end{aligned}
$$

that is, (7.5.1)(b) holds. Hence, (g, h) is a solution to the P.E. with charge c, and, by (7.5.14) and (7.5.15), the function h satisfies (7.5.13). The latter condition, (7.5.13), and Corollary 7.5.6(d) give the uniqueness of (g, h). $\square$

7.5.8 Remark: Finite X. If the state space X is a *finite* set, in which case the stochastic kernel P is a square matrix, it is well known that the **limiting matrix**

$$\widehat{P} := \lim_{n \to \infty} \frac{1}{n} \sum_{t=0}^{n-1} P^t \quad \text{(componentwise)} \tag{7.5.19}$$

exists, and that $I - P + \widehat{P}$ is nonsingular; its inverse

$$Z := (I - P + \widehat{P})^{-1}$$

is called the **fundamental matrix** associated to P. Moreover, the **deviation matrix** associated to P (or **Drazin inverse** of $I - P$), defined by

$$H := \lim_{N \to \infty} \frac{1}{N} \sum_{n=1}^{N} \sum_{t=0}^{n-1} (P - \widehat{P})^t (I - \widehat{P}), \qquad (7.5.20)$$

can also be written as

$$H = \lim_{N \to \infty} \frac{1}{N} \sum_{n=1}^{N} \sum_{t=0}^{n-1} (P^t - \widehat{P}) \qquad (7.5.21)$$

and satisfies that

$$H = Z(I - \widehat{P}) = (I - P + \widehat{P})^{-1}(I - \widehat{P}). \qquad (7.5.22)$$

Further, the pair (g, h) of vectors given by

$$g := \widehat{P}c, \quad \text{and} \quad h := Hc \qquad (7.5.23)$$

is a solution to the P.E. for P with charge c, and it is precisely of the form given by Theorem 7.5.7(ii). In fact, in a suitable theoretical setting, all of the expressions (7.5.19)–(7.5.23) have a well-defined meaning in a much more general context than the finite-state case. (See Hernández-Lerma and Lasserre [6] for details.) □

B. The unichain P.E.

Suppose that P has an i.p.m. μ. Then by the *Individual Ergodic Theorem* (Yosida [1]) for any given function u in $L_1(\mu) \equiv L_1(X, \mathcal{B}(X), \mu)$ there is a function u^* in $L_1(\mu)$ such that

$$\text{(a)} \quad u^* = \lim_{n \to \infty} \frac{1}{n} \sum_{t=0}^{n-1} P^t u \quad \mu\text{-a.e., and (b)} \quad \int u^* d\mu = \int u d\mu. \quad (7.5.24)$$

On the other hand, the *Mean Ergodic Theorem* ensures that the convergence in (a) holds in $L_1(\mu)$, that is, for every u in $L_1(\mu)$,

$$\text{(a)} \quad u^* = \lim_{n \to \infty} \frac{1}{n} \sum_{t=0}^{n-1} P^t u \quad \text{in} \quad L_1(\mu), \quad \text{and (b)} \quad Pu^* = u^*. \quad (7.5.25)$$

Further, *if μ is the unique i.p.m. for P*, then u^* is a constant μ-a.e. and in fact, by (7.5.24)(b),

$$u^* = \int u d\mu \quad \mu\text{-a.e.} \qquad (7.5.26)$$

Consider now the *unichain P.E.*, so that P has a *unique* i.p.m. μ, and let (g, h) be a solution to the P.E. with charge c. Moreover, assume that:

$$c \text{ is in } L_1(\mu), \text{ and } h \text{ satisfies } (7.5.9). \qquad (7.5.27)$$

Then, by (7.5.26) and Corollary 7.5.6(b), we see that $g = c^*$, that is,

$$g = \mu(c) := \int c d\mu \quad \mu\text{-a.e.} \qquad (7.5.28)$$

This gives an "explicit" expression for g in the unichain case. However, we wish to distinguish (7.5.28) from the case where $g = \mu(c)$ *holds everywhere*.

7.5.9 Definition. The P.E. (7.5.1) is called **strictly unichain** if it is unichain and $g(x) = \mu(c)$ for all $x \in X$, where μ denotes the unique i.p.m. of P.

We show next that the P.E. is strictly unichain when the kernel P is w-geometrically ergodic (Definition 7.3.9).

7.5.10 Theorem. *Suppose that P is w-geometrically ergodic with limiting p.m. μ Then:*

(a) *Each function in $\mathbb{B}_w(X)$ is μ-integrable, i.e., $\mathbb{B}_w(X) \subset L_1(\mu)$.*

(b) *For any function u in $\mathbb{B}_w(X)$, $\limsup_n \|P^n u\|_w \leq |\mu(u)|$. Hence u satisfies (7.5.9), that is,*

$$P^n u / n \to 0$$

in the w-norm and, of course, pointwise.

(c) *μ is the unique i.p.m. of P.*

(d) *If $u \in \mathbb{B}_w(X)$ is invariant (or harmonic) with respect to P (i.e., $Pu = u$), then*

$$u(x) = \mu(u) \quad \text{for all} \quad x \in X.$$

(e) *Let c be an arbitrary function in $\mathbb{B}_w(X)$. Then **(i)** a pair (g, h) of functions in $\mathbb{B}_w(X)$ is the unique solution of the strictly unichain P.E. with charge c and*

$$\mu(h) = 0 \qquad (7.5.29)$$

if and only if $g(x) = \mu(c)$ for all x and

$$h = \sum_{t=0}^{\infty} P^t[c - \mu(c)] = \lim_{n \to \infty} \sum_{t=0}^{n-1} P^t[c - \mu(c)] \quad \text{in } w\text{-norm.} \quad (7.5.30)$$

*Moreover, **(ii)** for any two solutions (g_1, h_1) and (g_2, h_2) of the strictly unichain P.E. with charge c we have $g_1 = g_2 = \mu(c)$ and h_1, h_2 differ at most by an additive constant; in fact,*

$$h_1 - h_2 = \mu(h_1 - h_2).$$

Proof. (a) By definition (7.2.1) of the w-norm, for any function u in $\mathbb{B}_w(X)$ we have:

$$\int |u| d\mu \le \|u\|_w \int w d\mu = \|u\|_w \|\mu\|_w < \infty,$$

where the last equality is due to (7.2.4).

(b) This follows from (7.3.8) and the inequality

$$|P^t u(x)| \le |P^t u(x) - \mu(u)| + |\mu(u)|.$$

(c) This fact was already proved in the paragraph after Definition 7.3.9.

(d) If $u = Pu$, then $u = P^t u$ for all $t = 0, 1, \ldots$. Thus, the desired conclusion follows from (7.3.8).

(e) The statement (ii) follows from (d) and Corollary 7.5.6(d). Similarly, statement (i) follows from Theorem 7.5.7 if we can show that the functions in (7.5.14) and (7.5.30) coincide. In turn, to prove the latter, first note that *if a sequence $\{S_n\}$ converges in (some) norm to S, then the sequence of Cesàro sums*

$$\frac{1}{N} \sum_{n=1}^{N} S_n$$

also converges in (the same) norm to S. Now, let S_n be the sequence in (7.5.30), i.e.,

$$S_n := \sum_{t=0}^{n-1} P^t[c - \mu(c)],$$

and use (7.3.8) to show that $\{S_n\}$ is a Cauchy sequence in $\mathbb{B}_w(X)$, and, therefore, it converges to a function, say, $S := h$ in $\mathbb{B}_w(X)$. Finally, observe that (7.5.14) with $g = \mu(c)$ and the w-norm, is precisely the limit of the Cesàro sums of $\{S_n\}$, so that the function in (7.5.14) coincides with $S = h$. □

7.5.11 Remark. In the context of Theorem 7.5.10 we have the following:

(a) By part (ii) of Theorem 7.5.10(e), *for any two solutions* (g, h_1), (g, h_2) of the strictly unichain P.E. we have $h_1 = h_2 + k$ where k is the *constant* $\mu(h_1 - h_2)$. Thus if we wish to guarantee that the P.E. has a *unique* solution it suffices to have $k = 0$, which is precisely the role of condition (7.5.29). In general, to "fix" a unique h we only need to take *any solution* (g, h) of the P.E. and replace h by $\hat{h} = h - \mu(h)$. This is tacitly what we did in Theorems 7.5.10(e) and 7.5.7. Indeed, if we look at (7.5.15) we see that the "full form" of h is

$$
\begin{aligned}
h &= \lim_{N \to \infty} \frac{1}{N} \sum_{n=1}^{N} \sum_{t=0}^{n-1} P^t[c - \mu(c)] + \mu(h). \\
&= \sum_{t=0}^{\infty} P^t[c - \mu(c)] + \mu(h) \quad \text{[by (7.5.30)]}.
\end{aligned}
\tag{7.5.31}
$$

There are other ways one can "fix" a unique h. For instance, let $\bar{x}$ be an arbitrary fixed point in X and replace (7.5.29) by the condition $h(\bar{x}) = 0$. Then in the above notation we again get $\mu(h_1 - h_2) = 0$.

(b) We can replace the "convergence estimate" in Corollary 7.5.6(c) by the following estimate, which is obtained from (7.3.8): For all $n \geq 1$

$$\| \sum_{t=0}^{n-1} P^t(c - \mu(c))\|_w \leq \|c\|_w R(1 - \rho^n)/(1 - \rho) \tag{7.5.32}$$

$$\leq \|c\|_w R/(1 - \rho).$$

Observe that (7.5.32) *holds for any function c in* $\mathbb{B}_w(X)$. On the other hand, again from (7.3.8) [or (7.3.7)] one can see that the function h in (7.5.4) can be written in the form (7.5.22)–(7.5.23), where the "limiting kernel (matrix)" $\widehat{P}(B|x)$ is the limiting p.m. μ, that is, $\widehat{P}(\cdot|x) = \mu(\cdot)$ for all $x \in X$. In this case, the "fundamental kernel" $Z = (I - P + \widehat{P})^{-1}$ is given by

$$(I - P + \mu)^{-1} = \sum_{t=0}^{\infty}(P - \mu)^t = \sum_{t=0}^{\infty}(P^t - \mu). \quad \square$$

For future reference we note that the conclusion of Theorem 7.5.10(d) holds μ-*almost everywhere* if u is **subinvariant** (i.e., $u \leq Pu$) or **super-invariant** (i.e., $u \geq Pu$). That is:

7.5.12 Lemma. *Suppose that P is as in Theorem 7.5.10, and let u be a function in* $\mathbb{B}_w(X)$ *such that either*

$$\text{(a)} \quad u(x) \geq Pu(x) \quad \forall x, \quad \text{or} \quad \text{(b)} \quad u(x) \leq Pu(x) \quad \forall x.$$

Then $u(x) = \mu(u) = \inf_X u(x)$ μ-a.e. in case (a), and $u(x) = \mu(u) = \sup_X u(x)$ μ-a.e. in case (b). Hence, in either case, $u(x) = \mu(u)$ μ-a.e.

Proof. (a) Suppose that $u \geq Pu$. This inequality yields $u \geq P^t u$ for all $t \geq 0$, so that, by (7.3.8), $u \geq P^t u \downarrow \mu(u)$. Thus

$$u(x) \geq \int u d\mu \quad \text{for all} \quad x.$$

Hence, letting $u_i := \inf_x u(x)$, we obtain $u_i \geq \int u d\mu \geq u_i$, that is, $\int u d\mu = u_i$. Finally, since $\int (u - u_i)d\mu = 0$ and $u \geq u_i$, we conclude that $u = u_i = \mu(u)$ μ-a.e. The proof in case (b) is similar [or apply (a) to $-u$]. $\square$

C. Examples

In general, solving the P.E. may be a challenging problem. There are cases, however, in which one can obtain, or perhaps "guess", a solution by some iterative procedure. For instance, consider the Markov chain in Examples 7.4.3 and 7.5.2. In the latter example we mentioned that a solution to the P.E. with charge $c(0) := -\bar{q}$ and $c(i) := 1$ for all $i \geq 1$ is given by the pair (g, h) with

$$g(\cdot) \equiv 0 \quad \text{and} \quad h(i) = i - \bar{q} \quad \forall i = 0, 1, \ldots, \tag{7.5.33}$$

where

$$\bar{q} := \sum_{i=1}^{\infty} iq(i) < \infty. \tag{7.5.34}$$

The question is, how did we get the solution (7.5.33)? To see this, recall that in Example 7.4.3 is shown that, under (7.5.34), the chain has the i.p.m.

$$\mu(i) = \mu(0) \sum_{k=i}^{\infty} q(k) \quad \forall i \geq 0, \quad \text{with} \quad \mu(0) = 1/(1 + \bar{q}). \tag{7.5.35}$$

Therefore, the constant $g(\cdot) \equiv \mu(c)$ is 0 since (7.5.34) and (7.5.35) yield

$$\mu(c) = \sum_{i=0}^{\infty} c(i)\mu(i) = c(0)\mu(0) + \mu(0) \sum_{i=1}^{\infty} iq(i) = \mu(0)(-\bar{q} + \bar{q}) = 0.$$

Hence to solve the P.E. (7.5.1) it suffices to consider (7.5.1)(b), which becomes

$$h(i) = c(i) + Ph(i) \quad \text{for all} \quad i = 0, 1, \ldots. \tag{7.5.36}$$

To compute

$$Ph(i) = \sum_{j=0}^{\infty} h(j)p(j|i),$$

recall from Example 7.4.3 that the transition probabilities $P(j|i)$ are given by

$$p(i-1|i) = 1 \quad \forall i \geq 1, \quad \text{and} \quad p(j|0) = q(j) \quad \forall j \geq 0.$$

Consequently,

$$Ph(i) = h(i-1) \quad \text{for} \quad i \geq 1, \quad \text{and} \quad Ph(0) = \sum_{j=0}^{\infty} h(j)q(j). \tag{7.5.37}$$

Then, replacing the values (7.5.37) in (7.5.36), we obtain

$$h(i) = i - \bar{q} + Ph(0) \quad \forall i \geq 0, \tag{7.5.38}$$

which together with $g = 0$ yields a solution (g, h) to the P.E. (7.5.1). Further, in view of the remark after Definition 7.5.1, in (7.5.38) we may subtract the *constant* $Ph(0)$ and so we obtain the solution (7.5.33). It is also illustrative, on the other hand, to verify other conditions in the results of this section. For instance, as shown in Example 7.4.3, the corresponding Markov chain can be described by the equation

$$x_{t+1} = x_t - 1 + (z_t + 1)I_0(x_t), \quad t = 0, 1, \ldots.$$

Thus, writing the charge c as $c(i) = 1 - (1 + \bar{q})I_0(i)$ and using (7.5.38), a direct calculation shows that the sequence M_n in (7.5.6) satisfies

$$\begin{aligned} M_{n+1} &= M_n + c(x_n) + h(x_{n+1}) - h(x_n) \\ &= M_n + (z_n - \bar{q})I_0(x_n), \end{aligned}$$

from which one can immediately verify, say, the martingale condition in Theorem 7.5.5(c).

In the above example, finding a solution to the P.E. was very easy because of the special form of the charge and the transition probabilities. In more general situations, one can try to obtain an "estimate" of $g = \mu(c)$ and then use the sequence

$$\sum_{t=0}^{n-1} P^t[c - \mu(c)] \tag{7.5.39}$$

in (7.5.30) to "guess" a feasible function h; finally, one would need of course to check that (g, h) is indeed a solution of the P.E. The following example shows how this procedure is supposed to work—for ease of exposition we restrict ourselves to a *scalar* case, but it should be clear that similar calculations can be done for the multivariable linear system (7.4.10).

7.5.13 Example: Linear systems (d = 1). Consider the *scalar* $(d = 1)$ linear system

$$x_{t+1} = Fx_t + z_t, \tag{7.5.40}$$

under the hypotheses (a) and (b) in Example 7.4.4. In particular, in the present case the hypothesis (a) translates into

$$|F| < 1. \tag{7.5.41}$$

We assume of course that $F \neq 0$. In addition, we will assume that the i.i.d. disturbances z_t have zero mean (which greatly simplifies the calculations), and finite second moment:

$$E(z_0) = 0, \quad \text{and} \quad \sigma^2 := E(z_0^2) < \infty. \tag{7.5.42}$$

The *charge* $c(x)$ is supposed to be the quadratic function

$$c(x) = \gamma x^2, \quad x \in \mathbb{R}, \tag{7.5.43}$$

for some constant γ, and we take the *weight function* $w := w_2$ in (7.4.11)

Step 1. By (7.3.8), to "estimate" $\mu(c)$ it suffices to compute

$$E_x c(x_t) = \int c(y) P^t(dy|x), \quad \forall x_0 = x,$$

and then let $t \to \infty$. To do this observe that, for any $t \geq 1$,

$$E[c(x_t)|x_{t-1}] = \gamma[F^2 x_{t-1}^2 + E(z_{t-1}^2) + 2Fx_{t-1}E(z_{t-1})]$$
$$= F^2 c(x_{t-1}) + \gamma\sigma^2$$

Therefore,

$$E_x c(x_t) = F^2 E_x c(x_{t-1}) + \gamma\sigma^2 \quad \forall t \geq 1,$$

and iterating we get

$$E_x c(x_t) = F^{2t} c(x) + \gamma\sigma^2 \sum_{k=0}^{t-1} F^{2k},$$

i.e.,

$$E_x c(x_t) = F^{2t} c(x) + \gamma\sigma^2 (1 - F^{2t})/(1 - F^2), \quad t \geq 1. \tag{7.5.44}$$

Thus, by (7.5.41), letting $t \to \infty$ we obtain

$$\mu(c) = \gamma\sigma^2/(1 - F^2). \tag{7.5.45}$$

Observe that we can write (7.5.44) as

$$E_x c(x_t) = [c(x) - \mu(c)]F^{2t} + \mu(c), \quad \forall t \geq 0, \ x \in \mathbb{R}. \tag{7.5.46}$$

Step 2. Compute (7.5.39) and let $t \to \infty$. From (7.5.46),

$$\sum_{t=0}^{n-1} E_x c(x_t) = [c(x) - \mu(c)](1 - F^{2n})/(1 - F^2) + n\mu(c).$$

Hence, (7.5.39) becomes

$$\sum_{t=0}^{n-1}[E_x c(x_t) - \mu(c)] = [c(x) - \mu(c)](1 - F^{2n})/(1 - F^2),$$

and letting $n \to \infty$, (7.5.30) yields that

$$h(x) = [c(x) - \mu(c)]/(1 - F^2), \quad x \in \mathbb{R}, \tag{7.5.47}$$

which obviously satisfies (7.5.29).

In conclusion, by Theorem 7.5.10(e), the pair (g, h) with $g(\cdot) \equiv \mu(c)$ in (7.5.45), and h in (7.5.47), is the unique solution of the strictly unichain P.E. that satisfies the constraint (7.5.29). $\square$

7.5.14 Example: Iterated function systems. Consider the IFS of Example 7.4.7 in the special case in which $X = \mathbb{R}$, and $F_i(x) = L_i x + l_i$, $i = 1, 2, \ldots, N$. Equivalently, the IFS is the "linear system"

$$x_{t+1} = L_{z_t} x_t + l_{z_t}$$

with random coefficients. In analogy with (7.4.22), we shall assume that $L_1, \ldots, L_N$ satisfy the condition

$$\sum_{i=1}^{N} |L_i| q_i < 1. \tag{7.5.48}$$

Now define the numbers

$$\widehat{L} := E(L_{z_t}) = \sum_{i=1}^{N} L_i q_i, \quad \text{and} \quad \widehat{l} := E(l_{z_t}) = \sum_{i=1}^{N} l_i q_i,$$

and consider the charge $c(x) = x$ for all $x \in X$. To solve the strictly unichain P.E. with charge c we may proceed exactly as in Example 7.5.13. In fact,

$$E(x_t | x_{t-1}) = \widehat{L} x_{t-1} + \widehat{l} \quad \text{for all} \quad t \geq 1 \quad [\text{cf. (7.4.19)}],$$

so that

$$E_x(x_t) = \widehat{L} E_x(x_{t-1}) + \widehat{l} = \widehat{L}^t x + \widehat{l}(1 - \widehat{L}^t)/(1 - \widehat{L})$$

for all $t \geq 0$. Finally, as in (7.5.44)–(7.5.47), using (7.5.48) we obtain

$$\mu(c) = \widehat{l}/(1 - \widehat{L}), \quad \text{and} \quad h(x) = [x - \mu(c)]/(1 - \widehat{L}), \quad x \in X,$$

which is a solution to the P.E. with charge $c(x) = x$. □

Notes on §7.5

1. Most of the results in subsection A are from Hernández-Lerma and Lasserre [6], where other approaches to the multichain P.E. are studied; see also papers [4] and [7] by the same authors, and Glynn and Meyn [1]. The concept of c-canonical pairs (Definition 7.5.4) is an adaptation to non-controlled Markov chains of the *canonical triplets* studied in §5.2 for MCPs, and which will also appear in later chapters. The P.E. also arises in potential theory (Nummelin [2], Revuz [1], Syski [1]) and stochastic approximation (Metivier and Priouret [1]), for instance.

2. For the special case of Markov chains with a *countable* state space X (in particular finite, as in Remark 7.5.8) see, for instance, Hordijk and Spieksma [1], Makowski and Shwartz [1], and Puterman [1].

8
Discounted Dynamic Programming with Weighted Norms

8.1 Introduction

In this chapter we consider the infinite-horizon discounted cost problem for a Markov control model $(X, A, \{A(x)|x \in X\}, Q, c)$. We already studied this problem using *dynamic programming* and *linear programming* in Chapters 4 and 6, respectively. Here we use again dynamic programming, so it is important to state at the outset the differences between this chapter and Chapter 4.

The main difference lies in the assumptions. In Chapter 4 we considered *nonnegative* cost-per-stage functions c, virtually *without restriction in their "growth rate"*, and we allowed *non-compact* control-constraint sets $A(x)$. In contrast, the hypotheses in this chapter—see Assumptions 8.3.1, 8.3.2 and 8.3.3, or 8.5.1, 8.5.2 and 8.5.3—require the sets $A(x)$ to be *compact*, but the cost functions c are allowed to take positive and *negative* values, provided that they satisfy a certain *growth condition* (Assumption 8.3.2). The corresponding dynamic programming theorems (Theorems 4.2.3 and 8.3.6) turn out to be very similar, except that in the present context the dynamic programming operator is a *contraction operator* with respect to a weighted w-norm (Proposition 8.3.9), which yields the *w-geometric convergence* of the *value iteration* (VI) algorithm [see (8.3.15)]. The latter fact, w-geometric convergence of the VI functions, gives many interesting results—such as evaluation of *rolling horizon* procedures, criteria for *elimination of nonoptimal control actions*, and existence and detection of *forecast horizons*—which are practically impossible to get in a context as general as that of Chapter 4.

To summarize, except for the sign of c, this chapter considers assumptions less general than those of Chapter 4, but in return we obtain much more information from the dynamic programming techniques.

What follows is an outline of the contents of this chapter. Section 8.2 presents an abridged version of §2.2 and §2.3; namely, it summarizes some of the main underlying concepts for a *Markov control process* (MCP), which are a Markov control model, and the admissible control policies. Section 8.3 deals with discounted *Dynamic Programming* (DP). The main result here is the DP Theorem 8.3.6, which concerns the discounted-cost optimality equation (or DCOE) (8.3.4), the convergence of the VI functions to the value function, and the existence of discount-optimal policies. In §8.4 we make a further analysis of the VI functions. The main question dealt with is, *how "close" is a VI-policy to being optimal?* The results in §8.3 and §8.4 are obtained under a set of hypotheses (Assumptions 8.3.1 to 8.3.3) which in particular impose a form of "strong continuity" on the controlled process' transition law Q—see Assumption 8.3.1(c). However, in §8.5 we show that those results remain valid under a "weak continuity" (weak Feller-like) condition on Q provided we change accordingly the hypotheses on the control model. The basic idea is to change the "measurability" requirements in Assumptions 8.3.1, 8.3.2, 8.3.3 to "continuity" conditions in Assumptions 8.5.1, 8.5.2, 8.5.3. Finally, in §8.6 we present a couple of examples, and we conclude in §8.7 with some general remarks.

8.2   The control model and control policies

Let $\mathcal{M} := (X, A, \{A(x)|x \in X\}, Q, c)$ be a **Markov control model** (MCM), where X and A are Borel spaces which denote the *state space* and the *action* (or *control*) *set*, respectively. For every state $x \in X$, $A(x)$ is a nonempty Borel subset of A whose elements are the *feasible actions* (or *controls*) if the system is in the state x. The set of feasible state-action pairs namely

$$\mathbb{K} := \{(x, a)|x \in X, a \in A(x)\}, \tag{8.2.1}$$

is assumed to be a Borel subset of $X \times A$. Moreover, the *transition law*

$$Q = \{Q(B|x, a)|B \in \mathcal{B}(X), \quad (x, a) \in \mathbb{K}\}$$

is a stochastic kernel on X given $\mathbb{K}$, and, finally, the measurable function $c : \mathbb{K} \to \mathbb{R}$ denotes the *cost-per-stage* (or *one-stage cost*) function.

8.2.1 Definition. $\mathbb{F}$ denotes the set of measurable functions $f : X \to A$ such that $f(x)$ is in $A(x)$ for all $x \in X$, and Φ stands for the set of stochastic kernels φ on A given X for which $\varphi(A(x)|x) = 1$ for all $x \in X$. The functions in $\mathbb{F}$ are referred to as **decision functions** or **selectors** of the set-valued mapping $x \mapsto A(x)$.

A selector $f \in \mathbb{F}$ may be identified with the stochastic kernel $\varphi \in \Phi$ for which $\varphi(\cdot|x)$ is the Dirac measure at $f(x)$ for all $x \in X$. Hence, we have $\mathbb{F} \subset \Phi$.

We shall assume that $\mathbb{F}$ is nonempty, or, equivalently, that the set $\mathbb{K}$ in (8.2.1) contains the graph of a measurable function from X to A. This assumption ensures that the set of control policies, defined below, is nonempty.

8.2.2 Definition. (Control policies.) For every $t = 0, 1, \ldots$, let H_t be the family of admissible histories up to time t, that is, $H_0 := X$, and

$$H_t := \mathbb{K}^t \times X = \mathbb{K} \times H_{t-1} \quad \text{for} \quad t = 1, 2, \ldots.$$

An element of H_t, called a "t-history", is a vector of the form

$$h_t = (x_0, a_0, \ldots, x_{t-1}, a_{t-1}, x_t) \tag{8.2.2}$$

with $(x_i, a_i) \in \mathbb{K}$ for $i = 0, \ldots, t-1$, and $x_t \in X$. A (**randomized**) **control policy** is a sequence $\pi = \{\pi_t\}$ of stochastic kernels π_t on the control set A given H_t, that satisfy the constraint

$$\pi_t(A(x_t)|h_t) = 1 \quad \forall h_t \in H_t, \; t = 0, 1, \ldots. \tag{8.2.3}$$

The set of all control policies is denoted by Π. Moreover, a control policy $\pi = \{\pi_t\}$ is said to be a:

(a) **randomized Markov policy** if there is a sequence $\{\varphi_t\}$ of stochastic kernels $\varphi_t \in \Phi$ such that

$$\pi_t(\cdot|h_t) = \varphi_t(\cdot|x_t) \quad \forall h_t \in H_t, \; t = 0, 1, \ldots; \tag{8.2.4}$$

(b) **randomized stationary policy** if (8.2.4) holds for a stochastic kernel $\varphi \in \Phi$ independent of t, i.e.,

$$\pi_t(\cdot|h_t) = \varphi(\cdot|x_t) \quad \forall h_t \in H_t, \; t = 0, 1, \ldots;$$

(c) **deterministic** (or **pure**) **policy** if there is a sequence $\{g_t\}$ of measurable functions $g_t : H_t \to A$ such that, for every $h_t \in H_t$ and $t = 0, 1, \ldots$, we have $g_t(h_t) \in A(x_t)$ and $\pi_t(\cdot|h_t)$ is the Dirac measure concentrated at $g_t(h_t)$;

(d) **deterministic Markov policy** if there is a sequence $\{f_t\}$ of selectors $f_t \in \mathbb{F}$ such that $\pi_t(\cdot|h_t)$ is the Dirac measure at $f_t(x_t) \in A(x_t)$ for all $h_t \in H_t$ and $t = 0, 1, \ldots;$

(e) **deterministic stationary policy** if there is a selector $f \in \mathbb{F}$ such that $\pi_t(\cdot|h_t)$ is the Dirac measure at $f(x_t) \in A(x_t)$ for all $h_t \in H_t$ and $t = 0, 1, \ldots.$

We denote by Π_{RM} the family of randomized Markov policies, and by Π_{RS} the subfamily of randomized stationary policies. Similarly, $\Pi_D \supset \Pi_{DM} \supset \Pi_{DS}$ denote the family of deterministic policies, and the subfamilies of deterministic Markov and deterministic stationary policies, respectively.

8.2.3 Remark. (a) If π is in Π_{RS} and $\varphi \in \Phi$ is as in Definition 8.2.2(b), then we write π as φ^∞. Similarly, if π is a deterministic stationary policy as in Definition 8.2.2(e), we write π as f^∞.

(b) Let $\mathbb{F}$ and Φ be the sets in Definition 8.2.1, and let $\mathbb{K}$ be as in (8.2.1). If G is a function on $\mathbb{K}$ we write

$$G(x,\varphi) := \int_A G(x,a)\varphi(da|x) \quad \text{if } \varphi \text{ is in } \Phi,$$

which reduces to

$$G(x,f) := G(x, f(x)) \quad \text{for } f \text{ in } \mathbb{F}.$$

In particular, for the cost function c and the transition law Q we write

$$c(x,\varphi) := \int_A c(x,a)\varphi(da|x), \quad Q(\cdot|x,\varphi) := \int_A Q(\cdot|x,a)\varphi(da|x) \quad (8.2.5)$$

if φ is in Φ, and

$$c(x,f) := c(x, f(x)), \quad Q(\cdot|x,f) = Q(\cdot|x, f(x)) \quad \text{if} \quad f \in \mathbb{F}. \quad (8.2.6)$$

(c) Let $\pi = \{\pi_t\}$ be an arbitrary control policy, and ν an arbitrary "initial distribution" on $\mathcal{B}(X)$. Further, let $(\Omega, \mathcal{F})$ be the measurable space consisting of the (canonical) "sample space" $\Omega := (X \times A)^\infty$ and the corresponding product σ-algebra $\mathcal{F}$. Then π and ν determine a probability measure (p.m.) P_ν^π and a stochastic process $\{(x_t, a_t), \ t = 0, 1, \ldots\}$ on Ω, where x_t and a_t represent the state and the control action at time t, respectively. (See §2.2.) The expectation operator with respect to P_ν^π is denoted by E_ν^π. If ν is the Dirac measure concentrated at the initial state $x_0 = x$, we write P_ν^π and E_ν^π as P_x^π and E_x^π, respectively.

(d) The p.m. P_ν^π in part (c) has the following properties: For every $B \in \mathcal{B}(X)$ and $C \in \mathcal{B}(A)$, and every t-history $h_t \in H_t$ as in (8.2.2), $t = 0, 1, \ldots$,

$$P_\nu^\pi(x_0 \in B) = \nu(B) \tag{8.2.7}$$

$$P_\nu^\pi(a_t \in C|h_t) = \pi_t(C|h_t) \tag{8.2.8}$$

$$P_\nu^\pi(x_{t+1} \in B|h_t, a_t) = Q(B|x_t, a_t). \tag{8.2.9}$$

In particular, from (8.2.7)–(8.2.9) one can deduce that if $\pi = \{\varphi_t\} \in \Pi_{RM}$ is a randomized Markov policy, then the state process $\{x_t\}$ is a nonhomogeneous Markov chain with transition kernels $Q(\cdot|\cdot, \varphi_t)$, $t = 0, 1, \ldots$; that is, for every $B \in \mathcal{B}(X)$ and $t = 0, 1, \ldots$ [and using the notation in (b)],

$$P_\nu^\pi(x_{t+1} \in B|x_0, \ldots, x_t) = P_\nu^\pi(x_{t+1} \in B|x_t)$$
$$= Q(B|x_t, \varphi_t). \tag{8.2.10}$$

For a deterministic Markov policy $\pi = \{f_t\} \in \Pi_{DM}$, the corresponding transition kernels are $Q(\cdot|\cdot, f_t)$. Moreover, for *stationary* policies $\varphi^\infty \in \Pi_{RS}$ or $f^\infty \in \Pi_{DS}$, $\{x_t\}$ is a *time-homogeneous* Markov chain with transition kernels $Q(\cdot|\cdot, \varphi)$ and $Q(\cdot|\cdot, f)$, respectively. In the latter case, the n-step transition kernel is written as $Q^n(\cdot|\cdot, \varphi)$, i.e.,

$$Q^n(B|x, \varphi) := \text{Prob}(x_n \in B|x_0 = x), \qquad (8.2.11)$$

and similarly for $Q^n(\cdot|\cdot, f)$. [For a proof of (8.2.10) see Proposition 2.3.5.]
□

8.3   The optimality equation

As in the previous section, let $(X, A, \{A(x)|x \in X\}, Q, c)$ be a Markov control model, which is fixed throughout the remainder of this chapter, and consider the α-**discounted cost** (abbreviated α-DC, or simply DC)

$$V(\pi, x) := E_x^\pi \left[\sum_{t=0}^\infty \alpha^t c(x_t, a_t) \right], \quad \pi \in \Pi, \ x \in X, \qquad (8.3.1)$$

where $\alpha \in (0, 1)$ is a given discount factor. The corresponding α-**discount value function** (or α-**discount optimal cost function**) is

$$V^*(x) := \inf_\Pi V(\pi, x), \quad x \in X. \qquad (8.3.2)$$

As in Chapter 4, the main problem we are concerned with is the calculation of V^* and in finding an α-**discount optimal** policy, that is, a control policy π^* such that

$$V(\pi^*, x) = V^*(x) \quad \forall x \in X. \qquad (8.3.3)$$

In Theorem 4.2.3 we showed in particular that, under quite general hypotheses, V^* is the pointwise-minimal solution of the α-**discounted cost optimality equation** (α-DCOE)

$$V^*(x) = \min_{A(x)} \left[c(x, a) + \alpha \int_X V^*(y) Q(dy|x, a) \right] \quad \forall x \in X. \qquad (8.3.4)$$

In this section we shall prove a similar result (Theorem 8.3.6) but which cannot be obtained from Theorem 4.2.3 because now we will consider cost-per-stage functions c that can take *negative* values, whereas all of Chapter 4 deals with *nonnegative* cost functions. Moreover, for application in later sections, we wish to obtain "good" estimates of the rate of convergence of the α-value iteration functions to V^* [see (8.3.15)], which are virtually impossible to get in a context as general as that of Theorem 4.2.3.

Hence, in this chapter we impose three different hypotheses. The first one, Assumption 8.3.1, is about the usual compactness-continuity conditions for Markov control models [and in fact is the same as Condition 3.3.2(ε), (b), (c2)]. The second, Assumption 8.3.2, uses a weight function w to impose a growth condition on the cost function and, in addition, it will yield that the *dynamic programming* (DP) *operator* T_α in (8.3.17) is a *contraction* on the space $\mathbb{B}_w(X)$ introduced in §7.2.A, namely, the Banach space of measurable functions on X with a finite w-norm. Finally, the third hypothesis, Assumption 8.3.3, is a further continuity condition that combined with the previous assumptions will ensure the existence of "measurable minimizers" for T_α (see Proposition 8.3.9). In §8.5 we introduce a different set of hypotheses.

A. Assumptions

8.3.1 Assumption. For every state $x \in X$:

(a) The control-constraint set $A(x)$ is compact;

(b) The cost-per-stage $c(x, a)$ is lower semicontinuous (l.s.c) in $a \in A(x)$; and

(c) The function $u'(x, a) := \int u(y)Q(dy|x, a)$ is continuous in $a \in A(x)$ for every function u in $\mathbb{B}(X)$, where $\mathbb{B}(X)$ denotes the Banach space of real-valued bounded measurable functions u on X, with the sup norm $\|u\| := \sup_X |u(x)|$ (see §7.2.A).

Remark. Assumption 8.3.1(c) is *equivalent* to the apparently weaker condition

(c') $u'(x, a)$ is *l.s.c.* in $a \in A(x)$ for every *nonnegative* function u in $\mathbb{B}(X)$.

Indeed, it is obvious that (c) implies (c'). Conversely, suppose that (c') holds and let u be an *arbitrary* function in $\mathbb{B}(X)$. Then $u + \|u\|$ is nonnegative, and, therefore, by (c'), the function

$$\int (u(y) + \|u\|)Q(dy|x, a) = u'(x, a) + \|u\|$$

is l.s.c. in $a \in A(x)$, which implies that $u'(x, a)$ is l.s.c. in $a \in A(x)$. In other words, $u'(x, a)$ is l.s.c. in $a \in A(x)$ for *any* function u in $\mathbb{B}(X)$. Moreover, applying the latter fact to $-u$, we can see that $u'(x, a)$ is also *upper semicontinuous* (u.s.c.) in $a \in A(x)$. Hence, as $u'(x, \cdot)$ is both l.s.c. and u.s.c., it is in fact *continuous*, and (c) follows. □

8.3.2 Assumption. There exist nonnegative constants $\bar{c}$ and β, with $1 \leq \beta < 1/\alpha$, and a weight function $w \geq 1$ on X such that for every state $x \in X$:

(a) $\sup_{A(x)} |c(x,a)| \le \bar{c}w(x)$; and

(b) $\sup_{A(x)} \int w(y)Q(dy|x,a) \le \beta w(x)$. (8.3.5)

Further, we suppose that w satisfies:

8.3.3 Assumption. For every state $x \in X$, the function $w'(x,a) := \int w(y)Q(dy|x,a)$ is continuous in $a \in A(x)$.

Concerning the requirement $\beta \ge 1$ in Assumption 8.3.2, see Note 4 at the end of this section.

An obvious sufficient condition for Assumption 8.3.2 is that c is *bounded*; that is, there is a constant $\bar{c}$ such that $|c(x,a)| \le \bar{c}$ for all (x,a) in $\mathbb{K}$. In this case, the function $w \ge 1$ can be taken to be bounded. Another sufficient condition is given in Remark 8.3.5(a) and also in Note 1 at the end of the section. On the other hand, we have:

8.3.4 Proposition. *Assumption 8.3.2 implies that*

$$C(x) := \sum_{t=0}^{\infty} \alpha^t c_t(x) < \infty \quad \text{for every} \quad x \in X, \qquad (8.3.6)$$

where $c_0(x) := \sup_{A(x)} |c(x,a)|$ and

$$c_t(x) := \sup_{A(x)} \int c_{t-1}(y)Q(dy|x,a) \quad \text{for} \quad t = 1, 2, \ldots \qquad (8.3.7)$$

are assumed to be measurable functions. Conversely, if (i) $C \ge 1$, and (ii) the inequality in (8.3.6) is satisfied for some α_0 with $\alpha < \alpha_0 < 1$, then Assumption 8.3.2 holds with

$$\bar{c} := 1 \quad w(x) := C(x), \quad \text{and} \quad \beta := \alpha_0^{-1}. \qquad (8.3.8)$$

Proof. If Assumption 8.3.2 holds, a straightforward induction argument shows that

$$c_t(x) \le \bar{c}\beta^t w(x) \quad \forall t = 0, 1, \ldots, x \in X. \qquad (8.3.9)$$

Hence, as $0 \le \alpha\beta < 1$, we get

$$C(x) \le \bar{c}w(x)/(1 - \alpha\beta) < \infty \quad \text{for every} \quad x \in X; \qquad (8.3.10)$$

that is, (8.3.6) holds.

Conversely, by (ii) and (8.3.7),

$$\int C(y)Q(dy|x,a) = \sum_{t=0}^{\infty} \alpha^t \int c_t(y)Q(dy|x,a)$$

$$\le \sum_{t=0}^{\infty} \alpha_0^t c_{t+1}(x)$$

$$= \alpha_0^{-1}[C(x) - c_0(x)],$$

i.e.,

$$\int C(y)Q(dy|x,a) \leq \alpha_0^{-1}C(x).$$

This inequality combined with the conditions (i), (ii), yields the desired conclusion. $\Box$

8.3.5 Remark. (a) In several places we shall consider inequalities of the form (7.3.9) or (7.3.10), so it is important to note that many results in this chapter remain valid if part (b) in Assumption 8.3.2 is replaced by an inequality of the form (8.3.11) below, where $\gamma > 0$ *is not necessarily* ≥ 1, in contrast to (8.3.5) where $\beta \geq 1$. More precisely, we have:

Assumption 8.3.2 is satisfied if there exists a real-valued measurable function $w \geq 1$ on X and positive constants m, γ and b such that $\alpha\gamma < 1$ and, for every state $x \in X$,

(i) $\sup_{A(x)} |c(x,a)| \leq mw'(x)$, *and*

(ii) $\sup_{A(x)} \int w'(y)Q(dy|x,a) \leq \gamma w'(x) + b$ (8.3.11)

Indeed, let $C(x)$ and $c_t(x)$ be as in (8.3.6) and (8.3.7), respectively, except that c_C is redefined as

$$c_0(x) := 1 + \sup_{A(x)} |c(x,a)|.$$

Observe that, by (i), we have $c_0 \leq Mw'$ with $M := 1 + m$. Then instead of (8.3.9) we obtain

$$c_t(x) \leq Mw'(x)\gamma^t + Mb\sum_{j=0}^{t-1}\gamma^j \quad \forall t = 1,2,\ldots,$$

and so (8.3.10) becomes

$$C(x) \leq Mw'(x)/(1-\alpha\gamma) + Mb\alpha/(1-\alpha)(1-\alpha\gamma).$$

Therefore, by the "converse" of Proposition 8.3.4, there exist constants $\bar{c}$ and β, and a function $w(\cdot) \geq 1$ that satisfy Assumption 8.3.2.

(b) In analogy with (7.2.8) [equivalently, (7.2.7)] we may define the w-**norm** of the transition law Q as

$$\|Q\|_w := \sup w(x)^{-1}\int_X w(y)Q(dy|x,a),$$

where the sup is over all feasible state-action pairs (x,a) in $\mathbb{K}$. Hence, under (8.3.5) we have $\|Q\|_w \leq \beta < \infty$, and so Proposition 7.2.5 is applicable in the present "controlled" context. $\Box$

B. The discounted-cost optimality equation

To state our main result in this section, we first recall from §4.2 the definition of the α-**value iteration** (or α-VI) **functions**

$$v_n(x) := \min_{A(x)} \left[c(x,a) + \alpha \int_X v_{n-1}(y) Q(dy|x,a) \right] \qquad (8.3.12)$$

for all $n \geq 1$ and $x \in X$, with $v_0(\cdot) \equiv 0$. For every $n = 1, 2, \ldots, v_n$ is the *optimal n-stage cost* [see (3.4.11)], i.e.,

$$v_n(x) = \inf_\Pi V_n(\pi, x) \quad \forall x \in X, \qquad (8.3.13)$$

where

$$V_n(\pi, x) := E_x^\pi \left[\sum_{t=0}^{n-1} \alpha^t c(x_t, a_t) \right]. \qquad (8.3.14)$$

The following theorem states among other things that the sequence $\{v_n\}$ converges *geometrically* in the w-norm to V^* [see (8.3.2)].

8.3.6 Theorem. *Suppose that Assumptions 8.3.1, 8.3.2 and 8.3.3 hold. Let β be the constant in (8.3.5), and define $\gamma := \alpha\beta$. Then:*

(a) *The α-discount value function V^* is the unique solution of the α-DCOE (8.3.4) in the space $\mathbb{B}_w(X)$, and*

$$\|v_n - V^*\|_w \leq \bar{c}\gamma^n/(1-\gamma), \quad n = 1, 2, \ldots, \qquad (8.3.15)$$

where $\bar{c}$ is the constant in Assumption 8.3.2(a).

(b) *There exists a selector $f_* \in \mathbb{F}$ such that $f_*(x) \in A(x)$ attains the minimum in (8.3.4) for every state x, that is [using the notation in (8.2.6)],*

$$V^*(x) = c(x, f_*) + \alpha \int V^*(y) Q(dy|x, f_*) \quad \forall x \in X, \qquad (8.3.16)$$

and the deterministic stationary policy $f_*^\infty \in \Pi_{DS}$ is α-discount optimal; conversely, if $f_*^\infty \in \Pi_{DS}$ is α-discount optimal, then it satisfies (8.3.16).

(c) *A policy π^* is α-discount optimal if and only if the corresponding cost function $V(\pi^*, \cdot)$ satisfies the α-DCOE.*

(d) *If an α-discount optimal policy exists, then there exists a <u>deterministic stationary</u> policy which is α-discount optimal.*

Except for the w-geometric convergence in (8.3.15), Theorem 8.3.6 looks very much the same as Theorem 4.2.3, and, consequently, it is important to note that none of these two results can be obtained from the other. First of all, it is obvious that Theorem 8.3.6 does not yield Theorem 4.2.3 because the latter allows *noncompact* control-constraint sets $A(x)$, in contrast to the compactness required in Assumption 8.3.1(a). On the other hand, Theorem 4.2.3 deals exclusively with *nonnegative* cost functions c, and so it cannot be used to obtain Theorem 8.3.6. However, if we restrict ourselves to cost functions $c \geq 0$, then Theorem 8.3.6 turns out to be a strengthened form of Theorem 4.2.3. (Later on we will need to consider cost functions with *negative* values.)

The remainder of this section is devoted to prove Theorem 8.3.6, which requires several preliminary results. First we shall introduce the DP (dynamic programming) operator T_α.

C. The dynamic programming operator

Given a measurable function $u : X \to \mathbb{R}$ and $0 \leq \alpha \leq 1$, we denote by $T_\alpha u$ the function given by

$$T_\alpha u(x) := \inf_{A(x)} \left[c(x,a) + \alpha \int_X u(y)Q(dy|x,a) \right], \quad x \in X, \qquad (8.3.17)$$

whenever the integral is well defined. If the infimum in (8.3.17) is actually attained at some action $a \in A(x)$ for every $x \in X$, then we shall write "min" in lieu of "inf". For $0 < \alpha < 1$, we shall prove that T_α is a contraction operator on the space $\mathbb{B}_w(X)$ (Proposition 8.3.9), which uses part (a) in the following lemma; the Fatou-like results in part (b) are also used below.

8.3.7 Lemma. *Suppose that Assumptions 8.3.3 and 8.3.1(c) hold. Then:*

(a) *The function $u'(x,a) := \int u(y)Q(dy|x,a)$ is <u>continuous</u> in $a \in A(x)$ for every $x \in X$ and every function u in $\mathbb{B}_w(X)$.*

(b) *(**An extension of Fatou's Lemma.**) Let $\{u_n\}$ be a bounded sequence in $\mathbb{B}_w(X)$, that is, there is a constant K such that $\|u_n\|_w \leq K$ for all n, and define*

$$u^I(x) := \liminf_{n\to\infty} u_n(x), \quad and \quad u^S(x) := \limsup_{n\to\infty} u_n(x).$$

Then for any state $x \in X$ and any sequence $\{a^n\}$ in $A(x)$ such that $a^n \to a$ in $A(x)$, we have

$$\liminf_{n\to\infty} \int u_n(y)Q(dy|x,a^n) \geq \int u^I(y)Q(dy|x,a), \qquad (8.3.18)$$

and

$$\limsup_{n\to\infty} \int u_n(y)Q(dy|x,a^n) \le \int u^S(y)Q(dy|x,a). \qquad (8.3.19)$$

Hence, if $u_n \to u$ (that is, $u^I = u^S$), then

$$\lim_{n\to\infty} \int u_n(y)Q(dy|x,a^n) = \int u(y)Q(dy|x,a). \qquad (8.3.20)$$

Proof. (a) Let u be a function in $\mathbb{B}_w(X)$, so that $|u(x)| \le mw(x)$ for all $x \in X$, where $m := \|u\|_w$. Then $u_m := u + mw$ is a *nonnegative* function in $\mathbb{B}_w(X)$, and so it is the limit of a nondecreasing sequence of measurable bounded functions $u^k \in \mathbb{B}(X)$. Now fix $x \in X$ and let $\{a^n\}$ be a sequence in $A(x)$ converging to $a \in A(x)$. Then, as $u^k \uparrow u_m$, Assumption 8.3.1(c) yields, for every k,

$$\liminf_{n\to\infty} \int u_m(y)Q(dy|x,a^n) \ge \liminf_{n\to\infty} \int u^k(y)Q(dy|x,a^n)$$
$$= \int u^k(y)Q(dy|x,a).$$

Hence, letting $k \to \infty$, monotone convergence yields that

$$\liminf_{n\to\infty} \int u_m(y)Q(dy|x,a^n) \ge \int u_m(y)Q(dy|x,a)$$

and, therefore, $\int u_m(y)Q(dy|x,\cdot)$ is l.s.c. on $A(x)$, which implies that $u'(x,\cdot)$ is l.s.c. on $A(x)$. In other words, $u'(x,\cdot)$ is l.s.c. on $A(x)$ for every function u in $\mathbb{B}_w(X)$. Hence, if we now apply the latter fact to $-u$ in lieu of u, we see that $u'(x,\cdot)$ is also u.s.c. Thus $u'(x,\cdot)$ is continuous on $A(x)$.

(b) Write u^I as $u^I(x) := \lim_{k\to\infty} U_k(x)$, where

$$U_k(x) := \inf_{n\ge k} u_n(x) \uparrow u^I(x) \quad \text{as} \quad k \to \infty.$$

Then, for all $n \ge k$,

$$\int u_n(y)Q(dy|x,a^n) \ge \int U_k(y)Q(dy|x,a^n),$$

and, as U_k is in $\mathbb{B}_w(X)$ (in fact, $\|U_k\|_w \le K$ for all k), part (a) yields

$$\liminf_{n\to\infty} \int u_n(y)Q(dy|x,a^n) \ge \int U_k(y)Q(dy|x,a). \qquad (8.3.21)$$

Thus, letting $k \to \infty$, we obtain (8.3.18) from (8.3.21) and monotone convergence. The proof of (8.3.19) is similar and is left to the reader. $\square$

We will also need the following lemma, whose parts (a) and (b) are the same as the "measurable selection theorem" in Proposition D.5 (Appendix D). Recall that the set-valued mapping (or multifunction) $x \mapsto A(x)$ from X to A is said to be **upper semicontinuous** (u.s.c.) if $\{x \in X | A(x) \cap F \neq \emptyset\}$ is a closed subset of X for every closed set $F \subset A$. [Equivalently, $x \mapsto A(x)$ is **u.s.c.** if $\{x \in X | A(x) \subset G\}$ is an open set in X for every open set $G \subset A$.]

8.3.8 Lemma. *Let $\mathbb{K}$ and $A(x)$ be as in (8.2.1) and Assumption 8.3.1(a), respectively, and let $v : \mathbb{K} \to \mathbb{R}$ be a given measurable function. Define*

$$v^*(x) := \inf_{A(x)} v(x,a), \quad x \in X. \qquad (8.3.22a)$$

(a) *If $v(x, \cdot)$ is l.s.c. on $A(x)$ for every $x \in X$, then there exists a selector $f \in \mathbb{F}$ such that $f(x) \in A(x)$ attains the minimum in (8.3.22a) for all $x \in X$, that is [using the notation in Remark 8.2.3(b)],*

$$v^*(x) = v(x,f) \quad \forall x \in X, \qquad (8.3.22b)$$

and v^* is a measurable function.

(b) *If the set-valued mapping $x \mapsto A(x)$ is u.s.c. and v is l.s.c. and bounded below on $\mathbb{K}$, then there exists a selector $f \in \mathbb{F}$ that satisfies (8.3.22b) and, moreover, v^* is l.s.c. and bounded below on X.*

(c) *Suppose that $x \mapsto A(x)$ is u.s.c, v is l.s.c. and, further,*

$$\sup_{A(x)} |v(x,a)| \leq kw(x) \quad \forall x \in X,$$

where k is a constant and $w(\cdot) \geq 1$ is a continuous function on X. Then there is a selector $f \in \mathbb{F}$ that satisfies (8.3.22b), v^* is l.s.c., and

$$|v^*(x)| \leq kw(x) \quad \forall x \in X; \qquad (8.3.22c)$$

that is, v^* is a l.s.c. function in the space $\mathbb{B}_w(X)$ and its w-norm satisfies $\|v^*\|_w \leq k$.

Proof. For the proof of parts (a) and (b) see Rieder [1] or Schäl [1]. To prove (c), apply (b) to the nonnegative l.s.c. function

$$u(x,a) := v(x,a) + kw(x). \quad \square$$

8.3.9 Proposition. *For $0 < \alpha < 1$, suppose that Assumptions 8.3.1, 8.3.2, and 8.3.3 hold, and let T_α be the map defined by (8.3.17). Then:*

(a) *T_α is a contraction operator on $\mathbb{B}_w(X)$, with modulus $\gamma := \alpha\beta < 1$; that is, T_α maps $\mathbb{B}_w(X)$ into itself and*

$$\|T_\alpha u - T_\alpha u'\|_w \leq \gamma \|u - u'\|_w \quad \forall u, u' \text{ in } \mathbb{B}_w(X); \qquad (8.3.23)$$

(b) *For every function u in $\mathbb{B}_w(X)$ there is a selector $f \equiv f_u$ in $\mathbb{F}$ such that*

$$T_\alpha u(x) = c(x, f) + \alpha \int_X u(y)Q(dy|x, f) \quad \forall x \in X. \qquad (8.3.24)$$

Part (b) holds for any $0 \leq \alpha \leq 1$.

Proof. Let u be an arbitrary function in $\mathbb{B}_w(X)$. Then, by Lemma 8.3.7(a) and Assumption 8.3.1(b), the function

$$v(x, a) := c(x, a) + \alpha \int u(y)Q(dy|x, a)$$

is l.s.c. in $a \in A(x)$ for every $x \in X$. Hence, Lemma 8.3.8(a) yields that $T_\alpha u$ is a measurable function and that there exists $f \in \mathbb{F}$ that satisfies (8.3.24). It is also clear that $T_\alpha u$ has a finite w-norm since, by Assumption 8.3.2,

$$
\begin{aligned}
|v(x, a)| &\leq \bar{c}w(x) + \alpha \|u\|_w \int w(y)Q(dy|x, a) \\
&\leq (\bar{c} + (\alpha\beta)\|u\|_w)w(x) \quad \forall x \in X,\ a \in A(x).
\end{aligned}
$$

Moreover, it is obvious that T_α is a monotone operator ($u \leq u'$ implies $T_\alpha u \leq T_\alpha u'$). Thus, in view of Proposition 7.2.9, to complete the proof it suffices to show that (7.2.20) holds. This, however, is immediate since (8.3.17) and (8.3.5) yield, for any real number r,

$$T_\alpha(u + rw)(x) \leq T_\alpha u(x) + (\alpha\beta)rw(x) \quad \forall x \in X.$$

That is, T_α is a contraction on $\mathbb{B}_w(X)$ with modulus $\gamma := \alpha\beta$. $\square$

Before presenting the proof of Theorem 8.3.6, observe that using T_α we may rewrite (8.3.4) and (8.3.12) as

$$V^* = T_\alpha V^* \qquad (8.3.25)$$

and

$$v_n = T_\alpha v_{n-1} = T_\alpha^n v_0 \quad \forall n = 0, 1, \ldots, \text{ with } v_0 = 0, \qquad (8.3.26)$$

respectively.

D. Proof of Theorem 8.3.6.

(a) By Proposition 8.3.9 and Banach's Fixed Point Theorem (Proposition 7.2.8), T_α has a unique fixed point u^* in $\mathbb{B}_w(X)$, i.e.,

$$T_\alpha u^* = u^* \qquad (8.3.27)$$

and

$$\|T_\alpha^n u - u^*\|_w \leq \gamma^n \|u - u^*\|_w \quad \forall u \in \mathbb{B}_w(X),\ n = 0, 1, \ldots. \qquad (8.3.28)$$

Hence, to prove part (a) we need to show that

(a$_1$) V^* is in $\mathbb{B}_w(X)$, with w-norm $\|V^*\|_w \le \bar{c}/(1-\gamma)$, and

(a$_2$) $V^* = u^*$.

In this case, (8.3.15) will follow from (8.3.26) and (8.3.28) with $u \equiv 0$.

To prove (a$_1$), let $\pi \in \Pi$ be an arbitrary policy and let $x \in X$ be an arbitrary initial state. Then for all $t = 0, 1, \ldots$

$$E_x^\pi w(x_t) \le \beta^t w(x), \tag{8.3.29}$$

and

$$|E_x^\pi c(x_t, a_t)| \le \bar{c}\beta^t w(x). \tag{8.3.30}$$

Indeed, (8.3.29) is trivial for $t = 0$. Now, if $t \ge 1$, it follows from (8.2.9) that

$$
\begin{aligned}
E_x^\pi[w(x_t)|h_{t-1}, a_{t-1}] &= \int w(y)Q(dy|x_{t-1}, a_{t-1}) \\
&\le \beta w(x_{t-1}) \quad \text{by (8.3.5)}.
\end{aligned}
\tag{8.3.31}
$$

Therefore $E_x^\pi w(x_t) \le \beta E_x^\pi w(x_{t-1})$, which iterated yields (8.3.29). Similarly, to get (8.3.30) observe that Assumption 8.3.2(a) yields

$$|c(x_t, a_t)| \le \bar{c}w(x_t) \quad \forall t = 0, 1, \ldots,$$

so that, by (8.3.29),

$$E_x^\pi |c(x_t, a_t)| \le \bar{c}\beta^t w(x). \tag{8.3.32}$$

Finally, note that (a$_1$) follows from (8.3.30) since a direct calculation gives

$$|V(\pi, x)| \le \sum_{t=0}^{\infty} \alpha^t E_x^\pi |c(x_t, a_t)| \le \bar{c}w(x)/(1-\gamma), \tag{8.3.33}$$

with $\gamma := \alpha\beta$. Thus, as $\pi \in \Pi$ and $x \in X$ were arbitrary,

$$|V^*(x)| \le \bar{c}w(x)/(1-\gamma). \tag{8.3.34}$$

To prove (a$_2$) let us first note that

$$\lim_{t\to\infty} \alpha^t E_x^\pi u(x_t) = 0 \quad \forall \pi \in \Pi, \, x \in X, \, u \in \mathbb{B}_w(X). \tag{8.3.35}$$

Indeed, by definition of w-norm and (8.3.29),

$$E_x^\pi |u(x_t)| \le \|u\|_w E_x^\pi w(x_t) \le \|u\|_w \beta^t w(x),$$

and (8.3.35) follows. Let us now consider the equality $u^* = T_\alpha u^*$ in (8.3.27). By Proposition 8.3.9(b), there exists a selector $f \in \mathbb{F}$ such that

$$u^*(x) = c(x, f) + \alpha \int u^*(y)Q(dy|x, f) \quad \forall x \in X. \tag{8.3.36}$$

Iteration of (8.3.36) yields

$$u^*(x) = E_x^f \sum_{t=0}^{n-1} \alpha^t c(x_t, f) + \alpha^n E_x^f u^*(x_n) \quad \forall n = 1, 2, \ldots,$$

and letting $n \to \infty$ we get, by (8.3.35),

$$u^*(x) = E_x^f \left[\sum_{t=0}^{\infty} \alpha^t c(x_t, f) \right] = V(f, x).$$

Thus, by definition of V^* [see (8.3.2)], we see that

$$u^*(x) \geq V^*(x). \tag{8.3.37}$$

To get the reverse inequality, note that (8.3.27) implies that

$$u^*(x) \leq c(x, a) + \alpha \int u^*(y) Q(dy|x, a) \quad \forall (x, a) \in \mathbb{K}. \tag{8.3.38}$$

Hence, for any policy $\pi \in \Pi$ and initial state $x \in X$, (8.2.9) and (8.3.38) yield

$$\alpha^t E_x^\pi [c(x_t, a_t) + \alpha u^*(x_{t+1}) - u^*(x_t)|h_t, a_t] \geq 0 \quad \forall t = 0, 1, \ldots.$$

Therefore, taking expectation E_x^π and summing over $t = 0, \ldots, n - 1$, we obtain

$$u^*(x) \leq E_x^\pi \left[\sum_{t=0}^{n-1} \alpha^t c(x_t, a_t) \right] + \alpha^n E_x^\pi u^*(x_n) \quad \forall n = 1, 2, \ldots.$$

Finally, letting $n \to \infty$ in the latter inequality and using (8.3.35), it follows that

$$u^*(x) \leq V(\pi, x),$$

so that, as π and x were arbitrary, we conclude that $u^*(x) \leq V^*(x)$ for all $x \in X$. This inequality and (8.3.37) yield (a$_2$).

(b) The existence of a selector $f_* \in \mathbb{F}$ that satisfies (8.3.16) follows from part (a) and Proposition 8.3.9(b). Conversely, *for any deterministic stationary policy f_*^∞, the corresponding α-discounted cost satisfies*

$$V(f_*^\infty, x) = c(x, f_*) + \alpha \int V(f_*^\infty, y) Q(dy|x, f_*) \quad \forall x \in X. \tag{8.3.39}$$

(See Remark 8.3.10.) Hence if f_*^∞ is α-discount optimal, we have $V(f_*^\infty, \cdot) = V^*(\cdot)$ so (8.3.39) yields (8.3.16).

Finally, part (c) is obvious, whereas (d) follows from (c) and (b). $\square$

8.3.10 Remark. Concerning (8.3.39), there are at least two ways in which one can show that, for any deterministic stationary policy $f^\infty \in \Pi_{DS}$,

$$V(f^\infty, x) = c(x, f) + \alpha \int_X V(f^\infty, y)Q(dy|x, f) \quad \forall x \in X. \qquad (8.3.40)$$

The first one is to expand the right-hand side of (8.3.1), with $\pi = f^\infty$, to obtain

$$V(f^\infty, x) = c(x, f) + \alpha E_x^f \left[\sum_{t=1}^\infty \alpha^{t-1} c(x_t, f) \right].$$

Then (8.3.40) follows from the Markov property (8.2.10) and using the definition (8.3.1) again, with $\pi = f^\infty$. The second way is via a *fixed-point argument* as in Proposition 8.3.9(a). Namely, for any given deterministic stationary policy f^∞, define an operator

$$R_f : \mathbb{B}_w(X) \to \mathbb{B}_w(X), \quad u \mapsto R_f u,$$

by

$$(R_f u)(x) := c(x, f) + \alpha \int_X u(y)Q(dy|x, f), \quad x \in X. \qquad (8.3.41)$$

Then using Proposition 7.2.9 it can be verified that R_f is a contraction operator on $\mathbb{B}_w(X)$ with modulus $\gamma := \alpha\beta$, and, therefore, R_f has a unique fixed point u_f in $\mathbb{B}_w(X)$, i.e.,

$$u_f = R_f u_f. \qquad (8.3.42)$$

From this equation and (8.3.41) we have then that u_f is the unique solution in $\mathbb{B}_w(X)$ of the equation

$$u_f(x) = c(x, f) + \alpha \int u_f(y)Q(dy|x, f), \quad x \in X. \qquad (8.3.43)$$

Moreover, iteration of (8.3.42) or (8.3.43) yields

$$u_f(x) = R_f^n u_f(x) = E_x^f \left[\sum_{t=0}^{n-1} \alpha^t c(x_t, f) \right] + \alpha^n E_x^f u_f(x_n)$$

for all $x \in X$ and $n = 1, 2, \ldots$. Finally, letting $n \to \infty$, we see from (8.3.35) and (3.3.1) that $u_f(x) = V(f^\infty, x)$ for all $x \in X$, and so (8.3.43) is the same as (8.3.40). $\square$

Notes on §8.3

1. Except for the fact that the cost-per-stage c is allowed to take *negative* values, Assumption 8.3.2 and (8.3.6) are, respectively, the same as conditions (b) and (c) in Proposition 4.3.1. However, *there is a misprint in Proposition 4.3.1(b): the inequality $1 \le k \le 1/\alpha$ should be $1 \le k < 1/\alpha$.*

2. Assumption 8.3.2 was introduced by Wessels [1], and it has been used by other authors, including Piunovski [1], and Wakuta [1]. On the other hand, van Nunen and Wessels [1] show that Assumption 8.3.2 is implied by the following condition introduced by Lippman [1]:

(L) There is a measurable function $w_0 \geq 1$ on X, a positive integer m, and positive constants b and M such that for all $(x, a) \in \mathbb{K}$:

$$|c(x, a)| \leq M w_0(x)^m$$

and

$$\int_X w_0^n(y) Q(dy | x, a) \leq [w_0(x) + b]^n \quad \text{for} \quad n = 1, \dots, m.$$

3. Also the condition (8.3.6) has been used by several authors; see, for instance, Bensoussan [1], Bhattacharya and Majumdar [1], Cavazos-Cadena [1].

4. If in Assumption 8.3.2 we allow β to be *less than* 1, then (8.3.29) would yield a *contradiction*. Namely, letting $t \to \infty$ in (8.3.29) we get $1 \leq 0$, since $w \geq 1$.

5. Let $\mathbb{K}$ be as in (8.2.1). Then for every pair (x, a) in $\mathbb{K}$ there is a decision function $f \in \mathbb{F}$ such that $a = f(x)$. (See Rieder [1], Example 2.6.) This fact and Proposition 8.3.9(b) yield that we can rewrite $T_\alpha u$ in (8.3.17) as

$$T_\alpha u(x) = \inf_{\mathbb{F}} \left[c(x, f) + \alpha \int u(y) Q(dy | x, f) \right], \quad x \in X, \tag{8.3.44}$$

for every function u in $\mathbb{B}_w(X)$.

8.4 Further analysis of value iteration

For each $n = 1, 2, \dots$, let $\mathbb{F}_n$ be the family of selectors $f \in \mathbb{F}$ for which $f(x) \in A(x)$ attains the minimum in (8.3.12); that is, $f \in \mathbb{F}_n$ if and only if

$$v_n(x) = c(x, f) + \alpha \int_X v_{n-1}(y) Q(dy | x, f) \quad \forall x \in X. \tag{8.4.1}$$

A deterministic Markov policy $\pi = \{f_n\} \in \Pi_{DM}$ such that f_n is in $\mathbb{F}_n$ for all $n = 1, 2, \dots$ is called an α-**value iteration** (α-VI) **policy**. The selector $f_0 \in \mathbb{F}$ may be arbitrarily chosen. On the other hand, the family of selectors f_* that satisfy (8.3.16) is denoted by $\mathbb{F}_*$. Thus, in view of Theorem 8.3.6(b), f_* belongs to $\mathbb{F}_*$ if and only if the deterministic stationary policy $f_*^\infty \in \Pi_{DS}$ is α-discount optimal. By (8.3.15), one would expect that $\mathbb{F}_n$ is "close" to $\mathbb{F}_*$ for all n sufficiently large. The question is, "close" in what sense? To deal with this question, we first consider the notion of "asymptotic discount optimality" (already introduced in §4.6). Then we give an estimate for the difference

$$V(f_n^\infty, \cdot) - V^*(\cdot),$$

which is used to introduce *rolling horizon* policies, and, finally, we consider the problem of existence and detection of *forecast horizons*, which requires in particular a *criterion to eliminate nonoptimal actions*.

Throughout this section, Assumptions 8.3.1, 8.3.2 and 8.3.3 are supposed to be true.

A. Asymptotic discount optimality

Let $D : \mathbb{K} \to \mathbb{R}$ be the α-**discount discrepancy function** defined by

$$D(x, a) := c(x, a) + \alpha \int_X V^*(y)Q(dy|x, a) - V^*(x). \qquad (8.4.2)$$

From the α-DCOE (8.3.4) we can see that D is a *nonnegative* function and that (8.3.4) can be rewritten as

$$\min_{A(x)} D(x, a) = 0 \quad \forall x \in X. \qquad (8.4.3)$$

Furthermore, by Theorem 8.3.6(b), a deterministic stationary policy f_*^∞ is α-discount optimal if and only if [using the notation in Remark 8.2.3(b)]

$$D(x, f_*) = 0 \quad \forall x \in X. \qquad (8.4.4)$$

Motivated by (8.4.4), we introduce the following concept.

8.4.1 Definition. A deterministic Markov policy $\pi = \{f_n\}$ is called **pointwise asymptotically discount optimal** (pointwise-ADO) if, for every state $x \in X$,

$$D(x, f_n) \to 0 \quad \text{as} \quad n \to \infty. \qquad (8.4.5)$$

8.4.2 Proposition. Let $\pi = \{f_n\}$ be an α-VI policy, that is, f_n is in $\mathbb{F}_n$ for all $n = 1, 2, \ldots$, and $f_0 \in \mathbb{F}$ is an arbitrary selector. Then π is a pointwise-ADO policy; in fact, for every $x \in X$ and $n = 1, 2, \ldots$,

$$0 \le D(x, f_n) \le 2\bar{c}\gamma^n w(x)/(1 - \gamma) \to 0, \quad \text{with } \gamma := \alpha\beta, \qquad (8.4.6)$$

where $\bar{c}, \beta$ and $w(\cdot)$ are as in Assumption 8.3.2.

Proof. By (8.4.2) and (8.4.1)

$$D(x, f_n) = c(x, f_n) + \alpha \int V^*(y)Q(dy|x, f_n) - V^*(x) \qquad (8.4.7)$$

$$= v_n(x) - V^*(x) + \alpha \int [V^*(y) - v_{n-1}(y)]Q(dy|x, f_n)$$

for all $x \in X$ and $n = 1, 2, \ldots$. Moreover, by (8.3.15),

$$|v_n(x) - V^*(x)| \leq \bar{c}\gamma^n w(x)/(1 - \gamma) \tag{8.4.8}$$

and, similarly, by (8.3.15) and (8.3.5),

$$\int |V^*(y) - v_{n-1}(y)|Q(dy|x, f_n) \leq \bar{c}\gamma^{n-1}(1 - \gamma)^{-1} \int w(y)Q(dy|x, f_n)$$
$$\leq \bar{c}\gamma^{n-1}(1 - \gamma)^{-1}\beta w(x).$$

Combining these inequalities with (8.4.7) we obtain (8.4.6). □

B. Estimates of VI convergence

In view of (8.4.3) and (8.4.4), *we can interpret (8.4.6) as an estimate of how "close" is $f_n \in \mathbb{F}_n$ to being optimal.* Another estimate can be obtained if we consider the *deterministic stationary policy* $f_n^\infty = \{f_n, f_n, \ldots\}$, which uses the control action $a_t = f_n(x_t)$ for all $t = 0, 1, \ldots$, and compute the difference between the corresponding infinite-horizon discounted cost $V(f_n^\infty, \cdot)$ and the α-discount value function V^*. In this case we obtain the following.

8.4.3 Proposition. *Fix an arbitrary integer $n \geq 1$ and let f_n be a selector in $\mathbb{F}_n$. Then, for all $x \in X$,*

$$0 \leq V(f_n^\infty, x) - V^*(x) \leq 2\bar{c}\gamma^n w(x)/(1 - \gamma) \tag{8.4.9}$$

with $\gamma := \alpha\beta$. Multiplying by $w(x)^{-1}$ in (8.4.9) [and (8.4.6)] we obtain estimates that are uniform in the w-norm.

Proof. As f_n is *fixed*, let us write $\widehat{f} := f_n$, and $\widehat{v}(x) := V(\widehat{f}^\infty, x)$, so that we wish to estimate

$$0 \leq \widehat{v}(x) - V^*(x) = [\widehat{v}(x) - v_n(x)] + [v_n(x) - V^*(x)].$$

Thus, by the inequality (8.4.8), we will obtain (8.4.9) if we show that

$$|\widehat{v}(x) - v_n(x)| \leq \bar{c}\gamma^n w(x)/(1 - \gamma). \tag{8.4.10}$$

To prove the latter inequality, first note that [as in (8.3.40)] we have

$$\widehat{v}(x) = c(x, \widehat{f}) + \alpha \int \widehat{v}(y)Q(dy|x, \widehat{f}),$$

which together with (8.4.1) gives

$$\widehat{v}(x) - v_n(x) = \alpha \int [\widehat{v}(y) - v_{n-1}(y)]Q(dy|x, \widehat{f}),$$

and so

$$|\widehat{v}(x) - v_n(x)| \le \alpha \int |\widehat{v}(y) - v_{n-1}(y)| Q(dy|x, \widehat{f}). \qquad (8.4.11)$$

Iteration of (8.4.11) gives [recalling the notation (8.2.11) and that $v_0 \equiv 0$]

$$\begin{aligned} |\widehat{v}(x) - v_n(x)| &\le \alpha^n \int |\widehat{v}(y)| Q^n(dy|x, \widehat{f}) \\ &= \alpha^n E_x^{\widehat{f}^\infty} |\widehat{v}(x_n)|, \end{aligned}$$

so that

$$|\widehat{v}(x) - v_n(x)| \le \alpha^n \|\widehat{v}\|_w E_x^{\widehat{f}^\infty} w(x_n). \qquad (8.4.12)$$

From the latter inequality, together with (8.3.29) and (8.3.33), we obtain (8.4.10) and, as we already mentioned, (8.4.9) follows. $\square$

C. Rolling horizon procedures

The ideal goal in optimal control problems is, of course, to explicitly determine the optimal value function and an optimal control policy. Unfortunately, this goal is quite often very "difficult"—if perhaps not impossible—to obtain. Thus, there are many cases in which one prefers to use a *suboptimal* but more *practical* procedure, provided its global performance can be assessed and compared with that of an optimal policy. We shall now discuss one such procedure, which is of frequent use in engineering and economics applications, such as stabilization of control systems, production management, and economic growth and macroplanning problems.

In a **rolling horizon** (RH) procedure—also known as a *moving* or *receding* or *sliding* horizon procedure—we begin by fixing a positive integer N, which is called the *rolling horizon*, and proceed as follows:

Step 1. Set $k = 0$ and determine an optimal control policy $\pi_{k,N} = \{f_{k,t}, t = k, k+1, \ldots, k+N-1\}$ for the N-stage problem starting at time k; in other words, $\pi_{k,N}$ minimizes the N-stage cost [cf. (3.3.14)]

$$V_{k,N}(\pi, x) := E^\pi \left[\sum_{t=k}^{k+N-1} \alpha^{k-t} c(x_t, a_t) | x_k = x \right]. \qquad (8.4.13)$$

Define $\widehat{f}_k := f_{k,k}$, the first optimal decision function for the N-stage problem.

Step 2. Substitute k by $k+1$ and go back to step 1.

This procedure determines a control policy $\widehat{\pi} = \{\widehat{f}_0, \widehat{f}_1, \ldots\}$ for the original infinite-horizon problem, and to validate the procedure the question then is to find error bounds for the "degree of suboptimality" of the *RH* policy $\hat{\pi}$, measured by the difference $V(\hat{\pi}, \cdot) - V^*(\cdot)(\geq 0)$, where V^* is the optimal value function in (8.3.2). However, as our Markov control model $(X, A, \{A(x)|x \in X\}, Q, c)$ is *stationary*, in the sense that all of its components $X, A, \ldots$, are time-independent, we can see that, for any $k = 0, 1, \ldots$, minimizing $V_{k,N}$ in (8.4.13) is equivalent to minimizing the N-stage cost V_N given by (8.3.14), namely,

$$V_N(\pi, x) = E_x^\pi \left[\sum_{t=0}^{N-1} \alpha^t c(x_t, a_t) \right].$$

Therefore, by (8.3.13) and (8.4.1), $\hat{f}_k = f_N$ for all $k \geq 0$, where f_N is a selector in $\mathbb{F}_N$, and so the cost function $V(\hat{\pi}, \cdot)$ is the same as $V(f_N^\infty, \cdot)$. Consequently, Proposition 8.4.3 yields:

8.4.4 Corollary. For any rolling horizon N,

$$0 \leq V(\hat{\pi}, x) - V^*(x) \leq 2\bar{c}\gamma^N w(x)/(1 - \gamma) \quad \forall x \in X. \tag{8.4.14}$$

D. Forecast horizons and elimination of nonoptimal actions

There is another way of interpreting the "closeness" of $\mathbb{F}_n$ to $\mathbb{F}_*$: By Theorem 4.6.5 (see Note 4 at the end of this section), if $\pi = \{f_n\}$ is an α-VI policy, then there exists an optimal policy $f_* \in \mathbb{F}_*$ such that, for each state $x \in X$, $f_*(x)$ is an *accumulation point* of the sequence $\{f_n(x)\} \subset A(x)$; that is, for every $x \in X$, there is a subsequence $\{n_i\} \equiv \{n_i(x)\}$ of $\{n\}$ such that

$$f_{n_i}(x) \to f_*(x) \quad \text{as} \quad i \to \infty. \tag{8.4.15}$$

[Theorem 4.6.5 requires the action set A to be locally compact. However, in our present context this requirement can be omitted because of Assumption 8.3.1(a).]

In turn, (8.4.15) suggests that there might be some control problems for which, for all n sufficiently large, either

$$\mathbb{F}_n \subset \mathbb{F}_* \tag{8.4.16}$$

or

$$\mathbb{F}_n \supset \mathbb{F}_*, \tag{8.4.17}$$

or even

$$\mathbb{F}_n = \mathbb{F}_* \tag{8.4.18}$$

If there is a positive integer N^* such that (8.4.16) holds for all $n \geq N^*$, then N^* is called a **forecast horizon**. Since $f_n \in \mathbb{F}_n$ is the first optimal decision function for the n-stage problem [see (8.3.12)–(8.3.14)], the existence of a forecast horizon N^* means that, roughly speaking, N^* is a finite horizon that is far enough off that the data beyond it—namely, the "forecasts"—have no effect on the optimal decisions in the initial period. Unfortunately, the existence of forecast horizons requires strong assumptions (even in the case of a *finite* state space X—see Shapiro [1]) and, in fact, for general (Borel) state spaces we need the more restrictive notion in part (b) of the following definition.

8.4.5 Definition. (a) For every state $x \in X$, let $A_*(x)$ be the set of control actions $a \in A(x)$ for which the minimum is attained in (8.4.3) [or (8.3.4)]; that is a is in $A_*(x)$ if

$$V^*(x) = c(x, a) + \alpha \int V^*(y) Q(dy | x, a).$$

Similarly, for $n \geq 1$, $A_n(x)$ denotes the set of actions $a \in A(x)$ that attain the minimum in (8.3.12), i.e.,

$$v_n(x) = c(x, a) + \alpha \int v_{n-1}(y) Q(dy | x, a) \quad \text{if} \quad a \in A(x).$$

Hence a deterministic stationary policy f^∞ is α-optimal, that is, f is in $\mathbb{F}_*$, if and only if $f(x)$ is in $A_*(x)$ for all x; and, on the other hand, f is in $\mathbb{F}_n$ if and only if $f(x)$ is in $A_n(x)$ for all x.

(b) Let $x \in X$ be a given (initial) state, and let $\pi = \{f_n\}$ be an α-VI policy. Then a positive integer N is said to be a (x, π)-**forecast horizon** if $f_n(x)$ is in $A_*(x)$ for all $n \geq N$, i.e.,

$$V^*(x) = c(x, f_n) + \alpha \int_X V^*(y) Q(dy | x, f_n) \quad \forall n \geq N. \tag{8.4.19}$$

8.4.6 Proposition. (a) *If*

$$\text{the action set } A \text{ is finite,} \tag{8.4.20}$$

then for every initial state $x \in X$ and every α-VI policy $\pi = \{f_n\}$ there exists a (x, π)-forecast horizon N_1. If, in addition, the state space X is also finite, then N_1 is independent of x and π; in other words, N_1 is a forecast horizon in the sense that (8.4.16) holds for all $n \geq N_1$.

(b) *In addition to (8.4.20), suppose that there exists a <u>unique</u> α-optimal control policy f_*^∞; that is, $\mathbb{F}_*$ consists of the single selector $f_* \in \mathbb{F}$:*

$$\mathbb{F}_* = \{f_*\}. \tag{8.4.21}$$

Then for every initial state $x \in X$, there exists a positive integer $N_2 = N_2(x)$ such that $f_*(x)$ is in $A_n(x)$ for all $n \geq N_2$ [cf. (8.4.17)]; in other words,

$$v_n(x) = c(x, f_*) + \alpha \int_X v_{n-1}(y) Q(dy | x, f_*) \quad \forall n \geq N_2. \tag{8.4.22}$$

Thus, if X is finite, there exists $N \geq 1$ such that (8.4.17) and (8.4.18) hold for all $n \geq N$.

Proof. (a) Fix x and $\pi = \{f_n\}$, and suppose that (a) does not hold; that is, for every positive integer $N \geq 1$ there exists $n \geq N$ such that, instead of (8.4.19), we have

$$V^*(x) < c(x, f_n) + \alpha \int V^*(y) Q(dy | x, f_n).$$

Equivalently, there is a subsequence $\{m\}$ of $\{n\}$ such that

$$V^*(x) < c(x, f_m) + \alpha \int V^*(y) Q(dy | x, f_m) \quad \forall m,$$

and, furthermore, as π is an α-VI policy,

$$v_m(x) = c(x, f_m) + \alpha \int v_{m-1}(y) Q(dy | x, f_m) \quad \forall m.$$

On the other hand, as $A(x)$ *is a finite set* [by (8.4.20)], there is a further subsequence $\{m_i\}$ of $\{m\}$ and a control action $a_x \in A(x)$ such that

$$f_{m_i}(x) = a_x \quad \forall i, \quad \text{and} \quad a_x \notin A_*(x).$$

In other words,

$$v_{m_i}(x) = c(x, a_x) + \alpha \int v_{m_i-1}(y) Q(dy | x, a_x) \quad \forall i, \tag{8.4.23}$$

and

$$V^*(x) < c(x, a_x) + \alpha \int V^*(y) Q(dy | x, a_x). \tag{8.4.24}$$

Thus, letting $i \to \infty$ in (8.4.23), from (8.3.15) and (8.3.20) we get

$$V^*(x) = c(x, a_x) + \alpha \int V^*(y) Q(dy | x, a_x), \tag{8.4.25}$$

which contradicts (8.4.24). This proves the first part of (a); that is, there exists a (x, π)-forecast horizon, say $N_1(x, \pi)$.

Now, *if A and X are both finite,* then there are finitely many α-VI policies π. Hence, $N_1 := \max_{x,\pi} N_1(x, \pi)$ defines a forecast horizon.

(b) Fix the initial state x and suppose that (8.4.20) and (8.4.21) both hold. If (8.4.22) is not satisfied, then [arguing as in the proof of (a)] there is a subsequence $\{n_i\}$ of $\{n\}$, and controls $a_{n_i} \in A_{n_i}(x)$ and $a_x \in A(x)$ such that $a_{n_i} = a_x$ for all i, and

$$a_x \neq f_*(x). \tag{8.4.26}$$

That is, for all i,

$$
\begin{aligned}
v_{n_i}(x) &= c(x, a_x) + \alpha \int v_{n_i-1}(y)Q(dy|x, a_x) \\
&< c(x, f_*) + \alpha \int v_{n_i-1}(y)Q(dy|x, f_*).
\end{aligned}
\tag{8.4.27}
$$

Therefore, letting $i \to \infty$, from (8.3.15) and (8.3.20) we obtain

$$
\begin{aligned}
V^*(x) &= c(x, a_x) + \alpha \int V^*(y)Q(dy|x, a_x) \\
&\leq c(x, f_*) + \alpha \int V^*(y)Q(dy|x, f_*) \\
&= V^*(x) \quad \text{since } f_*(x) \text{ is in } A_*(x).
\end{aligned}
\tag{8.4.28}
$$

This means that a_x belongs to $A_*(x)$, which, by (8.4.21), yields $a_x = f_*(x)$. As this contradicts (8.4.26), we conclude that (8.4.22) holds for some $N_2(x)$.

Finally, if X is a *finite* set, we get (8.4.17) for all $n \geq N_2 := \max_X N_2(x)$, and also (8.4.18) for all $n \geq \max\{N_1, N_2\}$, with N_1 as in the second part of (a). ⊐

Of course, for Proposition 8.4.6(a) to be of any practical use we need a method to "find", or at least "estimate" N_1; this is called the **detection** (of a forecast horizon) **problem**. To deal with it, we need in turn some criterion to eliminate **nonoptimal actions**, that is, actions which do not belong to $A_*(x)$. To do this, we will use the α-**VI discrepancy functions** $D_n : \mathbb{K} \to \mathbb{R}$ defined—in analogy with (8.4.2)—by

$$
D_n(x, a) := c(x, a) + \int_X v_{n-1}(y)Q(dy|x, a) - v_n(x),
\tag{8.4.29}
$$

for $n = 1, 2, \ldots$. Observe that, by (8.3.12), the functions D_n are *nonnegative* and that (8.3.12) can be rewritten as

$$
\min_{A(x)} D_n(x, a) = 0, \quad x \in X.
\tag{8.4.30}
$$

Further, by (8.3.15) and (8.3.20),

$$
\lim_{n \to \infty} D_n(x, a) = D(x, a).
\tag{8.4.31}
$$

We also have (with $\bar{c}$ and β as in Assumption 8.3.2, and $\gamma := \alpha\beta$):

8.4.7 Proposition. (Criterion for elimination of nonoptimal actions.) *An admissible control $a \in A(x)$ is nonoptimal in state x, that is, $a \notin A_*(x)$, if and only if there is a positive integer $n(a)$ such that*

$$
D_{n(a)}(x, a) \geq 2\bar{c}\gamma^{n(a)-1}w(x)/(1 - \gamma).
\tag{8.4.32}
$$

Proof. First we prove that, for any $n \geq 1$,

$$D_{n+1}(x,a) \geq D_n(x,a) - 2\bar{c}\gamma^n w(x). \tag{8.4.33}$$

To see this, use (8.4.29) to obtain

$$D_{n+1}(x,a) - D_n(x,a) = \alpha \int [v_n(y) - v_{n-1}(y)]Q(dy|x,a) - [v_{n+1}(x) - v_n(x)]. \tag{8.4.34}$$

On the other hand, as in (8.4.11), one can show that

$$|v_n(x) - v_{n-1}(x)| \leq \bar{c}\gamma^{n-1}w(x) \quad \forall n \geq 1. \tag{8.4.35}$$

From the latter inequality, and using (8.3.5), we get

$$\alpha \int [v_n(y) - v_{n-1}(y)]Q(dy|x,a) \geq -\bar{c}\gamma^n w(x) \quad \forall n. \tag{8.4.36}$$

Finally, from (8.4.34)–(8.4.36), a straightforward calculation yields (8.4.33). Now, iteration of (8.4.33) yields, for all $m, n \geq 1$,

$$D_{n+m}(x,a) \geq D_n(x,a) - 2\bar{c}\gamma^n w(x) \sum_{k=0}^{m} \gamma^k;$$

hence,

$$D_{n+m}(x,a) \geq D_n(x,a) - 2\bar{c}\gamma^n w(x)/(1-\gamma) \quad \forall n, m \geq 1. \tag{8.4.37}$$

We are now ready to prove the proposition itself.

Suppose that (8.4.32) holds, and take $n := n(a)$ in (8.4.37). This gives

$$D_{n+m}(x,a) \geq 2\bar{c}\gamma^{n-1}w(x) \quad \text{for all} \quad m \geq 1,$$

so that

$$D(x,a) = \lim_{m \to \infty} D_{n+m}(x,a) \geq 2\bar{c}\gamma^{n-1}w(x) > 0;$$

hence, a is not in $A_*(x)$. Conversely, if (8.4.32) does not hold, then

$$0 \leq D_n(x,a) < 2\bar{c}\gamma^{n-1}w(x)/(1-\gamma) \to 0 \quad \text{as} \quad n \to \infty.$$

Therefore, $D(x,a) = 0$ [by (8.4.31)], which means $a \in A_*(x)$. $\square$

If $a \in A(x)$ belongs to $A_*(x)$, so that $D(x,a) = 0$, then in view of (8.4.31) we may define $n(a) := 0$. Then Proposition 8.4.7 yields the following.

8.4.8 Corollary. *Suppose that (8.4.20) holds, that is, A is a finite set, and let $x \in X$ be a given initial state. For every admissible action $a \in A(x)$ let $n(a)$ be as in Proposition 8.4.7 if a is not in $A_*(x)$, and $n(a) := 0$ otherwise. Moreover, define*

$$N^* := \max\{n(a)|a \in A(x)\}. \tag{8.4.38}$$

Then N^* is finite and it is a (x, π)-forecast horizon for every α-VI policy π.

Proof. Let $\pi = \{f_n\}$ be an arbitrary α-VI policy, and let $a := f_n(x)$. If a is *not* in $A_*(x)$, then as $A(x) \subset A$ is a *finite* set, it is clear that $n(a)$ is also finite, and so is N^*. Further, if $n \geq N^*$, then a is necessarily in $A_*(x)$ because, otherwise, $D_n(x, a) = 0$ would contradict (8.4.32). □

Finally, if both conditions (8.4.20) and (8.4.21) hold, then Proposition 8.4.7 and Corollary 8.4.8 yield the following **(on-line) algorithm to detect N^* in (8.4.38) and the optimal selector f_* in (8.4.21)** for a given initial state $x \in X$:

Initialization. Let $n = 0$, and define $\widehat{A}_0 := A(x)$.

If $\widehat{A}_0$ has a single element, say a_*, then stop: $a_* = f_*(x)$. Otherwise, go to step $n = 1$.

Step n: For every a in $\widehat{A}_{n-1}$ compute $D_n(x, a)$. If

$$D_n(x, a) \geq 2\bar{c}\gamma^{n-1}w(x)/(1 - \gamma),$$

then eliminate a from $\widehat{A}_{n-1}$, and define

$$\widehat{A}_n := \{a \in A(x) | D_n(x, a) < 2\bar{c}\gamma^{n-1}w(x)/(1 - \gamma)\}.$$

If $\widehat{A}_n$ consists of a single element a_*, stop: $a_* = f_*(x)$. Otherwise, go to step $n + 1$.

Under the conditions (8.4.20) and (8.4.21), this algorithm is ensured to stop after a finite number of steps, and the stopping time N^* will be a (x, π)-forecast horizon for any α-VI policy π and any given initial state x. Furthermore, if the state space X is *finite*, then repeating the algorithm for every $x \in X$, one can get a forecast horizon in the sense of (8.4.16).

Notes on §8.4

1. For further properties and several characterizations of asymptotic discount optimality, see §§4.5, 4.6.

2. Proposition 8.4.3 and its application to rolling horizon (RH) policies are from Hernández-Lerma and Lasserre [8], [9], although the former reference deals only with *bounded* cost functions. These references and also, for instance, Alden and Smith [1] consider nonhomogeneous MCPs. For applications of RH policies in economics see Easley and Spulber [1], and Johansen [1]; for the stabilization of control systems see Kleinman [1] and Kwon *et al.* [1].

3. The results in subsection D, on forecast horizons and elimination of nonoptimal policies, are basically an extension of Bes and Lasserre [1] and Hernández-Lerma and Lasserre [10], which deal with *bounded* costs. For related references and applications see the survey by Bes and Sethi [1], and Rempala [1]. Perhaps the first paper on forecast horizons for (finite-state, finite-action) Markov control problems was the work of Shapiro [1], which soon afterwards was improved by Hinderer and Hübner [1]. In mathematical economics, "turnpike theorems" refer to results on asymptotic properties of optimal paths of capital accumulation of economic growth (see McKenzie [1]), and, by extension, forecast horizons are also known as *turnpike-planning horizons*.

4. Theorem 4.6.5 referred to at the beginning of §8.4.D is based on the following result by M. Schäl [1]:

Let $A(x)$ and $\mathbb{K}$ be as in Assumption 8.3.1(a) and (8.2.1), respectively, and let $\{f_n\}$ be a sequence in $\mathbb{F}$. Then there exists a selector $f \in \mathbb{F}$ such that, for each state $x \in X$, $f(x) \in A(x)$ is an accumulation point of the sequence $\{f_n(x)\}$.

8.5 The weakly continuous case

From the point of view of applications, perhaps the most restrictive of the hypotheses in §8.3.A is the "strong continuity" condition in Assumption 8.3.1(c), which requires the function

$$u'(x,a) := \int_X u(y)Q(dy|x,a), \quad (x,a) \in \mathbb{K}, \qquad (8.5.1)$$

to be continuous in $a \in A(x)$ for every $x \in X$ and every bounded *measurable* function $u \in \mathbb{B}(X)$. This condition, which is similar to the *strong Feller property* (Definition 7.3.1), is *equivalent* to require:

$$Q(B|x,a) = \text{Prob}(x_{t+1} \in B|x_t = x,\, a_t = a) \qquad (8.5.2)$$

is continuous in $a \in A(x)$ for every $x \in X$, every Borel set $B \subset X$, and $t = 0, 1, \ldots$, and it is certainly too much to ask for a large class of control models. In this section we replace Assumption 8.3.1(c) by the "weak continuity" (*weak Feller*-like) condition in Assumption 8.5.1(c) but then to obtain the corresponding results in §§8.3 and 8.4 we need to pay a price; namely, we have to *strengthen* other parts of Assumptions 8.3.1, 8.3.2, 8.3.3. The reason for this strenghtening will be apparent below.

In this section we replace Assumptions 8.3.1–8.3.3 by the following Assumptions 8.5.1–8.5.3.

8.5.1 Assumption.

(a) $A(x)$ is compact for every $x \in X$, and the set-valued mapping $x \mapsto A(x)$ is *u.s.c.*;

(b) The cost-per-stage c *is l.s.c. on* $\mathbb{K}$; and

(c) Q is *weakly continuous* on $\mathbb{K}$, that is, the function $u'(x,a)$ in (8.5.1) is continuous on $\mathbb{K}$ for every bounded *continuous* function u on X.

8.5.2 Assumption. This is the same as Assumption 8.3.2 except that the function $w \geq 1$ is required to be *continuous*.

8.5.3 Assumption. The function $w'(x,a)$ in Assumption 8.3.3 is *continuous* on $\mathbb{K}$.

Under this new set of assumptions (being basically a strengthening of Assumptions 8.3.1, 8.3.2, 8.3.3) *Theorem 8.3.6 remains valid*, but *in addition* we get that

$$V^* \quad \text{is a l.s.c. function in} \quad \mathbb{B}_w(X). \qquad (8.5.3)$$

To obtain (8.5.3) we need to make some changes in the proof of Theorem 8.3.6. First we introduce the following notation.

8.5.4 Definition. $\mathbb{L}(X)$ denotes the family of l.s.c. functions on X, and $\mathbb{L}_w(X)$ stands for the subfamily of l.s.c. functions that also belong to $\mathbb{B}_w(X)$, i.e.,

$$\mathbb{L}_w(X) := \mathbb{L}(X) \cap \mathbb{B}_w(X).$$

Similarly, $\mathbb{C}(X) \subset \mathbb{L}(X)$ denotes the subfamily of continuous functions on X, and $\mathbb{C}_w(X) := \mathbb{C}(X) \cap \mathbb{B}_w(X)$ is the subfamily of continuous functions in $\mathbb{B}_w(X)$. The family of continuous bounded functions on X is denoted by $C_b(X)$.

To prove (8.5.3) we need the following lemma, whose parts (a) and (b) correspond to Lemma 8.3.7(a) and Proposition 8.3.9, respectively. On the other hand, Lemma 8.5.5(c) states that convergence in w-norm preserves lower semicontinuity.

8.5.5 Lemma. Suppose that Assumptions 8.5.1, 8.5.2 and 8.5.3 are satisfied. Then:

(a) The function u' in (8.5.1) is continuous on $\mathbb{K}$ whenever u is in $\mathbb{L}_w(X)$;

(b) Proposition 8.3.9 remains valid if $\mathbb{B}_w(X)$ is replaced by $\mathbb{L}_w(X)$;

(c) If $\{v_n\}$ is a sequence in $\mathbb{L}_w(X)$ that converges in w-norm to a function v, then v is in $\mathbb{L}_w(X)$.

Proof. (a) The proof of this part is essentially the same as the proof of Lemma 8.3.7(a) with the obvious changes. Let u be a function in $\mathbb{L}_w(X)$ and define u_m as in the proof of Lemma 8.3.7(a). Then u_m is a nonnegative l.s.c. function and, therefore, there is a nondecreasing sequence of *continuous* bounded functions $u^k \in C_b(X)$ such that $u^k \uparrow u$. Now let (x^n, a^n) be a sequence in $\mathbb{K}$ converging to $(x,a) \in \mathbb{K}$. Then, Assumption 8.5.1(c) yields that, for every k,

$$\liminf_{n \to \infty} \int u_m(y) Q(dy | x^n, a^n) \quad \geq \quad \liminf_{n \to \infty} \int u^k(y) Q(dy | x^n, a^n)$$

$$= \int u^k(y)Q(dy|x,a).$$

Hence, letting $k \to \infty$ we get that $\int u_m(y)Q(dy|\cdot)$ is l.s.c., which together with Assumption 8.5.3 gives that u' is l.s.c. on $\mathbb{K}$. That is, u' is l.s.c. on $\mathbb{K}$ for every function u in $\mathbb{L}_w(X)$. If we now apply the latter fact to $-u$ we see that u' is also u.s.c. on $\mathbb{K}$.

(b) Since T_α satisfies the contraction property (8.3.23), to complete the proof of part (b) it only remains to show that

(b$_1$) $T_\alpha u$ is a *l.s.c.* function in $\mathbb{B}_w(X)$ for every u in $\mathbb{L}_w(X)$, and that

(b$_2$) for every $u \in \mathbb{L}_w(X)$, there exists $f \in \mathbb{F}$ that satisfies (8.3.24).

To prove (b$_1$) and (b$_2$), let u be a function in $\mathbb{L}_w(X)$ and note that, by part (a) and Assumption 8.5.1(c), the function within brackets in (8.3.7), namely,

$$c(x,a) + \alpha u'(x,a) =: v(x,a),$$

is l.s.c., and, in fact, under the present assumptions, v satisfies all of the hypotheses of Lemma 8.3.8(c). Hence, the latter lemma yields (b$_1$) and (b$_2$).

(c) Clearly, v is in $\mathbb{B}_w(X)$ if $\{v_n\}$ is a sequence in $\mathbb{L}_w(X)$ and v_n converges to v in w-norm. [Recall that $\mathbb{B}_w(X)$ *is a Banach space*—see Proposition 7.2.1.] Thus, it only remains to show that v is *l.s.c.* To prove this observe that

$$v(x) = [v(x) - v_n(x)] + v_n(x) \geq -\|v_n - v\|_w w(x) + v_n(x)$$

for all $x \in X$ and n. Therefore, if $x^k \to x$, the lower semicontinuity of each v_n and the continuity of w (Assumption 8.5.2) yield

$$\liminf_{k \to \infty} v(x^k) \geq -\|v_n - v\|_w w(x) + v_n(x) \quad \forall n.$$

Finally, letting $n \to \infty$ we obtain $\liminf_k v(x^k) \geq v(x)$; that is, v is l.s.c. $\square$

We can now see that (8.5.3) is a direct consequence of Lemma 8.5.5(b), (c) and (8.3.15). Namely, by Lemma 8.5.5(b) and a trivial induction argument, the α-VI functions v_n in (8.3.26) [or (8.3.12)] belong to $\mathbb{L}_w(X)$, which combined with Lemma 8.5.5(c) and (8.3.15) yields that V^* is in $\mathbb{L}_w(X)$; that is, (8.5.3) holds.

The previous paragraphs illustrate a situation already mentioned in §§4.2 and 3.3: in addition to the features of the particular control problem we are dealing with, *the choice of hypotheses basically depends on whether one wishes to* (or *can*) *work in a class of <u>lower semicontinuous</u> functions* (as is the case under Assumptions 8.5.1, 8.5.2, 8.5.3) *or in a class of <u>measurable</u> functions* (Assumptions 8.3.1, 8.3.2, 8.3.3).

8.6 Examples

In this section we present a couple of examples of Markov control models that satisfy the assumptions in §8.3 and §8.5. These examples are intended to illustrate how one can proceed in similar cases.

When considering an X-valued controlled process $\{x_t\}$ of the form

$$x_{t+1} = F(x_t, a_t, z_t), \quad t = 0, 1, \ldots, \tag{8.6.1}$$

we always suppose the following:

8.6.1 Assumption.

(a) The disturbance sequence $\{z_t\}$ consists of i.i.d. random variables with values in a Borel space Z, and $\{z_t\}$ is independent of the initial state x_0. The common distribution of the z_t is denoted by G.

(b) $F : \mathbb{K} \times Z \to X$ is a given measurable function, where $\mathbb{K} \subset X \times A$ is the set defined in (8.2.1).

Let $\tau = \{a_t\}$ be an arbitrary control policy (Definition 8.2.2). Then, by Assumption 8.6.1(a), the variables (x_t, a_t) and z_t are independent for each $t = 0, 1, \ldots$. Thus the controlled process' transition law Q is given by

$$
\begin{aligned}
Q(B|x,a) &:= \mathrm{Prob}(x_{t+1} \in B | x_t = x, \ a_t = a) \\
&= \int_Z I_B[F(x, a, z)] G(dz)
\end{aligned} \tag{8.6.2}
$$

for every $B \in \mathcal{B}(X)$, $(x,a) \in \mathbb{K}$, and $t = 0, 1, \ldots$. The expression (8.6.2) is of course a particular case of the integral

$$u'(x,a) := \int_X u(y) Q(dy|x,a) = E[u(x_{t+1}) | x_t = x, \ a_t = a] \tag{8.6.3}$$

when u is the indicator function I_B. In general, we may use (8.6.1) and Assumption 8.6.1(a) to write (8.6.3) as

$$u'(x,a) = \int_Z u[F(x, a, z)] G(dz). \tag{8.6.4}$$

8.6.2 Example: An inventory system. Consider an inventory-production system in which the state variable x_t, the control action a_t, and the disturbance z_t, for every $t = 0, 1, \ldots$, have the following meaning:

- x_t denotes the stock level at the beginning of period t;

- a_t is the amount of product ordered (and immediately supplied) at the beginning of period t;

- z_t denotes the product's demand during period t.

Using the standard notation $r^+ := \max(r, 0)$, we assume that the stock level evolves according to the equation [see (8.6.1)]

$$x_{t+1} = (x_t + a_t - z_t)^+, \quad t = 0, 1, \ldots, \tag{8.6.5}$$

for some given initial stock level x_0, so that the state space is the half-line $X := [0, \infty)$. The production variables a_t are supposed to take values in the interval $A := [0, \theta]$, for some given constant $\theta > 0$, irrespective of the stock level—that is, the control-constraint sets $A(x)$ satisfy

$$A(x) = A \quad \forall x \in X. \tag{8.6.6}$$

In addition, we suppose that the demand process $\{z_t\}$ satisfies Assumption 8.6.1 with $Z := [0, \infty)$, so that z_t is nonnegative for each t, and that the demand distribution G has the following properties:

8.6.3 (a) G has a continuous bounded density g [i.e., $G(dz) = g(z)dz$];
(b) G has a finite mean value $\bar{z}$, i.e.,

$$\bar{z} := E(z_0) = \int_0^\infty zG(dz) < \infty.$$

Finally, to complete the description of the control model (X, A, Q, c), where Q is given by (8.6.5) and (8.6.2), we shall consider a cost-per-stage function c that represents a *net cost* of the form

production cost + maintenance (or holding) cost − sales revenue

given by

$$c(x, a) := p \cdot a + m \cdot (x + a) - s \cdot E\min(x + a, z_0), \tag{8.6.7}$$

where p, m, and s are positive constants. The unit production p and the unit maintenance cost m do not exceed the unit sale price, i.e.,

$$p, m \leq s. \tag{8.6.8}$$

We shall now proceed to verify the assumptions in §8.3 and §8.5.

Verification of Assumption 8.3.1 and 8.5.1. It is clear that parts (a) and (b) are satisfied. In particular, since

$$E\min(x + a, z_0) = (x + a)[1 - G(x + a)] + \int_0^{x+a} zG(dz), \tag{8.6.9}$$

the cost function $c(x, a)$ is *continuous* on $\mathbb{K} := X \times A$. On the other hand, from (8.6.5), (8.6.3)–(8.6.4), and the property 8.6.3(a) we get

$$\begin{aligned}
u'(x, a) &= \int_0^\infty u[(x + a - z)^+]g(z)dz \\
&= u(0)[1 - G(x + a)] + \int_0^{x+a} u(x + a - z)g(z)dz.
\end{aligned} \tag{8.6.10}$$

Thus, an elementary change of variables in the latter integral yields

$$u'(x,a) = u(0)[1 - G(x+a)] + \int_0^{x+a} u(z)g(x+a-z)dz,$$

and so we see that $u'(x,a)$ is *continuous* in $(x,a) \in \mathbb{K}$ for every bounded *measurable* function u on X. This implies part (c) in both Assumptions 8.3.1 and 8.5.1.

Verification of Assumptions 8.3.2 and 8.5.2. It suffices to find a continuous weight function w that satisfies the conditions (i) and (ii) in Remark 8.3.5(a). To do this, let us first consider the moment generating function ψ of the variable $\theta - z_0$,

$$\psi(r) := E \exp[r(\theta - z_0)], \quad \text{for} \quad r \geq 0.$$

As $\psi(0) = 1$ and ψ is continuous, for each $\varepsilon > 0$ there is a positive number $\hat{r}$ such that

$$\psi(\hat{r}) \leq 1 + \varepsilon. \tag{8.6.11}$$

Define

$$w(x) := \exp[\hat{r}(x + 2\bar{z})], \quad x \in X. \tag{8.6.12}$$

Then, from (8.6.10) with $u = w$,

$$w'(x,a) = w(0)[1 - G(x+a)] + w(x) \int_0^{x+a} \exp[\hat{r}(a-z)]G(dz), \tag{8.6.13}$$

so that, since $1 - G(x+a) \leq 1$ and $\hat{r}(a-z) \leq \hat{r}(\theta - z)$ for all $a \in A$, we get

$$w'(x,a) \leq w(0) + \psi(\hat{r})w(x) \leq \beta w(x) + b \quad \forall x \in X, \tag{8.6.14}$$

with

$$\beta := 1 + \varepsilon \quad \text{and} \quad b := w(0).$$

On the other hand, a straightforward calculation using (8.6.8) and (8.6.9) shows that $\sup_A |c(x,a)| \leq s \cdot (x + 2\bar{z})$ for all $x \in X$, and, therefore,

$$\sup_A |c(x,a)| \leq mw(x) \tag{8.6.15}$$

for some constant m sufficiently large. Hence, as the function w in (8.6.12) is *continuous*, we see from (8.6.14) and (8.6.15) that the conditions (i) and (ii) in Remark 8.3.5(a) are both satisfied for any discount factor α such that $\beta < 1/\alpha$.

Verification of Assumptions 8.3.3 and 8.5.3. This follows from (8.6.13). □

Example 8.6.2 is due essentially to Vega-Amaya [1]—see also Hernández-Lerma and Vega-Amaya [1]. For an inventory example illustrating the results in §8.4.D see Bes and Lasserre [1]. Many additional references on inventory theory are given in §1.3 and §3.7.

8.6.4 Example: A queueing system. Consider a control system of the form

$$x_{t+1} = (x_t + a_t \eta_t - \xi_t)^+, \quad t = 0, 1, \ldots, \tag{8.6.16}$$

with state space $X = [0, \infty)$. This system is related to the inventory model in Example 8.6.2 [in (8.6.5) take $\eta_t \equiv 1$], and also to Example 7.4.2 [compare (8.6.16) and (7.4.8)]. In fact, as noted in the last paragraph of Example 7.4.2, a model of the form (7.4.8) [or (8.6.16)] can have several interesting interpretations. Here, we interpret (8.6.16) as modelling a single-server queueing system (of general type $GI/GI/1$) with controllable service rates. Thus, x_t and η_t denote, respectively, the waiting time and a "base" service time of the tth customer ($t = 0, 1, \ldots$), whereas ξ_t stands for the interarrival time between the tth and $(t+1)^{th}$ customers. The control variable a_t denotes the reciprocal of the service rate for the tth customer.

We shall suppose the following:

8.6.5 Assumptions on (8.6.16).

(a) The action (or control) set $A = A(x)$ for all $x \in X$ is a compact subset of an interval $(0, \theta]$ for some (finite) number θ.

(b) $\{\eta_t\}$ and $\{\xi_t\}$ are independent sequences of i.i.d. random variables.

(c) η_0 and ξ_0 have continuous bounded densities g_1 and g_2, respectively.

(d) The random variable $z := \theta \eta_0 - \xi_0$ has a (finite) negative mean and a moment generating function $\psi(r) := E(e^{rz})$ that is finite for some $\bar{r} > 0$, that is

$$\text{(i)} \;\; E(z) < 0, \quad \text{and} \quad \text{(ii)} \;\; \psi(\bar{r}) < \infty. \tag{8.6.17}$$

By Assumption 8.6.5(d), we have $\psi(0) = 1$ and $\psi'(0) = E(z) < 0$. Hence there is a positive number $r \leq \bar{r}$ such that

$$\psi(r) < 1.$$

For such a number r, we define the *continuous* weight function

$$w(x) := e^{rx}, \quad x \in X. \tag{8.6.18}$$

Moreover, we shall suppose that the associated cost-per-stage function $c(x, a)$ satisfies the Assumption 8.5.1(b) [hence 8.3.1(b)] and 8.3.2(a); that is, c is l.s.c. on $\mathbb{K} := X \times A$ and, for some constant $\bar{c}$,

$$\sup_A |c(x, a)| \leq \bar{c} w(x) \quad \forall x \in X. \tag{8.6.19}$$

Observe that this is not a restrictive condition. Namely, as A is compact, the condition (8.6.19) is bound to be satisfied, for $\bar{c}$ sufficiently large, for all the typical—say, polynomial—cost functions that appear in applications.

We shall now verify that the given queueing system satisfies the other assumptions in §8.3 and §8.5.

Verification of Assumptions 8.3.1–8.3.3, and 8.5.1–8.5.3. In view of the previous paragraphs, to complete the verification of Assumptions 8.3.1 and 8.5.1 it suffices to check that

8.6.6 Condition. *The function $u'(x,a) := \int u(y)Q(dy|x,a)$ is continuous and bounded on $\mathbb{K}$ for every* measurable *bounded function u on X.*

This requires some preliminary calculations.

For every $a \in A$, let $z_a := a\eta_0 - \xi_0$. Then, by Assumptions 8.6.5(b),(c), the probability distribution function of z_a is given, for every real number y, by

$$
\begin{aligned}
P(z_a \leq y) &= \int_0^\infty P(z_a \leq y | \eta_0 = s) g_1(s) ds \\
&= \int_0^\infty \int_{as-y}^\infty g_2(t) dt\, g_1(s) ds.
\end{aligned}
\tag{8.6.20}
$$

Hence denoting by g_a the density of z_a, we get

$$
g_a(y) = \int_{y/a}^\infty g_1(s) g_2(as - y) ds,
\tag{8.6.21}
$$

which, by Assumption 8.6.5(c) is a *bounded* function, *continuous in both variables $a \in A$ and $y \in \mathbb{R}$.* Observe that the latter property, *continuity in a and y*, is also satisfied by the distribution function in (8.6.21). This implies the condition 8.6.6 since, by (8.6.16) and (8.6.3),

$$
\begin{aligned}
u'(x,a) &= Eu[(x + z_a)^+] \\
&= u(0)P(x + z_a \leq 0) + \int_{-x}^\infty u(x + y) g_a(y) dy
\end{aligned}
\tag{8.6.22}
$$

and the latter integral can be written as

$$
\int_{-x}^\infty u(x + y) g_a(y) dy = \int_0^\infty u(y) g_a(y - x) dy.
$$

This completes the verification of Assumptions 8.3.1 and 8.5.1.

On the other hand, Assumptions 8.3.3 and 8.5.3 can also be deduced from (8.6.22) since replacing u by the weight function w [see (8.6.18)] yields

$$
w'(x,a) = P(z_a \leq -x) + w(x) \int_{-x}^\infty e^{ry} g_a(y) dy,
\tag{8.6.23}
$$

which is *continuous on* $\mathbb{K} = X \times A$.

Finally, to verify (8.3.5) observe that, by the Assumptions 8.6.5(a),(d),

$$z_a = a\eta_0 - \xi_0 \leq \theta\eta_0 - \xi_0 = z \quad \forall a \in A,$$

so that the integral in (8.6.23) satisfies

$$
\begin{aligned}
\int_{-x}^{\infty} e^{ry} g_a(y) dy &\leq \int_{-\infty}^{\infty} e^{ry} g_a(y) dy \\
&= E \exp(rz_a) \quad\quad\quad\quad (8.6.24) \\
&\leq \psi(r) \quad \forall(x, a) \in \mathbb{K}.
\end{aligned}
$$

Thus, as $P(z_a \leq -x) \leq 1 \leq w(x)$ for all (x, a) in $\mathbb{K}$, (8.6.23) implies that (8.3.5) holds with $\beta := 1 + \psi(r)$. The latter fact and (8.6.19) show that Assumptions 8.3.2 and 8.5.2 are satisfied for every discount factor $\alpha < 1/\beta$.

In conclusion, all of the results in §8.3 and §8.5 are applicable to the queueing system (8.6.16). In fact, many results in §8.4 are also applicable since the compactness of A in Assumption 8.6.5(a) includes, for instance, the condition (8.4.20). $\square$

Example 8.6.4 comes from Gordienko and Hernández-Lerma [1].

8.7 Further remarks

This chapter introduced a weighted-norm approach to discounted cost Markov control problems. A comparison with the hypotheses and results in Chapters 4 and 6 shows that each particular context or solution technique has its own merits. For instance, the results in §8.4 are virtually impossible to obtain in the setting of Chapters 4 and 6, but then §8.4 (and §8.3) requires more restrictive assumptions.

There are other ways to study the discounted problem. In particular, in later chapters we will see that it can be studied by a "direct approach", or as a "transient" control problem, or using finite-dimensional linear programming approximations.

9
The Expected Total Cost Criterion

9.1 Introduction

Let $\mathcal{M} = (X, A, \{A(x)|x \in X\}, Q, c)$ be the Markov control model (MCM) in §8.2. In this chapter we study the *expected total cost* (ETC) criterion defined as

$$V_1(\pi, x) := E_x^\pi \left[\sum_{t=0}^\infty c(x_t, a_t) \right] \quad \text{for } \pi \in \Pi, \ x \in X, \qquad (9.1.1)$$

so the corresponding (optimal) *value function* is

$$V_1^*(x) := \inf_\Pi V_1(\pi, x), \quad x \in X. \qquad (9.1.2)$$

As usual, the main problems we are concerned with are: (i) to "characterize" V_1^*—for instance, as the solution of a certain "optimality (or dynamic programming) equation", and (ii) to determine conditions for the existence of *ETC-optimal* policies, that is, policies π^* for which

$$V_1^*(x) = V_1(\pi^*, x) \quad \forall x \in X. \qquad (9.1.3)$$

The ETC criterion was probably the earliest infinite-horizon control problem studied in the literature, going back at least to the 1920s; see, for example, Ramsey [1]. It is obviously very demanding from the technical viewpoint because simply for $V_1(\pi, x)$ to be finite valued, or even to be well defined, we need to impose very strong assumptions on the MCM $\mathcal{M}$. In later chapters we will study other optimality criteria—such as "overtaking optimality"—that are less restrictive but which at the same time maintain some of the features of the ETC criterion.

The remainder of the chapter is organized as follows. For the sake of completeness and ease of reference, §9.2 summarizes some facts on extended real numbers and "quasi-integrability." In §9.3 we consider several theoretical questions, including the measurability of $V_1(\pi, \cdot)$ and $V_1^*(\cdot)$. Moreover, letting

$$V_\alpha(\pi, x) := E_x^\pi \left[\sum_{t=0}^{\infty} \alpha^t c(x_t, a_t) \right] \qquad (9.1.4)$$

be the α-discounted cost $(0 < \alpha < 1)$ in (8.3.1), we show that $V_\alpha(\pi, \cdot)$ converges to $V_1(\pi, \cdot)$ as $\alpha \uparrow 1$. In §9.4 we study the *sufficiency problem* in which the basic issue is to show that in (9.1.2) we may replace the set Π of *all* policies by the smaller set Π_{RM} of *randomized Markov* policies (Definition 8.2.2); in this case we say that Π_{RM} is a "complete" set of policies for the ETC criterion. The sufficiency problem requires the introduction of the ETC-expected occupation measures, which is a concept important in itself.

Section 9.5 gives conditions for the value function V_1^* to be a solution of the ETC-optimality equation, as well as conditions for a policy to be ETC-optimal. Finally, in §9.6 we study *transient* MCMs. This is an important class of models for which all the hypotheses in §9.3 and §9.5 are satisfied, and, therefore, one can make a very detailed analysis of many optimality-related questions.

9.2   Preliminaries

This section contains background material. The reader may skip the section and refer to it as needed.

A. Extended real numbers

In the set $\overline{\mathbb{R}} := \mathbb{R} \cup \{+\infty\} \cup \{-\infty\}$ of *extended real numbers*, we adopt the usual rules of arithmetic (where $\infty := +\infty$):

$$r + \infty = \infty + r = \infty, \quad \text{and} \quad r - \infty = -\infty + r = -\infty \quad \forall r \in \mathbb{R};$$
$$\infty + \infty = \infty, \quad -\infty - \infty = -\infty;$$
$$\qquad (9.2.1)$$

Observe that $\infty - \infty$ *is not defined*. Further,

$$r \cdot \infty = \infty \cdot r = \begin{cases} \infty & \text{if } r \in \overline{\mathbb{R}}, \ r > 0 \\ 0 & \text{if } r = 0 \\ -\infty & \text{if } r \in \overline{\mathbb{R}}, \ r < 0. \end{cases} \qquad (9.2.2)$$

The *positive* and *negative* parts of an extended real number r are defined as

$$r^+ := \max(r, 0) \quad \text{and} \quad r^- := \max(-r, 0), \tag{9.2.3}$$

respectively, and satisfy

$$r = r^+ - r^- \quad \text{and} \quad |r| = r^+ + r^-.$$

Let $\{r_n\}$ be a sequence in $\overline{\mathbb{R}}$ that may contain one of the numbers $+\infty$, $-\infty$, but not both. Then the "partial sum" $s_N := \sum_{n=0}^{N} r_n$ is well defined for each $N = 1, 2, \ldots$, and we say that *the series* $\sum_{n=0}^{\infty} r_n$ *converges (in* $\overline{\mathbb{R}}$) *to* $r \in \overline{\mathbb{R}}$ if the limit $\lim_{N \to \infty} s_N$ exists (in $\overline{\mathbb{R}}$) and equals r. For example, if

$$r_n \geq 0 \quad \text{for all} \quad n = 0, 1, \ldots, \tag{9.2.4}$$

then the series $\sum r_n$ converges in $\overline{\mathbb{R}}$ (the limit may be $+\infty$). On the other hand, if $\{r_n\}$ is such that

$$\sum_{n=0}^{\infty} r_n^+ < \infty \quad or \quad \sum_{n=0}^{\infty} r_n^- < \infty, \tag{9.2.5}$$

then the series $\sum r_n$ converges in $\overline{\mathbb{R}}$ to

$$\sum_{n=0}^{\infty} r_n = \sum_{n=0}^{\infty} r_n^+ - \sum_{n=0}^{\infty} r_n^-. \tag{9.2.6}$$

Moreover, if a series $\sum r_n$ converges in $\overline{\mathbb{R}}$, then

$$\left(\sum_{n=0}^{\infty} r_n \right)^{\pm} \leq \sum_{n=0}^{\infty} r_n^{\pm}. \tag{9.2.7}$$

This follows from the fact (easily verified by induction) that

$$\left(\sum_{n=0}^{N} r_n \right)^{\pm} \leq \sum_{n=0}^{N} r_n^{\pm} \quad \forall N = 0, 1, \ldots.$$

The following elementary proposition will be used to relate the α-discounted cost V_α in (9.1.4) with the expected total cost V_1 [see Proposition 9.3.3(b)].

9.2.1 Proposition. *Let $\{r_n\}$ be a sequence in $\overline{\mathbb{R}}$ and α (a "discount factor") in $(0, 1]$. If $\{r_n\}$ satisfies (9.2.4) or (9.2.5), then the series $\sum \alpha^n r_n$ converges in $\overline{\mathbb{R}}$ for every α in $(0, 1]$, and*

$$\lim_{\alpha \uparrow 1} \sum_{n=0}^{\infty} \alpha^n r_n = \sum_{n=0}^{\infty} r_n. \tag{9.2.8}$$

Proof. If the proposition is true under (9.2.4), then [by (9.2.5) and (9.2.6)] it is also true under (9.2.5). Now, to prove the proposition assuming (9.2.4) it suffices to note that $\alpha^n r_n \geq 0$ for all n and, further, the partial sums

$$\sum_{n=0}^{N} \alpha^n r_n$$

are *nondecreasing* in both variables $\alpha \in (0, 1]$ and $N = 0, 1, \ldots$. Therefore (by Remark 9.6.13), we can interchange the following limits

$$
\begin{aligned}
\lim_{\alpha \uparrow 1} \sum_{n=0}^{\infty} \alpha^n r_n &= \lim_{\alpha \uparrow 1} \lim_{N \to \infty} \sum_{n=0}^{N} \alpha^n r_n \\
&= \lim_{N \to \infty} \lim_{\alpha \uparrow 1} \sum_{n=0}^{N} \alpha^n r_n \\
&= \lim_{N \to \infty} \sum_{n=0}^{N} r_n,
\end{aligned}
$$

and (9.2.8) follows. $\square$

B. Integrability

Let $(\Omega, \mathcal{F}, P)$ be a probability space, and $\overline{\mathbb{R}}$ the set of extended real numbers. A random variable $\xi : \Omega \to \overline{\mathbb{R}}$ is said to be **integrable** (with respect to P) if

$$E(\xi^+) < \infty \quad \text{and} \quad E(\xi^-) < \infty.$$

In this case, the *expectation* (or *expected value*) of ξ is the real number

$$E(\xi) := E(\xi^+) - E(\xi^-). \tag{9.2.9}$$

On the other hand, ξ is called **quasi-integrable** if

$$E(\xi^+) < \infty \quad \text{or} \quad E(\xi^-) < \infty. \tag{9.2.10}$$

The *expectation* of a quasi-integrable random variable ξ is again defined by (9.2.9), which is now an *extended real number*.

A *nonnegative* random variable ξ is quasi-integrable [as $E(\xi^-) = 0 < \infty$]. A random variable ξ is integrable if and only if $E|\xi| < \infty$.

For a proof of parts (a) to (e) of the following proposition see, for instance, Neveu [1, p.41]; for a proof of (f) see Hinderer [1, p.146].

9.2.2 Proposition. *Let ξ and ξ_n $(n = 1, 2, \ldots)$ be quasi-integrable random variables. Then:*

(a) $E(k\xi) = kE(\xi)$ *for every finite constant k.*

(b) $E(\xi_1 + \xi_2) = E(\xi_1) + E(\xi_2)$ if $\xi_1 + \xi_2$ is defined [see (9.2.1)] and if ξ_1^+ and ξ_2^+ (or ξ_1^- and ξ_2^-) are integrable.

(c) $\xi_1 \le \xi_2$ implies $E(\xi_1) \le E(\xi_2)$.

(d) $\xi_n \uparrow \xi$ implies $E(\xi_n) \uparrow E(\xi)$ if ξ_n^- is integrable for at least one n.

(e) $\xi_n \downarrow \xi$ implies $E(\xi_n) \downarrow E(\xi)$ if ξ_n^+ is integrable for at least one n.

(f) Suppose that $\sum_n E(\xi_n^+) < \infty$ or $\sum_n E(\xi_n^-) < \infty$. Then:

(f.1) $\sum_n E(\xi_n)$ converges (in $\overline{\mathbb{R}}$) to $\sum_n E(\xi_n^+) - \sum_n E(\xi_n^-)$.

(f.2) $\sum_n \xi_n$ converges almost surely to a quasi-integrable random variable.

(f.3) $E(\sum_n \xi_n) = \sum_n E(\xi_n)$.

(f.4) $E(\sum_n \xi_n)^\pm \le \sum_n E(\xi_n^\pm)$.

9.3   The expected total cost

Let $\mathcal{M} = (X, A, \{A(x)|x \in X\}, Q, c)$ be the Markov control model in §8.2, and consider the **expected total cost** (ETC)

$$V_1(\pi, x) := E_x^\pi \left[\sum_{t=0}^\infty c(x_t, a_t) \right] \tag{9.3.1}$$

when using the policy π, given the initial state $x_0 = x$. The corresponding (optimal) **value function** is

$$V_1^*(x) := \inf_\Pi V_1(\pi, x), \quad x \in X. \tag{9.3.2}$$

To abbreviate, sometimes we shall write (9.3.1) as

$$V_1(\pi, x) = E_x^\pi \left(\sum_{t=0}^\infty c_t \right), \quad \text{with} \quad c_t := c(x_t, a_t). \tag{9.3.3}$$

The first step in our study of the ETC criterion will be to consider the following basic theoretical issues.

9.3.1 Questions

(a) Given a policy π, is $V_1(\pi, \cdot) : X \to \mathbb{R}$ (or $\overline{\mathbb{R}}$) a *measurable* function? Similarly,

(b) Is $V_1^* : X \to \mathbb{R}$ (or $\overline{\mathbb{R}}$) a *measurable* function?

(c) For each policy π and each initial state x, let $J_0(\pi, x) := 0$ and

$$J_n(\pi, x) := E_x^\pi \left[\sum_{t=0}^{n-1} c(x_t, a_t) \right], \quad n = 1, 2, \ldots, \quad (9.3.4)$$

be the **n-stage expected total cost**. The corresponding **optimal n-stage cost** is $J_0^*(\cdot) := 0$ and, for $n = 1, 2, \ldots,$

$$J_n^*(x) := \inf_\Pi J_n(\pi, x), \quad x \in X. \quad (9.3.5)$$

The question is, as $n \to \infty$, does J_n^* converge to V_1^*? In other words, we would like to find conditions under which

$$\lim_{n \to \infty} J_n^*(x) = V_1^*(x) \quad \forall x \in X. \quad (9.3.6)$$

(d) Does the α-discounted cost $V_\alpha(\pi, x)$ converge to $V_1(\pi, x)$ as $\alpha \uparrow 1$?

(e) To obtain the optimal value function V_1^* in (9.3.2), is it "sufficient" to minimize $V_1(\cdot, x)$ over a *subset* Π' of Π? If this happens to be true, we then say that Π' is a **sufficient** set of policies for the ETC problem.

In this section we give conditions under which each of the questions (a) to (d) has an affirmative answer. Question (e) is postponed to the next section since it requires some preliminary concepts and results.

First, concerning Question 9.3.1(a), we need to ensure that the series $\sum c_t$ in (9.3.3) is (at least) quasi-integrable with respect to P_x^π. Hence [as in (9.2.10)], we consider the expectations of *nonnegative series*

$$V_1^{(+)}(\pi, x) := E_x^\pi \left(\sum_{t=0}^\infty c_t^+ \right), \quad V_1^{(-)}(\pi, x) := E_x^\pi \left(\sum_{t=0}^\infty c_t^- \right), \quad (9.3.7)$$

and we suppose the following:

9.3.2 Assumption. For each $x \in X$

$$\sup_\Pi V_1^{(-)}(\pi, x) < \infty. \quad (9.3.8)$$

Observe that (9.3.8) trivially holds if the cost-per-stage $c(x, a)$ is *nonnegative*, in which case $c^- = 0$. On the other hand, under Assumption 9.3.2 we obtain, in particular, an *affirmative answer to Questions 9.3.1(a) and (d)*:

9.3.3 Proposition. *If Assumption 9.3.2 is satisfied, then:*

(a) $x \mapsto V_1(\pi, x)$ *is an extended real-valued measurable function on* X *for every policy* π, *and similarly for* $x \mapsto V_\alpha(\pi, x)$ *for each discount factor* α *in* $(0, 1)$.

Moreover,

(b) $\lim_{\alpha \uparrow 1} V_\alpha(\pi, x) = V_1(\pi, x)$ *for each* $\pi \in \Pi$ *and* $x \in X$, *and*

(c) $V_1^*(x) > -\infty$ *for each* $x \in X$.

Proof. (a) For each policy π and initial state x, the condition (9.3.8) implies that $V_1^{(-)}(\pi, x) < \infty$, with $V_1^{(-)}$ as in (9.3.7). Hence, as

$$V_1(\pi, x) = V_1^{(+)}(\pi, x) - V_1^{(-)}(\pi, x) \tag{9.3.9}$$

and similarly for $V_\alpha(\pi, x)$, part (a) follows from Proposition 9.2.2(f) and the properties (8.2.7) and (8.2.9) of the p.m. P_ν^π with $\nu = \delta_x$, the Dirac measure concentrated at $x_0 = x$.

(b) This follows from Proposition 9.2.1. Incidentally, observe that since

$$V_\alpha^*(\cdot) := \inf_\Pi V_\alpha(\pi, \cdot) \le V_\alpha(\pi', \cdot) \quad \forall \pi',$$

taking the limit $\alpha \uparrow 1$ and using (b) and the definition (9.3.2) of V_1^*, we obtain

$$\limsup_{\alpha \uparrow 1} V_\alpha^*(x) \le V_1^*(x) \quad \forall x \in X. \tag{9.3.10}$$

(c) As $V_1^{(+)} \ge 0$, (9.3.9) yields $V_1(\pi, x) \ge -V_1^{(-)}(\pi, x)$. Thus, taking the infimum over all π and using (9.3.8) we obtain part (c). $\square$

Concerning Question 9.3.1(c), we show below [Proposition 9.3.5(a)] that Assumption 9.3.2 yields

$$\limsup_{n \to \infty} J_n^*(x) \le V_1^*(x) \quad \forall x \in X. \tag{9.3.11}$$

However, to get (9.3.6) we will introduce an additional assumption, which uses the following **notation**: $V_1^n(\pi, x)$ denotes the expected total cost from time n onwards when using the policy π, given the initial state $x_0 = x$, that is,

$$V_1^n(\pi, x) := E_x^\pi \left[\sum_{t=n}^{\infty} c(x_t, a_t) \right], \quad n = 0, 1, \ldots. \tag{9.3.12}$$

Note that $V_1^0(\pi, x) = V_1(\pi, x)$. Moreover, from (9.3.3) and (9.3.4),

$$V_1(\pi, x) = J_n(\pi, x) + V_1^n(\pi, x), \quad n = 0, 1, \ldots. \tag{9.3.13}$$

In addition, Assumption 9.3.2 and Proposition 9.2.2(f) yield that

$$\lim_{n \to \infty} J_n(\pi, x) = V_1(\pi, x) \quad \forall \pi, x, \tag{9.3.14}$$

and, therefore, by (9.3.13),

$$\lim_{n\to\infty} V_1^n(\pi, x) = 0 \quad \forall \pi, x. \tag{9.3.15}$$

The following assumption requires (9.3.15) to be true in a stronger form.

9.3.4 Assumption. For each $x \in X$

$$\liminf_{n\to\infty} \sup_{\Pi} V_1^n(\pi, x) = 0. \tag{9.3.16}$$

9.3.5 Theorem. (a) *If Assumption 9.3.2 is satisfied, then (9.3.11) holds.*

(b) *If both Assumptions 9.3.2 and 9.3.4 are satisfied, then*

$$\liminf_{n\to\infty} J_n^*(x) \geq V_1^*(x) \quad \forall x \in X, \tag{9.3.17}$$

which combined with (9.3.11) yields (9.3.6), i.e.,

$$\lim_{n\to\infty} J_n^*(x) = V_1^*(x) \quad \forall x \in X.$$

Proof. (a) By definition (9.3.5) of J_n^*,

$$J_n^*(\cdot) \leq J_n(\pi, \cdot) \quad \forall n, \pi,$$

and taking $\limsup_n$ we get, by (9.3.14),

$$\limsup_{n\to\infty} J_n^*(\cdot) \leq V_1(\pi, x) \quad \forall \pi.$$

This inequality and (9.3.2) imply (9.3.11).

(b) From (9.3.13),

$$V_1(\pi, x) \leq J_n(\pi, x) + \sup_{\Pi} V_1^n(\pi, x)$$

so that, taking the infimum over all $\pi \in \Pi$,

$$V_1^*(x) \leq J_n^*(x) + \sup_{\Pi} V_1^n(\pi, x).$$

Finally, taking $\liminf_n$ and using (9.3.16) we obtain (9.3.17). $\square$

From the previous paragraphs we can obtain two (obvious) sufficient conditions for Question 9.3.1(b) to have an affirmative answer.

9.3.6 Proposition. *The value function V_1^* is measurable if (for instance) one of the two following conditions is satisfied:*

(1) *Assumption 9.3.2 holds and, further, there exists an ETC-optimal policy π^*, that is, π^* such that*

$$V_1^*(x) = V_1(\pi^*, x) \quad \forall x \in X. \tag{9.3.18}$$

(2) *Assumptions 9.3.2 and 9.3.4 both hold and the functions J_n^* are measurable.*

Proof. If (1) holds, then the measurability of V_1^* follows from Proposition 9.3.3(a).

On the other hand, under (2), the measurability of V_1^* follows from a well-known result in Real Analysis: *a pointwise limit of Borel-measurable functions is Borel-measurable.* (See, for instance, Ash [1], Theor. 1.5.4.) Indeed, if (2) holds, then V_1^* is measurable because, by Theorem 9.3.5(b) and (9.3.6), it is the pointwise limit of the measurable functions J_n^*. □

Of course, Proposition 9.3.6 answers one question [Question 9.3.1(b)], but simultaneously it raises another: *When are (1) or (2) satisfied?* This question is dealt with in §9.5 and §9.6. First, however, in the next section we consider Question 9.3.1(e).

Notes on §9.3

1. Most of the works on the expected total cost (ETC) criterion deal with Markov control processes (MCPs) in which: (i) the state space X is a *countable* set, and/or (ii) the MCP is either *positive* (that is, $c \geq 0$) or *negative* ($c \leq 0$). For extensive bibliographies on these two cases see, for instance, Altman [1], Bertsekas [1], or Puterman [1]. Among the few works dealing with *Borel state spaces* and not distinguishing between positive and negative MCPs, we can mention the papers by Quelle [1], Rieder [2], Schäl [1], and Hinderer's [1] monograph.

2. With respect to Question 9.3.1(b), it is well known that, in a very general context, the value function V_1^* is *universally measurable* (see, for instance, Hinderer [1]), which is a concept much weaker than measurability. To our knowledge, measurability of V_1^* typically requires restrictive conditions, such as (1) and (2) in Proposition 9.3.6.

3. The Markov control model $\mathcal{M} = (X, A, \{A(x)|x \in X\}, Q, c)$ is called **convergent** if it satisfies Assumption 9.3.2 and, *in addition*,

$$\sup_{\Pi} V_1^{(+)}(\pi, x) < \infty \quad \text{for each} \quad x \in X. \tag{9.3.19}$$

Equivalently, as $|c_t| = c_t^+ + c_t^-$, the MCM (Markov control model) $\mathcal{M}$ is convergent if

$$\sup_{\Pi} E_x^\pi \left(\sum_{t=0}^{\infty} |c_t| \right) < \infty \quad \text{for each} \quad x \in X. \tag{9.3.20}$$

In §9.6 we introduce a class of MCMs for which (9.3.20) holds. Although (9.3.20) is a strong condition, it still allows some "pathologies"—for instance, it does not guarantee (9.3.6), as shown by counterexamples in Puterman [1, §7.3.3], Strauch [1], van Hee et al. [1], etc. This is one of the reasons for introducing Assumption 9.3.4, which is similar to "tail conditions" used by Schäl [1], van Hee et al. [1], and many other authors. Other works (for instance, van Nunen and Wessels [2]) use "Lyapunov functions" instead of tail conditions.

Note that if (9.3.20) holds, then of course

$$\lim_{n\to\infty} \frac{1}{n} E_x^\pi \left(\sum_{t=0}^{n-1} |c_t| \right) = 0 \quad \forall \pi, x. \tag{9.3.21}$$

MCMs with this property are called **zero-average cost** models.

9.4 Occupation measures and the sufficiency problem

In this section we consider the "sufficiency problem" posed in Question 9.3.1(e); namely, is there a (proper) *subset* Π' of Π which is *sufficient* for the ETC problem? Here, "sufficient" means that, with V_1^* as in (9.3.2),

$$V_1^*(\cdot) = \inf_\Pi V_1(\pi, \cdot) = \inf_{\Pi'} V_1(\pi, \cdot). \tag{9.4.1}$$

In Theorem 9.4.5 we show that, under suitable assumptions, (9.4.1) holds with $\Pi' = \Pi_{RM}$, the family of *randomized Markov* policies [Definition 8.2.2(a)].

This result is important for at least two reasons: First, the minimization in (9.4.1) is greatly "simplified" in that it suffices to do it in the smaller set Π'. Second, (9.4.1) states that the *information* required to determine an optimal control action at each time t reduces to the *current state* x_t [see (8.2.4)], in contrast to a general policy which requires the *full history* h_t [see (8.2.3)].

We also show (Theorem 9.4.6) that if $c(x,a)$ is nonnegative, then the second equality in (9.4.1) is satisfied when Π and Π' are replaced by Π_{RS} and Π_{DS}, respectively, i.e.,

$$\inf_{\Pi_{RS}} V_1(\pi, \cdot) = \inf_{\Pi_{DS}} V_1(\pi, \cdot). \tag{9.4.2}$$

In other words, the family Π_{DS} of *deterministic stationary* policies is "sufficient" within the class Π_{RS} of *randomized stationary* policies when $c(x,a)$ is nonnegative.

To study the sufficiency problem we use the notion of *ETC-expected occupation measures*, which are similar to the α-*discount expected occupation measures* (or *state-action frequencies*) introduced in §6.3.

A. Expected occupation measures

In this subsection, $\pi \in \Pi$ denotes an arbitrary policy, and ν stands for an arbitrary initial distribution, that is, a p.m. (probability measure) on the state space X. The corresponding ETC is

$$V_1(\pi, \nu) := E_\nu^\pi \left[\sum_{t=0}^{\infty} c(x_t, a_t) \right]. \tag{9.4.3}$$

In particular, if ν is the Dirac measure δ_x concentrated at $x_0 = x$ we obtain (9.3.1). We suppose the following:

9.4.1 Assumption. Assumption 9.3.2 is satisfied and, in addition, (9.3.8) holds with ν in lieu of x.

The easiest way to obtain the *ETC-expected occupation measure* μ_ν^π is to replace the cost $c(x, a)$ in (9.4.3) by the indicator function $I_\Gamma(x, a)$ of a Borel set Γ in $X \times A$, and then let Γ vary in $\mathcal{B}(X \times A)$. This yields the measure

$$\mu_\nu^\pi(\Gamma) := \sum_{t=0}^{\infty} P_\nu^\pi[(x_t, a_t) \in \Gamma], \quad \Gamma \in \mathcal{B}(X \times A). \tag{9.4.4}$$

9.4.2 Remark. (a) Let $\{\lambda_n\}$ be a sequence of measures on some measurable space $(S, \mathcal{S})$. If the sequence is *increasing* ($\lambda_n \leq \lambda_{n+1} \forall n$) and *converges setwise* to λ ($\lambda_n(B) \to \lambda(B) \quad \forall B \in \mathcal{S}$), then the limit λ is a measure. (For a proof see, for instance, Doob [1], pp.30–31.) It follows that μ_ν^π in (9.4.4) is indeed a measure since it is the setwise limit of the increasing sequence of measures

$$\sum_{t=0}^{n} P_\nu^\pi[(x_t, a_t) \in \cdot], \quad n = 0, 1, \ldots. \tag{9.4.5}$$

(b) By (8.2.3) and the properties (8.2.7)–(8.2.9) of P_ν^π, the expected occupation measure μ_ν^π is defined on (the Borel subsets of) $X \times A$ but *it is concentrated on the set* $\mathbb{K}$ *in (8.2.1)*, i.e.,

$$\mu_\nu^\pi(\mathbb{K}^c) = 0, \quad \text{where} \quad \mathbb{K}^c := \text{complement of } \mathbb{K}. \tag{9.4.6}$$

Moreover, using again that μ_ν^π is the limit of the *increasing* sequence in (9.4.5), and writing $V_1(\pi, \nu)$ as the difference of its positive and negative parts [see (9.3.9) and (9.3.7)], it can be seen that we may rewrite (9.4.3) as

$$V_1(\pi, \nu) = \int_{X \times A} c(x, a) \mu_\nu^\pi(d(x, a)). \tag{9.4.7}$$

To see this, suppose that c is *nonnegative*, as is the case for c^+ and c^-. In addition, in (9.3.4) replace x with ν, and let γ_{n+1} be the measure in (9.3.5). Then, since $c \geq 0$ and $\gamma_n \leq \mu_\nu^\pi$, we get

$$J_n(\pi, \nu) = \int c \, d\gamma_n \leq \int c \, d\mu_\nu^\pi \quad \forall n.$$

Therefore, letting $n \to \infty$, (9.3.14) gives

$$V_1(\pi, \nu) \leq \int c d\mu_\nu^\pi.$$

To obtain the reverse inequality, let $\{u_k\}$ be a nondecreasing sequence of nonnegative simple functions such that $u_k \uparrow c$ pointwise. Hence

$$\int u_k d\gamma_n \leq V_1(\pi, \nu) \quad \forall k, n,$$

and the setwise convergence of γ_n to μ_ν^π gives

$$\int u_k d\mu_\nu^\pi \leq V_1(\pi, \nu) \quad \forall k.$$

Thus, letting $k \to \infty$, it follows from the Monotone Convergence Theorem that

$$\int c d\mu_\nu^\pi \leq V_1(\pi, \nu),$$

which completes the proof of (9.4.7) when c is nonnegative. Finally, replacing c with c^+ and c^- we obtain (9.4.7) for a general c.

Strictly speaking, we should write (9.4.7) as an integral over $\mathbb{K}$ instead of $X \times A$, since $c(x, a)$ is defined on $\mathbb{K}$ only. However, we can always measurably extend c to all of $X \times A$ for (9.4.7) to be well defined. For instance, we may take $c(x, a) := +\infty$ on the complement of $\mathbb{K}$, and then [by (9.4.6)] the convention (9.2.2) would yield that (9.4.7) consists of the integral over $\mathbb{K}$ plus a term equal to zero.

Also note that, for the given initial distribution ν, we may rewrite (9.3.8) as

$$\sup_\Pi \int c^-(x, a) \mu_\nu^\pi(d(x, a)) < \infty.$$

(c) If B and C are Borel sets in X and A, respectively, and if Γ is the measurable rectangle $B \times C$, then (9.4.4) becomes

$$\mu_\nu^\pi(B \times C) = \sum_{t=0}^{\infty} P_\nu^\pi(x_t \in B, a_t \in C). \tag{9.4.8}$$

In particular, if C is the action space A itself (i.e., $C = A$), then we obtain the **marginal** (or **projection**) $\hat{\mu}_\nu^\pi$ of μ_ν^π on X, i.e.,

$$\hat{\mu}_\nu^\pi(B) := \mu_\nu^\pi(B \times A) = \sum_{t=0}^{\infty} P_\nu^\pi(x_t \in B) \tag{9.4.9}$$

for all $B \in \mathcal{B}(X)$. $\square$

To deal with Question 9.3.1(e) it is convenient to rewrite (9.4.8) as

$$\mu_\nu^\pi(B \times C) = \sum_{t=0}^\infty \mu_{\nu,t}^\pi(B \times C), \qquad (9.4.10)$$

where

$$\mu_{\nu,t}^\pi(B \times C) := P_\nu^\pi(x_t \in B, a_t \in C) \qquad (9.4.11)$$

for $B \in \mathcal{B}(X)$, $C \in \mathcal{B}(A)$, $t = 0, 1, \ldots$. Similarly, we may rewrite (9.4.9) as

$$\widehat{\mu}_\nu^\pi(B) = \sum_{t=0}^\infty \widehat{\mu}_{\nu,t}^\pi(B), \quad B \in \mathcal{B}(X), \qquad (9.4.12)$$

where

$$\widehat{\mu}_{\nu,t}^\pi(B) := \mu_{\nu,t}^\pi(B \times A) = P_\nu^\pi(x_t \in B), \quad B \in \mathcal{B}(X), \qquad (9.4.13)$$

is the **marginal** of $\mu_{\nu,t}^\pi$ on X. We have, on the other hand:

9.4.3 Lemma.

(a) $\widehat{\mu}_{\nu,0}^\pi(\cdot) = \nu(\cdot)$.

(b) $\widehat{\mu}_{\nu,t}^\pi(\cdot) = \int_{X \times A} Q(\cdot|x,a)\mu_{\nu,t-1}^\pi(d(x,a)), \quad t = 1, 2, \ldots$.

(c) $\widehat{\mu}_\nu^\pi(\cdot) = \nu(\cdot) + \int_{X \times A} Q(\cdot|x,a)\mu_\nu^\pi(d(x,a))$.

Proof. (a) This follows from (9.4.13) with $t = 0$, and (8.2.7).

(b) First note that, for $t = 1, 2, \ldots$, we may write (8.2.9) as

$$\begin{aligned}
P_\nu^\pi(x_t \in B|h_{t-1}, a_{t-1}) &= E_\nu^\pi[I_B(x_t)|h_{t-1}, a_{t-1}] \\
&= Q(B|x_{t-1}, a_{t-1}).
\end{aligned}$$

So, taking expectation $E_\nu^\pi(\cdot)$ and using (9.4.13), we get, for any B in $\mathcal{B}(X)$,

$$\begin{aligned}
\widehat{\mu}_{\nu,t}^\pi(B) &= E_\nu^\pi[I_B(x_t)] \\
&= E_\nu^\pi[Q(B|x_{t-1}, a_{t-1})] \\
&= \int_{X \times A} Q(B|x,a)\mu_{\nu,t-1}^\pi(d(x,a)),
\end{aligned}$$

which proves (b).

(c) For any Borel set B in X, (9.4.12) and part (a) yield

$$\widehat{\mu}_\nu^\pi(B) = \nu(B) + \sum_{t=1}^\infty \widehat{\mu}_{\nu,t}^\pi(B).$$

Thus, using (b) and (9.4.10), we obtain (c). $\square$

B. The sufficiency problem

We need the following preliminary result stated in Proposition D.8(a); for a proof see, for instance, Dynkin and Yushkevich [1, pp. 88–89] or Hinderer [1, p. 89].

9.4.4 Lemma. *Let* $\mathbb{K}$ *and* Φ *be as in (8.2.1) and Definition 8.2.1, respectively. If* μ *is a p.m. on* $X \times A$ *concentrated on* $\mathbb{K}$, *then there exists a stochastic kernel* $\varphi \in \Phi$ *such that*

$$\mu(B \times C) = \int_B \varphi(C|x)\widehat{\mu}(dx) \quad \forall B \in \mathcal{B}(X), C \in \mathcal{B}(A),$$

where $\widehat{\mu}(\cdot) := \mu(\cdot \times A)$ *is the marginal (or projection) of* μ *on* X.

We next give an affirmative answer to Question 9.3.1(e) by showing that in (9.3.2) we may replace Π by the subset Π_{RM} of the randomized Markov policies [Definition 8.2.2(a)]; in other words, Π_{RM} is a *sufficient* set of policies for the ETC problem.

9.4.5 Theorem. *Let* π *be an arbitrary policy and* ν *an arbitrary initial distribution. Suppose that Assumption 9.4.1 is satisfied. Then there exists a randomized Markov policy* $\pi' = \{\varphi_t\} \in \Pi_{RM}$ *such that the ETC-expected occupation measures for* π *and* π' *coincide, i.e.,*

$$\mu_\nu^\pi = \mu_\nu^{\pi'}. \tag{9.4.14}$$

Hence

$$V_1(\pi, \nu) = V_1(\pi', \nu), \tag{9.4.15}$$

and

$$V_1^*(\nu) := \inf_\Pi V_1(\pi, \nu) = \inf_{\Pi_{RM}} V_1(\pi, \nu). \tag{9.4.16}$$

Proof. By (9.4.10), to prove (9.4.14) it suffices to find a randomized Markov policy $\pi' = \{\varphi_t, t = 0, 1, \ldots\}$ such that

$$\mu_{\nu,t}^\pi = \mu_{\nu,t}^{\pi'} \quad \forall t = 0, 1, \ldots, \tag{9.4.17}$$

where $\mu_{\nu,t}^\pi$ is the p.m. defined in (9.4.11). In turn, to prove (9.4.17) we may use Lemma 9.4.4 as follows: Since [by (8.2.3)] $\mu_{\nu,t}^\pi$ is a p.m. on $X \times A$ concentrated on $\mathbb{K}$, there exists a stochastic kernel φ_t in Φ such that (as in Lemma 9.4.4)

$$\mu_{\nu,t}^\pi(B \times C) = \int_B \varphi_t(C|x)\widehat{\mu}_{\nu,t}^\pi(dx) \tag{9.4.18}$$

for every $B \in \mathcal{B}(X)$, $C \in \mathcal{B}(A)$, and $t = 0, 1, \ldots$. Now let π' be the randomized Markov policy $\pi' := \{\varphi_0, \varphi_1, \ldots\}$, with φ_t as in (9.4.18). Then (9.4.17) trivially holds for $t = 0$ since, by (9.4.11) and Lemma 9.4.3(a), for all B in $\mathcal{B}(X)$ and C in $\mathcal{B}(A)$ we have

$$
\begin{aligned}
\mu_{\nu,0}^{\pi'}(B \times C) \quad &:= \quad P_\nu^{\pi'}(x_0 \in B, a_0 \in C) \\
&= \quad \int_B \varphi_0(C|x)\nu(dx) \\
&= \quad \mu_{\nu,0}^{\pi}(B \times C),
\end{aligned}
$$

where the last equality is due to (9.4.18) and Lemma 9.4.3(a) again. The proof now proceeds by induction: Suppose that (9.4.17) holds for some integer $t \geq 0$. Then, by Lemma 9.4.3(b), the marginal of $\mu_{\nu,t+1}^{\pi}$ on X satisfies that

$$
\begin{aligned}
\widehat{\mu}_{\nu,t+1}^{\pi}(\cdot) \quad &= \quad \int_X Q(\cdot|x,a)\mu_{\nu,t}^{\pi}(d(x,a)) \\
&= \quad \int_X Q(\cdot|x,a)\mu_{\nu,t}^{\pi'}(d(x,a)) \quad \text{[by the induction hypothesis]} \\
&= \quad \widehat{\mu}_{\nu,t+1}^{\pi'}(\cdot) \quad \text{[by Lemma 9.4.3(b)]};
\end{aligned}
$$

that is, the marginals (on X) of $\mu_{\nu,t+1}^{\pi}$ and $\mu_{\nu,t+1}^{\pi'}$ coincide. This implies that (9.4.17) holds for $t + 1$ because [by (9.4.11)]

$$
\begin{aligned}
\mu_{\nu,t+1}^{\pi'}(B \times C) \quad &:= \quad P_\nu^{\pi'}(x_{t+1} \in B, a_{t+1} \in C) \\
&= \quad \int_B \varphi_{t+1}(C|x)\widehat{\mu}_{t+1}^{\pi'}(dx) \\
&= \quad \int_B \varphi_{t+1}(C|x)\widehat{\mu}_{\nu,t+1}^{\pi}(dx) \\
&= \quad \mu_{\nu,t+1}^{\pi}(B \times C) \quad \text{[by (9.4.18)]}.
\end{aligned}
$$

This completes the proof of (9.4.17), which, as was already mentioned, gives (9.4.14). In turn, (9.4.14) and (9.4.7) give the equality (9.4.15), and also give (9.4.16) since π was an arbitrary policy. $\square$

In connection with the question of "sufficiency" of sets of policies, we next present an interesting result which states that, under appropriate assumptions, the ETC corresponding to a randomized stationary policy $\varphi^\infty \in \Pi_{RS}$ can always be "improved" (or "minorized") by the ETC of some *deterministic* stationary policy $f^\infty \in \Pi_{DS}$ in the sense that

$$
V_1(\varphi^\infty, \nu) \geq V_1(f^\infty, \nu),
$$

where ν is the given initial distribution. In other words, this fact would yield another affirmative answer to Question 9.3.1(e) when Π and Π' are replaced by Π_{RS} and Π_{DS}, respectively—see (9.4.21). The precise statement is as follows.

9.4.6 Theorem. *Suppose that the cost-per-stage function $c(x,a)$ is non-negative (so, that, in particular, Assumption 9.3.2 is satisfied). Further, let φ^∞ and ν be a given randomized stationary policy and a given initial distribution, respectively, such that*

(*) $V_1(\varphi^\infty, \cdot)$ *is a finite-valued function on X, integrable with respect to ν.*

Then there exists a deterministic stationary policy $f^\infty \in \Pi_{DS}$ such that

$$V_1(\varphi^\infty, x) \geq V_1(f^\infty, x) \quad \forall x \in X; \tag{9.4.19}$$

hence

$$V_1(\varphi^\infty, \nu) \geq V_1(f^\infty, \nu). \tag{9.4.20}$$

Moreover, if the condition (*) holds for every randomized stationary $\varphi^\infty \in \Pi_{RS}$, then

$$\inf_{\Pi_{RS}} V_1(\pi, \nu) = \inf_{\Pi_{DS}} V_1(\pi, \nu). \tag{9.4.21}$$

The proof of Theorem 9.4.6 requires two preliminary facts. The first one is the following lemma of Hinderer [1, Lemma 15.1], which is an extension of a result by Blackwell [1].

9.4.7 Lemma. *Let $\mathbb{K}$ and $\mathbb{F}, \Phi$ be as in (8.2.1) and Definition 8.2.1, respectively. If φ is a stochastic kernel in Φ and $v : \mathbb{K} \to \bar{\mathbb{R}}$ is a measurable function such that*

$$x \mapsto \int_A v^-(x,a)\varphi(da|x)$$

is a finite-valued map on X, then there exists a decision function $f \in \mathbb{F}$ that satisfies

$$\int_A v(x,a)\varphi(da|x) \geq v(x, f(x)) \quad \forall x \in X.$$

The second fact we need is simply another expression for the ETC V_1 in (9.3.1), which will also be useful in later sections. [The expression (9.4.23) corresponds to the special case $n = 1$ of Lemma 9.5.6(a).] We will use the following notation: Given a policy $\pi = \{\pi_t, \ n = 0, 1, \ldots\}$ the 1-*shift policy* $\pi^{(1)} = \{\pi_t^{(1)}, \ t = 0, 1, \ldots\}$ is defined as

$$\pi_0^{(1)}(\cdot|x_1) := \pi_1(\cdot|x_0, a_0, x_1),$$

and for $t = 1, 2, \ldots$,

$$\pi_t^{(1)}(\cdot|x_1, a_1, \ldots, x_{t+1}) := \pi_{t+1}(\cdot|x_0, a_0, x_1, a_1, \ldots, x_{t+1}).$$

In particular, if $\pi = \varphi^\infty$ is a randomized stationary policy, then [by Definition 8.2.2 (b)] the 1-shift policy is given by

$$\pi_t^{(1)}(\cdot|x_1, a_1, \ldots, x_{t+1}) = \varphi(\cdot|x_{t+1}) \quad \forall t = 0, 1, \ldots. \tag{9.4.22}$$

9.4.8 Lemma. *Suppose that Assumption 9.3.2 is satisfied. Then, for each policy* $\pi = \{\pi_t\}$ *and initial state* $x \in X$,

$$V_1(\pi, x) = \int_A \left[c(x, a) + \int_X V_1(\pi^{(1)}, y) Q(dy|x, a) \right] \pi_0(da|x). \qquad (9.4.23)$$

In particular, if $\pi = \varphi^\infty$ *is a randomized stationary policy, then [by (9.4.22)]*

$$V_1(\varphi^\infty, x) = \int_A \left[c(x, a) + \int_X V_1(\varphi^\infty, y) Q(dy|x, a) \right] \varphi(da|x). \qquad (9.4.24)$$

Proof. From (9.3.1),

$$V_1(\pi, x) = E_x^\pi[c(x_0, a_0)] + E_x^\pi \left[\sum_{t=1}^\infty c(x_t, a_t) \right].$$

Furthermore, by (8.2.7)–(8.2.9), the first term on the right-hand side equals

$$\int_A c(x, a) \pi_0(da|x),$$

and the second equals

$$E_x^\pi \left\{ E_x^\pi \left[\sum_{t=1}^\infty c(x_t, a_t) | x_0, a_0, x_1 \right] \right\} = E_x^\pi[V_1(\pi^{(1)}, x_1)]$$
$$= \int_A \int_X V_1(\pi^{(1)}, y) Q(dy|x, a) \pi_0(da|x).$$

Combining these expressions we obtain (9.4.23). □

We are now ready to prove Theorem 9.4.6.

Proof of Theorem 9.4.6. Rewrite (9.4.24) in the form

$$V_1(\varphi^\infty, x) = \int_A v(x, a) \varphi(da|x)$$

with

$$v(x, a) := c(x, a) + \int_X V_1(\varphi^\infty, y) Q(dy|x, a).$$

Therefore, by the hypothesis (*) and Lemma 9.4.7, there is a decision function $f \in \mathbb{F}$ such that, using the notation (8.2.6),

$$V_1(\varphi^\infty, x) \geq c(x, f) + \int_X V_1(\varphi^\infty, y) Q(dy|x, f).$$

Iteration of this inequality gives, for every $n = 1, 2, \ldots$ [with $Q^n(\cdot|x, f)$ as in (8.2.11) and $f^\infty \in \Pi_{DS}$ the deterministic stationary policy determined by $f \in \mathbb{F}$—see Remark 8.2.3(a)],

$$V_1(\varphi^\infty, x) \geq E_x^{f^\infty} \left[\sum_{t=0}^{n-1} c(x_t, f) \right] + \int_X V_1(\varphi^\infty, y) Q^n(dy|x, f).$$

Therefore, by (9.3.4) (with $\pi = f^\infty$) and the assumption that $c \geq 0$, we get

$$V_1(\varphi^\infty, x) \geq J_n(f^\infty, x) \qquad \forall n = 1, 2, \ldots,$$

and letting $n \to \infty$, we see that (9.4.19) follows from (9.3.14).

On the other hand, integration of both sides of (9.4.19) with respect to ν yields (9.4.20).

Finally, to prove (9.4.21), observe that (9.4.20) implies that

$$\inf_{\Pi_{RS}} V_1(\pi, \nu) \geq \inf_{\Pi_{DS}} V_1(\pi, \nu),$$

whereas the reverse inequality follows from the fact that Π_{DS} is contained in Π_{RS} (since $\mathbb{F}$ is contained in Φ—see the paragraph after Definition 8.2.1).
$\square$

Notes on §9.4

1. For MCPs with a countable state space X, Theorem 9.4.5 is due to Strauch [1], and Derman and Strauch [1]—see also Derman [1]. Theorem 9.4.6, on the other hand, is modelled after results by Kurano and Kawai [1] and González-Hernández and Hernández-Lerma [1] for discounted cost problems, but in fact it can be seen as a very special case of Hinderer's [1] Theorem 15.2, which deals with nonhomogeneous Markov control models. In our present context, Hinderer's result can be stated as follows:

Suppose that Assumption 9.4.1 is satisfied and that the cost-per-stage $c(x, a)$ *is* **nonnegative**. *Then for each policy* π *there exists a* **deterministic** *Markov policy* $\pi^d = \{f_t\}$ *such that*

$$V_1(\pi, x) \geq V_1(\pi^d, x) \quad \forall x \in X. \tag{9.4.25}$$

To prove this result the idea is that by Theorem 9.4.5 we may assume at the outset that π is a randomized Markov policy, say $\pi = \{\varphi_t\}$. Then, by (9.4.23) and Lemma 9.4.7, there exist $f_0 \in \mathbb{F}$ such that

$$\begin{aligned}
V_1(\pi, x) &= \int_A \left[c(x, a) + \int_X V_1(\pi^{(1)}, y) Q(dy|x, a) \right] \varphi_0(da|x) \\
&\geq c(x, f_0) + \int_X V_1(\pi^{(1)}, y) Q(dy|x, f_0).
\end{aligned}$$

Next, applying the same argument to $V_1(\pi^{(1)}, \cdot)$ one obtains $f_1 \in \mathbb{F}$, and continuing in this manner we get $\pi^d = \{f_0, f_1, \ldots\}$, which satisfies (9.4.25).

2. The ETC-expected occupation measure μ_ν^π in (9.4.4) corresponds to the case $\alpha = 1$ of the α-**discount** expected occupation measures (or state-action frequencies) in §6.3, namely,

$$\mu_\nu^{\pi, \alpha}(\Gamma) := \sum_{t=0}^\infty \alpha^t P_\nu^\pi[(x_t, a_t) \in \Gamma], \quad \Gamma \in \mathcal{B}(X \times A), \ 0 < \alpha < 1. \tag{9.4.26}$$

Of course, this measure is *finite*—with total measure $1/(1-\alpha)$—but μ_ν^π is not. Another important difference between the α-discount and the undiscounted ($\alpha = 1$) expected occupation measures is the following. In the former case ($0 < \alpha < 1$) the corresponding equation in Lemma 9.4.3(c) is

$$\widehat{\mu}_\nu^{\pi,\alpha}(\cdot) = \nu(\cdot) + \alpha \int_{X \times A} Q(\cdot|x,a)\mu_\nu^{\pi,\alpha}(d(x,a)), \qquad (9.4.27)$$

which *characterizes the family of α-discount expected occupation measures* (Theorem 6.3.7); that is, any such a measure satisfies (9.4.27), and *conversely*, if μ is a measure on $X \times A$, concentrated on $\mathbb{K}$, and such that μ satisfies (9.4.27), i.e.,

$$\widehat{\mu}(\cdot) = \nu(\cdot) + \alpha \int_{X \times A} Q(\cdot|x,a)\mu(d(x,a)),$$

then μ *is the α-discount expected occupation measure $\mu_\nu^{\pi,\alpha}$ corresponding to some policy π*. The latter statement (the converse) is *not* true in the undiscounted case, $\alpha = 1$, unless we impose additional assumptions on the transition law Q.

9.5 The optimality equation

This section has three main objectives. First, we give conditions under which the value function V_1^* [see (9.3.2)] satisfies the optimality equation (9.5.2), (9.5.3). Second, we present several optimality criteria similar to those in §4.5 for the discounted cost. Third, we use the optimality equation to produce optimality criteria when using deterministic stationary policies.

A. The optimality equation

Let $T \equiv T_1$ be the *dynamic programming operator* T_α in (8.3.17) when $\alpha = 1$; that is

$$Tu(x) := \inf_{A(x)} \left[c(x,a) + \int_X u(y)Q(dy|x,a) \right], \qquad x \in X. \qquad (9.5.1)$$

A measurable function u is said to satisfy (or to be a solution of) the **optimality equation** for the expected total cost (ETC) criterion if u is a fixed point of T, that is, $u = Tu$. In Theorem 9.5.3 below we show that, under suitable assumptions, the value function V_1^* satisfies the optimality equation so that

$$V_1^* = TV_1^*, \qquad (9.5.2)$$

or, more explicitly,

$$V_1^*(x) = \inf_{A(x)} \left[c(x,a) + \int_X V_1^*(y)Q(dy|x,a) \right], \qquad x \in X. \qquad (9.5.3)$$

Observe that (9.5.3), which is also referred to as the *dynamic programming equation* for the ETC criterion, does not have a unique solution (if it has one). In fact, if a function $u(\cdot)$ satisfies (9.5.3), then so does $u(\cdot) + k$ for any constant k. Thus, it is important to "characterize" V_1^* within a certain class of solutions of (9.5.3); one such characterization is given in Theorem 9.5.9.

To obtain (9.5.3) we impose several hypotheses (Assumption 9.5.2), which use the following definition.

9.5.1 Definition. $\mathcal{U}$ denotes the family of extended real-valued functions that are integrable with respect to $Q(\cdot|x, a)$ for each $x \in X$ and $a \in A(x)$.

Note that $\mathcal{U}$ is nonempty since it contains (at least) all the measurable bounded functions on X. In particular, the 0-stage cost $J_0^*(\cdot) := 0$ [see (9.3.5)] is in $\mathcal{U}$. Part (b) in the following assumption requires, among other things, that the optimal n-stage cost J_n^* belongs to $\mathcal{U}$ for all $n = 1, 2, \ldots$. This is known to be true under a variety of conditions—see, for instance, §3.3, or the assumptions in §8.3 and §8.5.

9.5.2 Assumption.

(a) The hypotheses of Theorem 9.3.5(b)—that is, Assumptions 9.3.2 and 9.3.4—are satisfied.

(b) For each $n = 1, 2, \ldots, J_n^*$ is in $\mathcal{U}$ and satisfies

$$J_n^*(x) = T J_{n-1}^*(x) \quad \forall x \in X. \tag{9.5.4}$$

(c) Either (i) the cost-per-stage function $c(x, a)$ is *nonnegative*, in which case the functions J_n^* form a *nondecreasing* sequence, or (ii) there is a function W in $\mathcal{U}$ such that

$$J_n^* \leq W \quad \forall n = 1, 2, \ldots. \tag{9.5.5}$$

Observe that (9.5.4) is simply the *value iteration* equation that we have already seen in several chapters; see, for instance, (8.3.12) or (8.3.26) for the discounted case.

Under Assumption 9.5.2, *which is supposed to hold throughout this section*, Theorem 9.3.5(b) and Proposition 9.3.6 yield (9.3.6), i.e.,

$$\lim_{n \to \infty} J_n^*(x) = V_1^*(x) \quad \forall x \in X, \tag{9.5.6}$$

and that V_1^* is a measurable function on X. Moreover, V_1^* is quasi-integrable with respect to the measure $Q(\cdot|x, a)$ for each $x \in x$ and $a \in A(x)$. Indeed, this is obvious if the condition (i) in Assumption 9.5.2(c) holds—in this case V_1^* is *nonnegative* and so the integral of its negative part is zero [see (9.2.9), or (9.3.7), (9.3.8)]. On the other hand, under the condition (ii), we have [by (9.5.5) and Proposition 9.3.3(c)]

$$-\infty < V_1^*(x) \leq W(x) \quad \forall x \in X; \tag{9.5.7}$$

hence, as W belongs to the set $\mathcal{U}$ (see Definition 9.5.1), (9.5.7) and Proposition 9.2.2(c) yield

$$-\infty \le \int_X V_1^*(y)Q(dy|x,a) < \infty \quad \forall x \in X, \ a \in A(x).$$

We also get the optimality equation (9.5.2), (9.5.3):

9.5.3 Theorem. *If Assumption 9.5.2 holds, then V_1^* satisfies (9.5.2), (9.5.3).*

Proof. We will show that

$$\text{(a) } V_1^* \ge TV_1^*, \quad \text{and} \quad \text{(b) } V_1^* \le TV_1^*. \tag{9.5.8}$$

To prove (a), choose an arbitrary policy $\pi \in \Pi$ and an arbitrary initial state $x \in X$. Then, by Lemma 9.4.8,

$$
\begin{aligned}
V_1(\pi, x) &= \int_A \left[c(x,a) + \int_X V_1(\pi^{(1)}, y)Q(dy|x,a) \right] \pi_0(da|x) \\
&\ge \int_A \left[c(x,a) + \int_X V_1^*(y)Q(dy|x,a) \right] \pi_0(da|x) \quad \text{[by (9.3.2)]} \\
&\ge TV_1^*(x) \quad \text{[by definition (9.5.1) of T]};
\end{aligned}
$$

that is,

$$V_1(\pi, x) \ge TV_1^*(x).$$

This inequality and (9.3.2) give (9.5.8)(a) since π and x were arbitrary.
  To obtain (9.5.8)(b) use (9.5.4) and (9.5.1) to write

$$J_n^*(x) \le c(x,a) + \int_X J_{n-1}^*(y)Q(dy|x,a) \quad \forall x \in X, \ a \in A(x). \tag{9.5.9}$$

Now consider the condition (i) in Assumption 9.5.2(c). In this case the sequence J_n^* is *nondecreasing* and, therefore, letting $n \to \infty$ in (9.5.9), we obtain [by (9.5.6) and the Monotone Convergence Theorem]

$$V_1^*(x) \le c(x,a) + \int V_1^*(y)Q(dy|x,a) \quad \forall x \in X, \ a \in A(x), \tag{9.5.10}$$

which implies (9.5.8)(b). On the other hand, under the condition (ii) in Assumption 9.5.2(c), we may take $\limsup_n$ in (9.5.9) and obtain (9.5.10) again, by (9.5.6) and Fatou's Lemma. This completes the proof. $\square$

B. Optimality criteria

Having the optimality equation (9.5.3) [or (9.5.2)] we can proceed to obtain several optimality criteria, which informally can be obtained taking a "discount factor" $\alpha = 1$ in Theorem 4.5.1 (on discounted cost problems). In particular, the concepts in the following definition correspond to the case "$\alpha = 1$" of those introduced in §4.5. [Alternatively, to obtain the discrepancy function D_1 in (9.5.11), we may take $\alpha = 1$ in (8.4.2).]

9.5.4 Definition.

(a) The **discrepancy function** for the ETC criterion is the nonnegative function D_1 on $\mathbb{K}$ [the set defined in (8.2.1)] given by

$$D_1(x, a) := c(x, a) + \int_X V_1^*(y) Q(dy|x, a) - V_1^*(x). \qquad (9.5.11)$$

(b) $\{M_n^*\}$ denotes the sequence defined by $M_0^* := V_1^*(x_0)$, and

$$M_n^* := \sum_{t=0}^{n-1} c(x_t, a_t) + V_1^*(x_n) \quad \text{for } n = 1, 2, \ldots. \qquad (9.5.12)$$

(c) Given a policy $\pi = \{\pi_t\}$ and an integer $n \geq 0$, the corresponding n-**shift policy** $\pi^{(n)} = \{\pi_t^{(n)}, \ t = 0, 1, \ldots\}$ is given [with h_t as in (8.2.2)] by

$$\pi_0^{(n)}(\cdot|x_n) := \pi_n(\cdot|h_n),$$

and for $t = 1, 2, \ldots,$

$$\pi_t^{(n)}(\cdot|x_n, a_n, \ldots, x_{n+t}) := \pi_{n+t}(\cdot|h_n, a_n, \ldots, x_{n+t}).$$

In particular, the 0-**shift policy** is the same as π, i.e., $\pi^{(0)} = \pi$.

The role of the discrepancy function D_1 is, of course, the same as in other Markov control problems. For instance, it gives an alternative way of writing the optimality equation (9.5.2)–(9.5.3) as

$$\inf_{A(x)} D_1(x, a) = 0 \qquad \forall x \in X, \qquad (9.5.13)$$

and, more importantly, it gives an "explicit" expression, (9.5.19) below, for the difference between the expected total cost $V_1(\pi, \cdot)$ and the value function $V_1^*(\cdot)$ for every policy π.

9.5.5 Theorem. (Optimality criteria.) *Suppose that Assumption 9.5.2 is satisfied, and let π be a policy such that $V_1(\pi, x) < \infty$ for each $x \in X$. Then the following statements (a) and (b) are equivalent, and also (c) and (d) are equivalent:*

(a) *π is ETC-optimal, i.e., $V_1(\pi, x) = V_1^*(x)$ $\forall x$.*

(b) *$V_1^n(\pi, x) = E_x^\pi V_1^*(x_n)$ $\forall n, x$ [with $V_1^n(\pi, x)$ as in (9.3.12)].*

(c) $E_x^\pi D_1(x_n, a_n) = 0 \quad \forall n, x.$

(d) $\{M_n^*, \sigma(h_n)\}$ is a P_x^π-martingale for all x [where $\sigma(h_n)$ is the σ-algebra generated by the n-history h_n in (8.2.2), $n = 0, 1, \ldots$]; that is, for every n, M_n^* is P_x^π-integrable, $\sigma(h_n)$-measurable and

$$E_x^\pi(M_{n+1}^* | h_n) = M_n^* \qquad (P_x^\pi\text{-a.s.}) \qquad (9.5.14)$$

If, in addition, V_1^* satisfies that

$$\liminf_{n \to \infty} E_x^\pi V_1^*(x_n) \geq 0 \quad \text{for each} \quad \pi \in \Pi \quad \text{and} \quad x \in X, \qquad (9.5.15)$$

then the four conditions (a) to (d) are equivalent.

Observe that (9.5.15) holds if, for instance, $c(x, a)$ is nonnegative.

The proof of Theorem 9.5.5 will follow from the next two lemmas, the first of which (Lemma 9.5.6) basically states that the expected total cost $V_1(\pi, x)$ in (9.3.1) and (9.3.13) can be written in other alternative forms.

9.5.6 Lemma. For each $\pi \in \Pi$, $x \in X$ for which $V_1(\pi, x) < \infty$, and $n = 0, 1, \ldots$:

(a) $V_1(\pi, x) = J_n(\pi, x) + E_x^\pi[V_1(\pi^{(n)}, x_n)]$, where $\pi^{(n)}$ is the n-shift policy corresponding to π;

(b) $V_1(\pi, x) = E_x^\pi(M_n^*) + [V_1^n(\pi, x) - E_x^\pi V_1^*(x_n)]$.

Moreover,

$$\limsup_{n \to \infty} E_x^\pi V_1^*(x_n) \leq 0 \quad \forall \pi, x; \qquad (9.5.16)$$

hence, if (9.5.15) holds, then

$$\lim_{n \to \infty} E_x^\pi V_1^*(x_n) = 0 \quad \forall \pi, x, \qquad (9.5.17)$$

and, furthermore,

$$V_1^n(\pi, x) = E_x^\pi V_1^*(x_n) + \sum_{t=n}^{\infty} E_x^\pi D_1(x_t, a_t) \quad \forall \pi, x, n. \qquad (9.5.18)$$

In particular, with $n = 0$,

$$V_1(\pi, x) = V_1^*(x) + \sum_{t=0}^{\infty} E_x^\pi D_1(x_t, a_t) \quad \forall \pi, x. \qquad (9.5.19)$$

Proof. (a) From (9.3.12), (8.2.8)–(8.2.9), and Definition 9.5.4(c) we see that

$$V_1^n(\pi, x) \quad := \quad E_x^\pi \left[\sum_{t=n}^{\infty} c(x_t, a_t) \right]$$

$$= E_x^\pi \left\{ E_x^\pi \left[\sum_{t=n}^\infty c(x_t, a_t) | h_n \right] \right\}$$

$$= E_x^\pi V_1(\pi^{(n)}, x_n),$$

i.e.,

$$V_1^n(\pi, x) = E_x^\pi V_1(\pi^{(n)}, x_n). \tag{9.5.20}$$

Thus, part (a) follows from (9.5.20) and (9.3.13).

(b) This is obtained by adding and subtracting $E_x^\pi V_1^*(x_n)$ in (9.3.13), and using the definition (9.5.12) of M_n^*.

To prove (9.5.16), note that (9.3.2) and (9.3.20) yield

$$V_1^n(\pi, x) \ge E_x^\pi V_1^*(x_n) \quad \forall \pi, x, n, \tag{9.5.21}$$

and so (9.5.16) follows from (9.3.15). Also note that (9.5.17) is a consequence of (9.5.16) and (9.5.15).

To prove (9.5.18) observe that (9.5.11) and (8.2.9) give, for every $t = 0, 1, \ldots,$

$$E_x^\pi[D_1(x_t, a_t) | h_t, a_t] = E_x^\pi[c(x_t, a_t) + V_1^*(x_{t+1}) - V_1^*(x_t) | h_t, a_t].$$

Taking the expectation E_x^π on both sides of the latter equation we get

$$E_x^\pi D_1(x_t, a_t) = E_x^\pi c(x_t, a_t) + E_x^\pi V_1^*(x_{t+1}) - E_x^\pi V_1^*(x_t),$$

and then

$$\sum_{t=n}^{N-1} E_x^\pi D_1(x_t, a_t) = \sum_{t=n}^{N-1} E_x^\pi c(x_t, a_t) + E_x^\pi V_1^*(x_N) - E_x^\pi V_1^*(x_n).$$

Finally, letting $N \to \infty$ and using (9.5.17) we obtain (9.5.18). $\square$

9.5.7 Remark. Suppose that the cost-per-stage $c(x, a)$ is *nonnegative*— as in part (i) of Assumption 9.5.2(c). Then (9.5.15) is obviously satisfied, and so the conditions (a) to (d) in Theorem 9.5.5 are all equivalent and, moreover, (9.5.17) and (9.5.18) hold. $\square$

9.5.8 Lemma. *For each policy π and initial state x for which $V_1(\pi, x) < \infty$, the sequence $\{M_n^*, \sigma(h_n)\}$ is a P_x^π-submartingale; that is, for every n, M_n^* is P_x^π-integrable, $\sigma(h_n)$-measurable and*

$$E_x^\pi(M_{n+1}^* | h_n) \ge M_n^* \quad (P_x^\pi\text{-a.s.}); \tag{9.5.22}$$

hence, for every $n = 0, 1, \ldots,$

$$E_x^\pi(M_{n+1}^*) \ge E_x^\pi(M_n^*) \ge E_x^\pi(M_0^*) = V_1^*(x). \tag{9.5.23}$$

Proof. By Definition 9.5.4(b) and Assumption 9.5.2, it is evident that M_n^* is P_x^π-integrable and $\sigma(h_n)$-measurable for every n, and, furthermore,

$$M_{n+1}^* = M_n^* + [c(x_n, a_n) + V_1^*(x_{n+1}) - V_1^*(x_n)]$$

Thus, in view of (9.5.11),

$$E_x^\pi(M_{n+1}^* | h_n) = M_n^* + E_x^\pi[D_1(x_n, a_n) | h_n], \tag{9.5.24}$$

which gives (9.5.22) since D_1 is a nonnegative function. $\square$

Using Lemmas 9.5.6 and 9.5.8 we can now easily prove Theorem 9.5.5:

Proof of Theorem 9.5.5.(a) $\Leftrightarrow$ (b). Suppose that (a) holds. Then, by (9.5.21), to prove (b) it suffices to show that

$$V_1^n(\pi, x) \le E_x^\pi V_1^*(x_n) \quad \forall n, x. \tag{9.5.25}$$

To obtain this inequality observe that, as π is optimal [i.e., $V_1(\pi, \cdot) = V_1^*(\cdot)$], Lemma 9.5.6(b) and (9.5.23) yield

$$V_1^*(x) \ge V_1^*(x) + [V_1^n(\pi, x) - E_x^\pi V_1^*(x_n)],$$

and (9.5.25) follows. This shows that (a) implies (b). The converse is obviously true: take $n = 0$ in (b).

(c) $\Leftrightarrow$ (d). As

$$E_x^\pi D_1(x_n, a_n) = E_x^\pi \{ E_x^\pi[D_1(x_n, a_n) | h_n] \},$$

the equivalence of (c) and (d) follows from (9.5.24)—recall that D_1 is nonnegative.

Finally, if (9.5.15) holds, then (9.5.18) shows that (b) and (c) [hence (a) to (d)] are equivalent. This completes the proof of Theorem 9.5.5. $\square$

The inequality (9.5.16) can be used to obtain a characterization of V_1^* as the pointwise "maximal" solution of the optimality equation within a certain subclass of functions in $\mathcal{U}$ (Definition 9.5.1). This is the essential content of the following result. (In Theorem 9.5.13 we give conditions for V_1^* to be the "unique" solution of the optimality equation.)

9.5.9 Theorem. (A "characterization" of V_1^*.) *Suppose that Assumption 9.5.2 is satisfied. Let $u \in \mathcal{U}$ be a function that satisfies the optimality equation (9.5.3) and the inequality (9.5.16), i.e.,*

$$u(x) = \inf_{A(x)} \left[c(x, a) + \int_X u(y) Q(dy | x, a) \right] \quad \forall x \in X. \tag{9.5.26}$$

and

$$\limsup_{n \to \infty} E_x^\pi u(x_n) \le 0 \quad \forall x \in X, \ \pi \in \Pi, \tag{9.5.27}$$

respectively. Then

$$u(\cdot) \leq V_1^*(\cdot). \tag{9.5.28}$$

Hence, if V_1^* belongs to the class $\mathcal{U}$, then V_1^* is the maximal function in $\mathcal{U}$ that satisfies (9.5.26) and (9.5.27).

Proof. We will show that if $u \in \mathcal{U}$ satisfies (9.5.26) and (9.5.27), then

$$u(x) \leq V_1(\pi, x) \quad \forall \pi \in \Pi, \ x \in X, \tag{9.5.29}$$

which, by (9.3.2), gives (9.5.28).

To prove (9.5.29), choose an arbitrary policy π and an arbitrary initial state x. Then, for every $t = 0, 1, \ldots$, (8.2.9) and (9.5.26) yield

$$
\begin{aligned}
E_x^\pi[u(x_{t+1})|h_t, a_t] &= \int u(y)Q(dy|x_t, a_t) \\
&\geq u(x_t) - c(x_t, a_t),
\end{aligned}
$$

so that taking expectation $E_x^\pi(\cdot)$ and rearranging terms we obtain

$$E_x^\pi u(x_{t+1}) - E_x^\pi u(x_t) + E_x^\pi c(x_t, a_t) \geq 0.$$

Thus, summing over $t = 0, \ldots, n-1$,

$$E_x^\pi u(x_n) - u(x) + J_n(\pi, x) \geq 0,$$

which we can rewrite as

$$u(x) \leq J_n(\pi, x) + E_x^\pi u(x_n).$$

As a result, taking $\lim \sup_n$, we obtain (9.5.29) from (9.5.27) and (9.3.14). $\square$

C. Deterministic stationary policies

We conclude this section with some remarks on the expected total cost (ETC) when using a deterministic stationary policy $f^\infty \in \Pi_{DS}$. [Recall Definitions 8.2.1 and 8.2.2(e).]

First note that replacing the policy φ^∞ in (9.4.24) by f^∞ we get that the ETC $V_1(f^\infty, \cdot)$ satisfies [using the notation (8.2.6)]

$$V_1(f^\infty, x) = c(x, f) + \int_X V_1(f^\infty, y)Q(dy|x, f), \quad x \in X. \tag{9.5.30}$$

In other words, the function $h(\cdot) := V_1(f^\infty, \cdot)$ satisfies

$$h(x) = c(x, f) + \int_X h(y)Q(dy|x, f), \quad x \in X. \tag{9.5.31}$$

Comparing this equation and (7.5.1) we see that (9.5.31) is the *Poisson equation* for the stochastic kernel P and the charge c given by

$$P(\cdot|x) := Q(\cdot|x, f) \quad \text{and} \quad c(\cdot) := c(\cdot, f), \tag{9.5.32}$$

respectively, with P-invariant (or P-harmonic) function

$$g(\cdot) := 0. \tag{9.5.33}$$

As a result, Theorem 7.5.5 gives:

9.5.10 Proposition. *The following conditions are equivalent for any deterministic stationary policy f^∞ for which $V_1(f^\infty, \cdot)$ is finite valued:*

(a) $h(\cdot) := V_1(f^\infty, \cdot)$ *satisfies the Poisson equation (9.5.31).*

(b) $h(x) = J_n(f^\infty, x) + E_x^{f^\infty} h(x_n) \quad \forall x \in X, \; n = 0, 1, \ldots.$ $\qquad(9.5.34)$

(c) *The sequence $\{M_n(f)\}$ with $M_0(f) := h(x_0)$ and*

$$M_n(f) := \sum_{t=0}^{n-1} c(x_t, f) + h(x_n) \quad \text{for} \quad n = 1, 2, \ldots,$$

is a $P_x^{f^\infty}$-martingale for every initial state x.

On the other hand, (9.5.34) and (9.3.14) imply that $h(\cdot) = V_1(f^\infty, \cdot)$ satisfies

$$\lim_{n\to\infty} E_x^{f^\infty} h(x_n) = 0 \qquad \forall x \in X. \tag{9.5.35}$$

This statement has the following obvious converse:

9.5.11 Proposition. *If h satisfies (9.5.31) and (9.5.35), then*

$$h(x) = V_1(f^\infty, x) \qquad \forall x \in X. \tag{9.5.36}$$

Proof. Iteration of (9.5.31) gives (9.5), and letting $n \to \infty$ we obtain (9.5.36). $\square$

The connection between Propositions 9.5.10, 9.5.11 and the optimality equation (9.5.3) is provided by the following result.

9.5.12 Theorem. *Suppose that Assumption 9.5.2 is satisfied. Then a deterministic stationary policy f^∞ such that $V_1(f^\infty, \cdot) < \infty$ is ETC-optimal if and only if $f(x) \in A(x)$ attains the minimum in (9.5.3) for all $x \in X$, i.e.,*

$$V_1^*(x) = c(x, f) + \int_X V_1^*(y) Q(dy|x, f) \quad \forall x \in X, \tag{9.5.37}$$

and [cf. (9.5.16)]

$$\limsup_{n\to\infty} E_x^{f^\infty} V_1^*(x_n) = 0 \qquad \forall x \in X. \tag{9.5.38}$$

Proof. Suppose that f^∞ is ETC-optimal, that is, $V_1(f^\infty, \cdot) = V_1^*(\cdot)$. Then (9.5.37) follows from Theorem 9.5.3 and the Poisson equation (9.5.30), whereas (9.5.38) is a consequence of (9.5.35).

Conversely, suppose that (9.5.37) and (9.5.38) are satisfied. Then, from (9.5.37) and (9.5.34), we have

$$V_1^*(x) = J_n(f^\infty, x) + E_x^{f^\infty} V_1^*(x_n) \quad \forall x, n,$$

and taking $\limsup_n$, (9.5.38) and (9.3.14) yield

$$V_1^*(x) = J(f^\infty, x) \quad \forall x.$$

That is, f^∞ is ETC-optimal. $\square$

The conditions (9.5.37) and (9.5.38) are known in the MCP literature as the *conserving* and the *equalizing* properties of f^∞, respectively. There are many elementary examples illustrating that indeed if one of these properties fails the optimality of f^∞ cannot be guaranteed—see, for instance, Bertsekas [1], Puterman [1], Cavazos-Cadena and Montes-de-Oca [1], Schweitzer [1].

Finally, we will combine Theorems 9.5.9 and 9.5.12 to characterize V_1^* as the unique solution of the optimality equation within a certain class of functions.

9.5.13 Theorem. (Uniqueness of V_1^*.) *Suppose that Assumption 9.5.2 is satisfied, and let u be a function in $\mathcal{U}$ such that*

(a) *u satisfies the optimality equation (9.5.3);*

(b) *there is a decision function $f \in \mathbb{F}$ such that*

$$u(x) = c(x, f) + \int u(y) Q(dy|x, f) \quad \forall x \in X;$$

(c) $\limsup_{n \to \infty} E_x^\pi u(x_n) = 0 \quad \forall \pi, x.$

Then

$$u = V_1^*. \tag{9.5.39}$$

In other words, V_1^* is the *unique solution* of the optimality equation (9.5.3) in the class of functions $u \in \mathcal{U}$ that satisfy (a), (b), (c).

Proof. By Theorem 9.5.9, the present hypotheses (a) and (c) yield that $u \leq V_1^*$. On the other hand, (b) and (c) yield—as in Proposition 9.5.11— that $u(\cdot) = V_1(f^\infty, \cdot)$; hence $u \geq V_1^*$. This completes the proof of (9.5.39). $\square$

Notes on §9.5

In this section we followed Hernández-Lerma, Carrasco and Pérez-Hernández [1], although the main results are already known in some form or other—see Hinderer [1], Quelle [1], Rieder [2], Schäl [1].

9.6 The transient case

In this section we study the class of so-called *transient* MCMs (Markov control models) introduced by Veinott [1] in the case of *finite* state spaces X and *finite* action sets A. This class contains the *discounted* models (see Proposition 9.6.3), and, under a suitable condition, it is contained in the class of *convergent* models (Proposition 9.6.4). Many authors have extended Veinott's definition and results to *countable* state space X but most of these extensions very much depend on the "countability"—or "discreteness"—of X. To our knowledge, the only author who has studied the transient case in *Borel state spaces* is Pliska [1] and here we will follow, and slightly generalize, his approach.

The section is divided into four parts. Part A presents the transient MCM and some related models. In part B we go back to the conditions (1) and (2) in Proposition 9.3.6, whereas in part C we show that to verify if a MCM is transient it suffices to consider deterministic stationary policies. Finally, in part D we show that the *policy iteration* algorithm converges. [The convergence of the *value iteration* algorithm will follow from Corollary 9.6.5 and Theorem 9.3.5(b).]

A. Transient models

In view of Theorem 9.4.5, we may—and will—restrict ourselves to deal with randomized Markov policies $\pi = \{\varphi_t, t = 0, 1, \ldots\}$ in Π_{RM}. [See Definitions 8.2.1 and 8.2.2(a).] We will also use the following *notational conventions*:

- Given a stochastic kernel φ in Φ, the kernel $Q(\cdot|x, \varphi)$ in (8.2.5) will be written as $Q_\varphi(\cdot|x)$, so that

$$Q_\varphi(\cdot|x) := \int_A Q(\cdot|x, a)\varphi(da|x), \quad x \in X. \tag{9.6.1}$$

- Let $\pi = \{\varphi_0, \varphi_1, \ldots\}$ be a randomized Markov policy. Sometimes, if there is no risk for confusions, we shall write Q_{φ_t} in (9.6.1) as Q_t, i.e.,

$$Q_t(\cdot|x) := \int_A Q(\cdot|x, a)\varphi_t(da|x), \quad x \in X. \tag{9.6.2}$$

Moreover, using (9.6.2) and the "product" (or "composition") formula (7.2.14), we define the *t-step transition kernels*

$$Q_\pi^t := Q_0 Q_1 \dots Q_{t-1} \quad \text{for} \quad t = 1, 2, \dots, \qquad (9.6.3)$$

with $Q_\pi^0(\cdot|x) := \delta_x(\cdot)$.

In particular, if $\pi = \{\varphi_t\}$ is a *randomized stationary* policy φ^∞ (that is, $\varphi_t = \varphi$ for all $t = 0, 1, \dots$), then Q_π^t reduces to the t-step transition kernel in (8.2.11), i.e.,

$$Q_\pi^t(\cdot|x) = Q^t(\cdot|x, \varphi) \equiv Q_\varphi^t(\cdot|x) \quad \text{if } \pi = \varphi^\infty, \qquad (9.6.4)$$

where we have used (9.6.1), and similarly if $\pi = f^\infty$ is a *deterministic stationary* policy.

Observe, on the other hand, that the kernel (9.6.3) coincides with the marginal measure $\hat{\mu}_{x,t}^\pi$ in (9.4.13), i.e.,

$$\hat{\mu}_{x,t}^\pi(\cdot) = Q_\pi^t(\cdot|x) \qquad \forall x \in X, t = 0, 1, \dots . \qquad (9.6.5)$$

Indeed, from Lemma 9.4.3(b) and equations (9.4.18) and (9.6.1) we have

$$\hat{\mu}_{x,t}^\pi(\cdot) = \int_X Q_{t-1}(\cdot|y) \hat{\mu}_{x,t-1}^\pi(dy) \qquad \forall t = 1, 2, \dots, \qquad (9.6.6)$$

with $\hat{\mu}_{x,0}^\pi = \delta_x$ [by Lemma 9.4.3(a)]. Then iteration of (9.6.6) gives (9.6.5).

Throughout the following, $w : X \to [1, \infty)$ denotes a given *weight function* such that

$$\|Q_\varphi\|_w < \infty \qquad \forall \varphi \in \Phi,$$

where $\| \cdot \|_w$ denotes the (operator) w-norm in (7.2.8)—see also (7.2.7) or (7.2.10). We will use this norm to define the transient case but the reader should keep in mind that, in principle, in (9.6.7) below we can replace the w-norm by an arbitrary operator norm, as in Pliska [1].

9.6.1 Definition. The Markov control model is said to be **transient** if there is a constant k such that

$$\left\| \sum_{t=0}^\infty Q_\pi^t \right\|_w \leq k \qquad \forall \pi \in \Pi_{RM}. \qquad (9.6.7)$$

The w-norm in (9.6.7) can be written in several equivalent forms, such as [by (7.2.8)]:

$$\left\| \sum_{t=0}^\infty Q_\pi^t \right\|_w = \sup_X w(x)^{-1} \sum_{t=0}^\infty \int_X w(y) Q_\pi^t(dy|x) \qquad (9.6.8)$$

$$= \sup_X w(x)^{-1} \sum_{t=0}^\infty E_x^\pi w(x_t) \quad \text{[by (9.6.3)]}$$

$$= \sup_X w(x)^{-1} \sum_{t=0}^{\infty} \int w(y) \widehat{\mu}_{x,t}^{\pi}(dy) \quad \text{[by (9.6.5)]}$$

$$= \sup_X w(x)^{-1} \int_X w(y) \widehat{\mu}_x^{\pi}(dy) \quad \text{[by (9.4.12)]}.$$

9.6.2 Remark. It is clear that for (9.6.7) to be true the transition kernel $Q(\cdot|x, a)$ has to be very "special" and, in general, it is convenient to think of it as being a **substochastic** kernel, that is, $Q(X|x, a) \leq 1$ for all $x \in X$ and $a \in A(x)$. This is precisely the case for the *discounted* model in Proposition 9.6.3, below. A related situation occurs for **absorbing** MCMs in which there exists a Borel set $X_0 \subset X$ such that

$$Q(X_0|x, a) = 1 \quad \text{and} \quad c(x, a) = 0 \quad \forall x \in X_0, \ a \in A(x), \tag{9.6.9}$$

and, in addition,

$$\sup_{\Pi, X} E_x^{\pi}(\tau_0) < \infty, \quad \text{where} \quad \tau_0 := \inf\{t \geq 1 | x_t \in X_0\}. \tag{9.6.10}$$

The idea is that the state process $\{x_t\}$ "lives" in the complement of X_0 but once it reaches X_0 [which occurs in a finite expected time, by (9.6.10)] then, by (9.6.9), it remains there forever—it is "absorbed" by X_0—at zero cost. If the state and action spaces are both *finite*, then the transient and the absorbing models—as well as the class of so-called "contracting" models— are all equivalent; see Kallenberg [1]. $\square$

We next show that a discounted model can be transformed into a transient model.

Consider the usual MCM $\mathcal{M} = (X, A, \{A(x)|x \in X\}, Q, c)$ and let $0 < \alpha < 1$ be a "discount factor". We suppose that the weight function w satisfies Assumption 8.3.2(b); that is, there exists a constant β such that $1 \leq \beta < 1/\alpha$ and [as in (8.3.5)]

$$\sup_{A(x)} \int_X w(y) Q(dy|x, a) \leq \beta w(x) \quad \forall x \in X. \tag{9.6.11}$$

9.6.3 Proposition. *Suppose that (9.6.11) holds and let $\widetilde{\mathcal{M}}$ be a MCM that is the same as $\mathcal{M}$ except that the transition law Q is replaced by*

$$\widetilde{Q}(\cdot|x, a) := \alpha Q(\cdot|x, a). \tag{9.6.12}$$

Then $\widetilde{\mathcal{M}}$ is transient; in fact, for all $\pi \in \Pi_{RM}$,

$$\left\| \sum_{t=0}^{\infty} \widetilde{Q}_{\pi}^t \right\|_w \leq k, \quad \text{with} \quad k := 1/(1 - \alpha\beta). \tag{9.6.13}$$

Proof. Let $\pi = \{\varphi_t\}$ be an arbitrary randomized Markov policy. Then, by (9.6.12) and (9.6.3),

$$\widetilde{Q}_\pi^t = \alpha^t Q_\pi^t \qquad \forall t = 0, 1, \ldots.$$

Thus, by (9.6.11),

$$
\begin{aligned}
\int w(y)\widetilde{Q}_\pi^t(dy|x) &= \alpha^t \int w(y)Q_\pi^t(dy|x) \\
&\leq (\alpha\beta)^t w(x) \qquad \forall x, t,
\end{aligned}
$$

so that

$$\sum_{t=0}^\infty \int w(y)\widetilde{Q}_\pi^t(dy|x) \leq kw(x) \quad \forall x, \text{ with } k := 1/(1 - \alpha\beta).$$

This inequality and (9.6.8) give (9.6.13). $\square$

In short, Proposition 9.6.3 states that a discounted model can be transformed into a transient model if *part (b) in Assumption 8.3.2* is satisfied. If instead of (b) we use *part (a)* of Assumption 8.3.2, then a transient model is *convergent*, that is, (9.3.20) holds. In other words:

9.6.4 Proposition. *Suppose that the MCM $\mathcal{M}$ is transient and that there is a constant $\overline{c} \geq 0$ for which*

$$\sup_{A(x)} |c(x, a)| \leq \overline{c}w(x) \quad \forall x \in X. \tag{9.6.14}$$

Then $\mathcal{M}$ is convergent. In fact, if k satisfies (9.6.7), then (by Theorem 9.4.5)

$$\sup_\Pi E_x^\pi \left(\sum_{t=0}^\infty |c_t| \right) \leq \overline{c}kw(x) \quad \forall x \in X; \tag{9.6.15}$$

hence, in particular, the expected total cost $V_1(\pi, \cdot)$ is a function in the space $\mathbb{B}_w(X)$ for all policy π.

Proof. Let $\pi = \{\varphi_t\}$ be an arbitrary randomized Markov policy, and x an arbitrary initial state. Then, for each $t = 0, 1, \ldots$, (9.6.3) [or (9.6.5)] yields

$$
\begin{aligned}
E_x^\pi |c(x_t, a_t)| &= \int_X |c(y, \varphi_t)| Q_\pi^t(dy|x) \\
&\leq \overline{c} \int_X w(y)Q_\pi^t(dy|x) \quad \text{[by (9.6.14)]},
\end{aligned}
$$

so that, by (9.6.8) and (9.6.7),

$$\sum_{t=0}^\infty E_x^\pi |c(x_t, a_t)| \leq \overline{c}kw(x).$$

This gives (9.6.15) since π and x were arbitrary. $\square$

As an obvious consequence of (9.6.15) we have:

9.6.5 Corollary. *Under the hypotheses of Proposition 9.6.4, part (a) in Assumption 9.5.2 is satisfied, and also part (c)(ii) with* $\mathcal{U} := \mathbb{B}_w(X)$ *and* $W(\cdot) := \bar{c}kw(\cdot)$.

We give below (subsection B) conditions ensuring Assumption 9.5.2(b). First, however, we will comment on the relation between transient MCMs and transient Markov chains, and we will also present a class of transient MCMs (Example 9.6.7).

9.6.6 Remark. (Relation between transient MCMs and transient Markov chains.) Let us suppose that the MCM $\mathcal{M}$ is transient and for every Borel set $B \subset X$ define the **occupation time** (as in the first paragraph of §7.3.A)

$$\eta_B := \sum_{t=1}^{\infty} I_B(x_t). \qquad (9.6.16)$$

The, for any policy $\pi \in \Pi_{RM}$ and initital state $x \in X$, we see from (9.6.5) that

$$E_x^\pi[I_B(x_t)] = P_x^\pi(x_t \in B) = Q_\pi^t(B|x),$$

so that the **expected occupation time** of B satisfies [by (9.6.16), (9.6.8) and (9.6.7)]

$$E_x^\pi(\eta_B) = \sum_{t=1}^{\infty} Q_\pi^t(B|x) \leq kw(x). \qquad (9.6.17)$$

Thus, extending Definition 7.3.2(b) to *Markov control processes*, it follows from (9.6.17) that if $w(\cdot)$ is bounded on the set B, then B is *uniformly transient* for *all* policy π; that is, if $\sup_B w(x) =: w_B < \infty$, then

$$\sup_\Pi E_x^\pi(\eta_B) \leq kw_B,$$

where we have used Theorem 9.4.5 to write Π in lieu of Π_{RM}. Moreover, if the weight function w is *bounded* [which might be the case when dealing with bounded costs $c(x,a)$], then the whole state space X is a transient set, and so the state process $\{x_t\}$ is transient for all policy π. $\square$

9.6.7 Example. We show that if the weight function w satisfies an inequality of the form (7.3.9) or (7.3.10), namely,

$$\int w(y)Q(dy|x,a) \leq \beta w(x) + bl(x) \qquad \forall x \in X,\ a \in A(x), \qquad (9.6.18)$$

then the MCM $\mathcal{M}$ is transient. In (9.6.18) we suppose that β and b are nonnegative constants with $\beta < 1$, and $0 \leq l(\cdot) \leq 1$ is a measurable function. In addition, we assume that there exists a nonnegative constant $\gamma < 1$

such that for every randomized Markov policy $\pi = \{\varphi_t\}$ [using the notation (9.6.2)]

$$Q_\pi^t l \leq \gamma^t \qquad \forall t = 1, 2, \ldots . \tag{9.6.19}$$

More explicitly, we can write (9.6.19) as

$$E_x^\pi[l(x_t)] \leq \gamma^t \quad \forall x \in X, t = 1, 2, \ldots . \tag{9.6.20}$$

To prove that $\mathcal{M}$ is transient we will show that (9.6.7) holds with

$$k := (1 - \beta)^{-1}[1 + b(1 - \gamma)^{-1}], \tag{9.6.21}$$

which follows from straightforward calculations. Indeed, note that integration of both sides of (9.6.18) with respect to $\varphi_j(\cdot|x)$ yields

$$\int w(y)Q(dy|x, \varphi_j) \leq \beta w(x) + bl(x) \quad \forall x \in X, j = 0, 1, \ldots,$$

which using (9.6.2) can be written in abbreviated form as

$$Q_j w \leq \beta w + bl.$$

Then from (9.6.3) and (9.6.18) we obtain

$$\begin{aligned} Q_\pi^0 w &= w, \\ Q_\pi^1 w &= Q_0 w \leq \beta w + bl, \end{aligned}$$

and for $t = 2, 3, \ldots$

$$\begin{aligned} Q_\pi^t w &\leq \beta^t w + \beta^{t-1} bl + b \sum_{j=1}^{t-1} \beta^{t-1-j} Q_\pi^j l \\ &\leq \beta^t w + \beta^{t-1} bl + b \sum_{j=1}^{t-1} \beta^{t-1-j} \gamma^j \quad \text{[by (9.6.19)]}. \end{aligned}$$

It follows that

$$\sum_{t=0}^{\infty} Q_\pi^t w \leq (1 - \beta)^{-1}[w + bl + b\gamma(1 - \gamma)^{-1}].$$

Finally, since $l \leq 1 \leq w$, we get

$$\sum_{t=0}^{\infty} Q_\pi^t w \leq kw$$

with k as in (9.6.21). Therefore, by (9.6.8), $\mathcal{M}$ is transient. $\square$

B. Optimality conditions

Let us suppose that the MCM $\mathcal{M}$ is as in Proposition 9.6.4. Then, by (9.6.15) and Corollary 9.6.5, one can see that *all of the results in the previous sections will be true* provided that Assumption 9.5.2(b) is satisfied. This is indeed the case because conditions such as, say, (c) in Theorem 9.5.3 trivially hold for all u in $\mathcal{U} := \mathbb{B}_w(X)$; namely,

$$|E_x^\pi u(x_n)| \le E_x^\pi |u(x_n)| \le \|u\|_w E_x^\pi w(x_n) \to 0 \text{ as } n \to \infty, \qquad (9.6.22)$$

by (9.6.8). In turn, as already shown in several chapters, Assumption 9.5.2(b) holds in a number of cases. In particular, as in this section we are dealing with MCMs with a weighted w-norm, it is natural to consider the analogue of Assumptions 8.3.1–8.3.3 or 8.5.1–8.5.3. To fix ideas, we will consider the former; that is, we suppose:

9.6.8 Assumption. In addition to the hypotheses of Proposition 9.6.4 we assume that for every state $x \in X$:

(a) $A(x)$ is compact;

(b) $c(x, a)$ is l.s.c. in $a \in A(x)$;

(c) the function $u'(x, a) := \int u(y) Q(dy|x, a)$ is continuous in $a \in A(x)$ for every bounded function u in $\mathbb{B}_w(X)$; and

(d) the function $w'(x, a) := \int w(y) Q(dy|x, a)$ is continuous in $a \in A(x)$.

Using Assumption 9.6.8 we get, in particular, that Assumption 9.5.2(b) holds with $\mathcal{U}$ being the space $\mathbb{B}_w(X)$, as shown in the following lemma.

9.6.9 Lemma. *If Assumption 9.6.8 is satisfied, then the optimal n-stage cost J_n^* belongs to the space $\mathbb{B}_w(X)$ and, moreover, $J_n^* = T J_{n-1}^*$ for all $n = 1, 2, \ldots$, with $J_0^*(\cdot) := 0$; that is,*

$$J_n^*(x) = \min_{A(x)} \left[c(x, a) + \int_X J_{n-1}^*(y) Q(dy|x, a) \right] \quad \forall x \in X. \qquad (9.6.23)$$

In addition, for every $n = 1, 2, \ldots$, there is a decision function $f_n \in \mathbb{F}$ such that $f_n(x) \in A(x)$ attains the minimum in (9.6.23) for each $x \in X$, i.e.,

$$J_n^*(x) = c(x, f_n) + \int_X J_{n-1}^*(y) Q(dy|x, f_n) \quad \forall x \in X. \qquad (9.6.24)$$

Proof. The proof follows from a direct induction argument, using Lemma 8.3.8(a) [as in the proof of Proposition 8.3.9(b)]. $\square$

From Lemma 9.6.9 and Corollary 9.6.5 we immediately deduce our main optimality result in this section. Namely:

9.6.10 Theorem. *Suppose that Assumption 9.6.8 holds. Then:*

(a) *The value function V_1^* belongs to the space $\mathbb{B}_w(X)$ and there is a decision function $f_* \in \mathbb{F}$ such that $f_*(x) \in A(x)$ attains the minimum in the right-hand side of the optimality equation (9.5.3), i.e.,*

$$V_1^*(x) = c(x, f_*) + \int_X V_1^*(y)Q(dy|x, f_*) \quad \forall x \in X, \qquad (9.6.25)$$

and the corresponding deterministic stationary policy f_*^∞ is ETC-optimal.

(b) *A deterministic stationary policy f_*^∞ is ETC-optimal if and only if the decision function $f_* \in \mathbb{F}$ satisfies (9.6.25).*

(c) *V_1^* is the unique solution of the optimality equation (9.5.3) in the space $\mathbb{B}_v(X)$.*

Proof. (a) From Lemma 9.6.9, Theorem 9.3.5(b) and Proposition 9.3.6(2), it follows that V_1^* is a measurable function, whereas the fact that it belongs to $\mathbb{B}_w(X)$ is a consequence of Proposition 9.6.4. The existence of a decision function f_* that satisfies (9.6.24) results from Theorem 9.5.3 and Lemma 8.3.8(a).

Further, from (9.6.22) and Theorem 9.5.12 we obtain the optimality of f_*^∞ as well as part (b).

Finally, part (c) follows from (a) together with (9.6.22) and Theorem 9.5.13. $\square$

One can also obtain other optimality results related to Theorem 9.6.10. For instance, for every n, let $f_n \in \mathbb{F}$ be as in (9.6.24). Then from the result by Schäl mentioned in Note 4 of §8.4 and the convergence in (9.5.6), one can deduce the existence of a decision function (or selector) $f_* \in \mathbb{F}$ such that, for each $x \in X$, $f_*(x) \in A(x)$ is an accumulation point of the sequence $\{f_n(x)\}$ and, furthermore, *the deterministic stationary policy f_*^∞ is ETC-optimal.* We will also show, on the other hand, that the *policy iteration algorithm* converges (see subsection D). But first we will prove an interesting result according to which *to verify (9.6.7) it suffices to consider deterministic stationary policies.*

C. Reduction to deterministic policies

9.6.11 Theorem. *Suppose that the MCM $\mathcal{M}$ and the weight function w satisfy the conditions (a), (b) and (d) of Assumption 9.6.8. In addition, suppose that there is a constant $k \geq 0$ such that [using the notation (9.6.4) with $f \in \mathbb{F}$ in lieu of $\varphi \in \Phi$]*

$$\left\| \sum_{t=0}^{\infty} Q_f^t \right\|_w \leq k \quad \forall f \in \mathbb{F}. \qquad (9.6.26)$$

Then (9.6.7) holds and so $\mathcal{M}$ is transient.

The proof of Theorem 9.6.11 is based on a clever idea of Veinott [1] (see also Pliska [1]), which consists in showing that there exists a deterministic stationary policy that *maximizes* the expected total "reward" function given by

$$W(\pi, x) := \sum_{t=0}^{\infty} Q_\pi^t w(x) = \sum_{t=0}^{\infty} E_x^\pi w(x_t), \quad \pi \in \Pi_{RM}, \ x \in X, \quad (9.6.27)$$

where the one-step "reward" is nothing less than the weight function w itself! The precise statement is as follows.

9.6.12 Lemma. *Under the hypotheses of Theorem 9.6.11, there exists a function u^* in $\mathbb{B}_w(X)$ and a decision function $f_* \in \mathbb{F}$ such that, for all $x \in X$,*

$$
\begin{aligned}
u^*(x) &= \max_{A(x)} \left[w(x) + \int_X u^*(y) Q(dy|x, a) \right] \\
&= w(x) + \int_X u^*(y) Q(dy|x, f_*);
\end{aligned}
\quad (9.6.28)
$$

hence

$$u^*(x) = W(f_*^\infty, x) = \sup_\Pi W(\pi, x) \quad \forall x \in X. \quad (9.6.29)$$

Proof. Let $R : \mathbb{B}_w(X) \to \mathbb{B}_w(X)$, $u \mapsto Ru$, be the operator defined as

$$Ru(x) := \max_{A(x)} \left[w(x) + \int_X u(y) Q(dy|x, a) \right], \quad a \in A(x). \quad (9.6.30)$$

If we define a function $v(x, a) := w(x) + \int u(y) Q(dy|x, a)$ and apply Lemma 8.3.8(a) to $-v$, we can see that for each $u \in \mathbb{B}_w(X)$ there is a decision function $f \in \mathbb{F}$ such that $f(x) \in A(x)$ attains the maximum in (9.6.29) for all $x \in X$, i.e.,

$$Ru(x) = w(x) + \int u(y) Q(dy|x, f) \quad \forall x \in X. \quad (9.6.31)$$

This will give the *second equality* in (9.6.28) if we show that u^* is indeed a function in $\mathbb{B}_w(X)$. To prove this we will use the *value iteration* approach.

Let $\{u_n\}$ be the sequence in $\mathbb{B}_w(X)$ given by $u_0 := 0$ and $u_n := Ru_{n-1}$ for $n \geq 1$. As R is a *monotone* operator ($u \geq u'$ implies $Ru \geq Ru'$), the sequence $\{u_n\}$ is nondecreasing, i.e.,

$$u_{n+1} \geq u_n \geq 0 \quad \forall n = 0, 1, \ldots. \quad (9.6.32)$$

We will next show that

$$u_n \leq kw \quad \forall n = 0, 1, \ldots, \quad (9.6.33)$$

where k is the constant in (9.6.26). Suppose that (9.6.33) it is not true; that is, there exists an integer $n \geq 1$ such that

$$u_{n-1} \leq kw \quad \text{but} \quad u_n \not\leq kw.$$

Now, as in (9.6.31), let $f \in \mathbb{F}$ be a decision function for which

$$u_n(x) = Ru_{n-1}(x) = w(x) + \int u_{n-1}(y)Q(dy|x, f) \quad \forall x \in X,$$

and consider a new sequence $\{v_j\}$ with $v_0 := u_{n-1}$ and

$$v_j(x) = w(x) + \int v_{j-1}(y)Q(dy|x, f), \quad j = 1, \ldots.$$

Thus, using (9.6.4) with f in lieu of φ, we can write $v_j = w + Q_f v_{j-1}$ so that

$$v_j = \sum_{t=0}^{j-1} Q_f^t w + Q_f^j v_0.$$

Observe that, by (9.6.22), $Q_f^j v_0 \to 0$ since $v_0 := u_{n-1}$ is in $\mathbb{B}_w(X)$. Therefore, as $j \to \infty$,

$$v_j \uparrow \sum_{t=0}^{\infty} Q_f^t w \leq kw \quad [\text{by (9.6.26)}],$$

which contradicts that $v_1 := u_n \not\leq kw$. This proves (9.6.33).

Now, from (9.6.32) and (9.6.33), there exists a function u^* in $\mathbb{B}_w(X)$ such that $u^* \leq kw$ and $u_n(x) \uparrow u^*(x)$ for all $x \in X$. But, on the other hand, we also have

$$u_n(x) = Ru_{n-1}(x) \uparrow u^*(x) \quad \text{for all} \quad x \in X;$$

hence (by Remark 9.6.13, below), u^* satisfies $u^*(x) = Ru^*(x)$, which is the same as the first equality in (9.6.28), and the second equality follows from a previous remark—see (9.6.31).

Finally, the first equality in (9.6.29) follows from the Poisson equation

$$u^*(x) = w(x) + \int u^*(y)Q(dy|x, f_*)$$

in (9.6.28) combined with (9.6.22) and Proposition 9.5.11 (with the obvious changes in notation), whereas the second equality in (9.6.29) can be obtained from the "optimality equation" (9.6.28) by standard arguments [see for instance the proof of (9.5.29)]. $\square$

Having Lemma 9.6.12, the proof of Theorem 9.6.11 is straightforward:

Proof of Theorem 9.6.11. As $u^* \leq kw$, (9.6.29) yields

$$W(\pi, x) \leq u^*(x) \leq kw(x) \quad \forall \pi \in \Pi_{RM}, \ x \in X.$$

This implies (9.6.7), by the definition (9.6.27) of $W(\pi, x)$. □

9.6.13 Remark. Let Y and Z be two arbitrary sets, and let v an extended-real-valued function on $Y \times Z$. Then it is easily seen that

$$\sup_{Y} \sup_{Z} v(y, z) = \sup_{Z} \sup_{Y} v(y, z).$$

It follows that if v_n is a nondecreasing sequence of functions on $\mathbb{K}$ [the set in (8.2.1)] such that $v_n \uparrow v^*$, so that $v^* = \sup_n v_n$, then

$$\lim_{n \to \infty} \sup_{A(x)} v_n(x, a) = \sup_{A(x)} \lim_{n \to \infty} v_n(x, a) = \sup_{A(x)} v^*(x, a) \quad \forall x \in X.$$

Similarly (replacing "sup" and "nondecreasing" by "inf" and "nonincreasing", respectively), if $v_n \downarrow v^*$ then

$$\lim_{n \to \infty} \inf_{A(x)} v_n(x, a) = \inf_{A(x)} v^*(x, a). \qquad □$$

D. The policy iteration algorithm

Suppose that Assumption 9.6.8 is satisfied and, for some integer $i \geq 0$, let g_i^∞ be a deterministic stationary policy. Then, by Proposition 9.6.4, the corresponding expected total cost

$$v_i(\cdot) := V_1(f_i^\infty, \cdot)$$

is a function in the space $\mathbb{B}_w(X)$, which, by (9.5.30), satisfies the Poisson equation

$$v_i(x) = c(x, f_i) + \int_X v_i(y) Q(dy|x, f_i) \quad \forall x \in X. \tag{9.6.34}$$

Now let T be the dynamic programming operator in (9.5.1), and let $f_{i+1} \in \mathbb{F}$ be a decision function such that

$$v_i(x) \geq Tv_i(x) = c(x, f_{i+1}) + \int v_i(y) Q(dy|x, f_{i+1}) \quad \forall x \in X. \tag{9.6.35}$$

Iteration of this inequality gives

$$v_i(x) \geq J_n(f_{i+1}^\infty, x) + \int v_i(y) Q^n(dy|x, f_{i+1})$$

and letting $n \to \infty$ we obtain [by (9.6.22) and (9.3.14)]

$$v_i(x) \geq v_{i+1}(x) \quad \forall x \in X, \tag{9.6.36}$$

where $v_{i+1}(\cdot) := V_1(f_{i+1}^\infty, \cdot)$. This algorithm, which starts with an arbitrary policy $f_0^\infty \in \Pi_{DS}$ and which at every step chooses the next policy f_{i+1}^∞

according to (9.6.35), is called the **policy iteration algorithm** (PIA)— also known as *Howard's policy improvement algorithm*. We say that *the PIA converges* if the nonincreasing sequence $\{v_i\}$ satisfies that

$$v_i(x) \downarrow V_1^*(x) \quad \forall x \in X. \tag{9.6.37}$$

We will next show that this is indeed the case.

9.6.14 Theorem. (Convergence of the PIA algorithm.) *Under Assumption 9.6.8, the PIA converges.*

Proof. First note that *if for some i the equality holds in (9.6.36) for all $x \in X$, then $v_i = V_1^*$ and the deterministic stationary policy f_i^∞ is ETC-optimal.* This is a consequence of Theorem 9.6.10(b), (c) because if the equality holds in (9.6.36), then [by (9.6.35)] v_i satisfies the optimality equation $v_i = Tv_i$. This proves the theorem in the case of equality in (9.6.36).

In the general case, (9.6.36) and (9.6.15) imply the existence of a function v^* in the space $\mathbb{B}_w(X)$ such that

$$v_i(x) \downarrow v^*(x) \quad \forall x \in X. \tag{9.6.38}$$

We wish to show that v^* satisfies the optimality equation $v^* = Tv^*$, which, as in the previous paragraph, will give $v^* = V_1^*$. Observe that (9.6.38) and $v_i \geq Tv_i$ [see (9.6.35)], combined with Remark 9.6.13 and Fatou's Lemma, yield

$$v^* \geq Tv^*. \tag{9.6.39}$$

To obtain the reverse inequality observe that for all x in X:

$$
\begin{aligned}
v^*(x) &\leq v_i(x) & \text{[by (9.6.38)]} \\
&= c(x, f_i) + \int v_i(y) Q(dy|x, f_i) & \text{[by (9.6.34)]} \\
&\leq c(x, f_i) + \int v_{i-1}(y) Q(dy|x, f_i) & \text{[by (9.6.36)]} \\
&= Tv_{i-1}(x) & \text{[by (9.6.35)].}
\end{aligned}
$$

Thus, by (9.5.1),

$$v^*(x) \leq c(x, a) + \int v_{i-1}(y) Q(dy|x, a) \quad \forall a \in A(x).$$

It follows that, letting $i \to \infty$ and using (9.6.38) again

$$v^*(x) \leq c(x, a) + \int v^*(y) Q(dy|x, a) \quad \forall x \in X, \ a \in A(x),$$

which implies that $v^* \leq Tv^*$. Therefore, in view of (9.6.39), $v^* = Tv^*$, which gives $v^* = V_1^*$ and (9.6.37) follows. $\square$

For the sequence $\{f_i\}$ in (9.6.34), (9.6.35), it also holds the remark in the paragraph following the proof of Theorem 9.6.10; namely:

9.6.15 Corollary. *Suppose that Assumption 9.6.8 is satisfied and let $\{f_i\}$ be the sequence of decision functions in (9.6.34), (9.6.35). Then there exists a decision function $f_* \in \mathbb{F}$ such that, for each $x \in X$, $f_*(x) \in A(x)$ is an accumulation point of $\{f_i(x)\}$ and, moreover, the deterministic stationary policy f_*^∞ is ETC-optimal.*

Notes on §9.6

1. This section comes from Hernández-Lerma, Carrasco and Pérez-Hernández [1], which also studies the "stability" of an ETC-optimal deterministic stationary policy.

2. As noted at the beginning of the section, transient MCMs were introduced by Veinott [1] in the case of *finite* state spaces, and they have been analyzed by several authors in the *countable* state case. Here we followed Pliska's [1] approach partly because, to begin with, he is—to the best of our knowledge—the only author who has considered transient models in "general" (that is, Borel) state spaces, and partly because it fits very nicely in the weighted-norm framework of previous (and later) chapters. One should keep in mind, however, that instead of using the w-norm in (9.6.7), he uses an *arbitrary* operator norm. This allows of course some flexibility in the selection of the norm, but one should be careful how this selection is done. For instance, a natural choice in some problems might be to use the **total variation norm** $\|Q\|_{TV}$ of Q, which is obtained from (7.2.7) replacing the w-norm $\|u\|_w := \sup |u(x)|/w(x)$ by the **sup norm** $\|u\| := \sup_X |u(x)|$, i.e.,

$$\|Q\|_{TV} := \sup\{\|Qu\| : \|u\| \le 1\}; \qquad (9.6.40)$$

alternatively, we can use (7.2.10) replacing $\|\mu\|_w$ by the *total variation norm* $\|\mu\|_{TV}$ of the measure μ [see (7.2.3)], i.e.,

$$\|Q\|_{TV} := \sup\{\|Q\mu\|_{TV} : \|\mu\|_{TV} \le 1\}. \qquad (9.6.41)$$

At any rate, if we use this norm rather that $\|\cdot\|_w$ in (9.6.7), then all of the results in this section remain valid but for MCMs with a **bounded** one-stage cost $c(x, a)$, i.e., for some constant $\bar{c}$

$$|c(x, a)| \le \bar{c} \qquad \forall x \in X, \ a \in A(x), \qquad (9.6.42)$$

which is the class of models considered by Pliska.

3. Theorem 9.6.11 might be very convenient for checking whether (9.6.7) holds. For example, Pliska [1, Theor. 3.2] shows that the following four conditions (a) to (d) are *equivalent* and that each of them implies (9.6.26):

(a) There exists a number k such that $\sum_{t=0}^{\infty} \|Q_f^t\|_w \le k$ for all $f \in \mathbb{F}$.

(b) For each $\gamma > 0$, there exists an integer N such that $\|Q_f^t\|_w \leq \gamma$ for all $t \geq N$ and all $f \in \mathbb{F}$.

(c) For each $\gamma > 0$, there exists an integer t such that $\|Q_f^t\|_w \leq \gamma$ for all $f \in \mathbb{F}$.

(d) There exist positive numbers γ and δ, with $\gamma < 1$, such that $\|Q_f^t\|_w \leq \delta \gamma^t$ for all $f \in \mathbb{F}$ and $t = 0, 1, \ldots$.

Incidentally, since Pliska considers only models that satisfy (9.6.42), in the proof of Theorem 9.6.11 and Lemma 9.6.12 one may take $w(\cdot) \equiv 1$ in (9.6.27) and (9.6.28). On the other hand, he shows that, when using a "general" operator norm $\|\cdot\|$, if the transition kernel Q is such that

$$\|Q_f\| \leq 1 \qquad \forall f \in \mathbb{F}, \qquad (9.6.43)$$

then the *five* conditions (a)–(d) *and* (9.6.26) are all equivalent. For the w-norm, (9.6.43) can also be written [by (7.2.8)] as

$$\int_X w(y) Q_f(dy|x) \leq w(x) \qquad \forall x \in X, \; f \in \mathbb{F}.$$

4. For a *general*—not necessarily transient—MCM, Schäl [1, §17] shows that the compactness condition (a) in Assumption 9.6.8 can be relaxed; that is, under suitable hypotheses, the control constraint sets $A(x)$ may be allowed to be *noncompact*. Alternatively, one may replace the compactness of $A(x)$ by inf-compactness of the cost function $c(x, a)$, as in Chapters 4, 5, 6. On the other hand, for a *general* MCM, Rieder [3] has given conditions for the convergence of the PIA [see (9.6.37)]; without the appropriate assumptions, it is well known that this convergence may not hold (for counterexamples see, for instance, Rieder [3], Bertsekas [1], or Puterman [1]).

10
Undiscounted Cost Criteria

10.1 Introduction

A. Undiscounted criteria

Infinite-horizon Markov control problems can be roughly classified as being "discounted" or "undiscounted". The former, which have been the main subject of Chapter 4 and Chapter 8, are basically well understood in the sense that their theory can be safely considered to be complete. This is not the case for undiscounted problems—in fact, to start with, "undiscounted" can have several different meanings.

For example, if in the α-discounted cost criterion [see, for instance, (9.1.4)] we take $\alpha = 1$, then the undiscounted problem concerns of course the **expected total cost** (ETC)

$$V_1(\pi, x) := E_x^\pi \left[\sum_{t=0}^{\infty} c(x_t, a_t) \right] = \lim_{n \to \infty} J_n(\pi, x), \qquad (10.1.1)$$

where

$$J_n(\pi, x) := E_x^\pi \left[\sum_{t=0}^{n-1} c(x_t, a_t) \right], \quad n = 1, 2, \ldots \qquad (10.1.2)$$

is the *n-stage expected total cost*. [See (9.3.14).] In this case a policy π^* is "optimal", or **ETC-optimal**, if [as in (9.1.3)]

$$V_1(\pi^*, x) \le V_1(\pi, x) \quad \forall \pi \in \Pi, \ x \in X. \qquad (10.1.3)$$

The ETC criterion, however, has at least two main drawbacks: (i) it might not be well defined for all policies, and (ii) it does not look at how the finite horizon cost $J_n(\pi, x)$ varies with n.

A common way of coping with drawback (i) is to consider the long-run expected average cost (AC), already studied in Chapter 5 and which is also considered in the present chapter from a different perspective. But, as noted in Chapter 5, the AC criterion has the inconvenience of being extremely underselective, for it ignores what occurs in virtually any finite period of time.

Thus, to cope with (i) and (ii), it might be more "convenient" to put the undiscounted problem in the form introduced by Ramsey [1]: A policy π^* is said to **overtake** a policy π if for every initial state x there exists an integer $N = N(\pi^*, \pi, x)$ such that

$$J_n(\pi^*, x) \le J_n(\pi, x) \quad \forall n \ge N. \tag{10.1.4}$$

Then a policy π^* is called **strongly overtaking optimal** (strongly O.O.), or optimal in the sense of Ramsey, if π^* overtakes any other policy π.

Under suitable conditions—for instance, if the sequences in (10.1.4) converge [as in (9.3.14)]—strong overtaking optimality is equivalent to compare policies with respect to the ETC criterion. In general, as is to be expected, strong overtaking optimality turns out to be extremely overselective—there are many well-known, elementary (finite-state) MCPs for which there is no strongly O.O. policy (see §10.9).

Hence, we have to go back to the original undiscounted problem and put it in a form weaker that Ramsey's. This was done by Gale [1] and von Weizsäcker [1] introducing the notion of weak overtaking optimality: A policy π^* is said to be **weakly overtaking optimal** (weakly O.O.) if for every policy π, initial state x, and $\varepsilon > 0$, there exists an integer $N = N(\pi^*, \pi, x, \varepsilon)$ such that

$$J_n(\pi^*, x) \le J_n(\pi, x) + \varepsilon \quad \forall n \ge N. \tag{10.1.5}$$

This notion of optimality seems to be more "reasonable" than (10.1.4).

But then again, if we are interested in the way $J_n(\pi, x)$ varies with n, it seems to be even more reasonable to compare $J_n(\pi, x)$ with the *optimal n-stage cost*

$$J_n^*(x) := \inf_{\Pi} J_n(\pi, x). \tag{10.1.6}$$

We thus arrive at Flynn's [1] **opportunity cost** of π, given the initial state x, which is defined as

$$OC(\pi, x) := \limsup_{n \to \infty} [J_n(\pi, x) - J_n^*(x)]. \tag{10.1.7}$$

A policy π^* is said to be **opportunity cost-optimal** (OC-optimal) if

$$OC(\pi^*, x) = \inf_{\Pi} OC(\pi, x) \quad \forall x \in X. \tag{10.1.8}$$

Subtracting $J_n^*(x)$ on both sides of (10.1.5), it follows immediately that OC optimality is weaker than weak overtaking optimality, i.e.,

$$\pi^* \text{ weakly O.O.} \Rightarrow \pi^* \text{ OC-optimal.} \qquad (10.1.9)$$

On the other hand, instead of comparing $J_n(\pi, x)$ with $J_n^*(x)$, we might compare it with $nJ^*(x)$, where $J^*(x)$ is the **optimal expected average cost** (AC), i.e.,

$$J^*(x) := \inf_{\Pi} J(\pi, x), \qquad (10.1.10)$$

and

$$J(\pi, x) := \limsup_{n \to \infty} J_n(\pi, x)/n \qquad (10.1.11)$$

is the *long-run expected average cost* (AC for short) when using π, given the initial state x. Thus, instead of (10.1.7), we have **Dutta's [1] criterion**

$$D(\pi, x) := \limsup_{n \to \infty} [J_n(\pi, x) - nJ^*(x)]. \qquad (10.1.12)$$

In the terminology of Gale [1] and Dutta [1], a policy π for which $D(\pi, \cdot)$ is finite is said to be a "good" policy.

A policy π^* is called **D-optimal**, or optimal in the sense of Dutta, if

$$D(\pi^*, x) = \inf_{\Pi} D(\pi, x) \quad \forall x \in X. \qquad (10.1.13)$$

Again, in analogy with (10.1.9), it follows directly from the definitions that D-optimality is weaker than weak overtaking optimality, i.e.,

$$\pi^* \text{ weakly O.O.} \Rightarrow \pi^* \text{ D-optimal.} \qquad (10.1.14)$$

B. AC criteria

In general, the converse of (10.1.14) and (10.1.9) does not hold. But, on the other hand, we do have that if π^* is such that $OC(\pi^*, \cdot)$ is finite valued, then

$$\pi^* \text{ OC-optimal} \Rightarrow \pi^* \text{ AC-optimal,} \qquad (10.1.15)$$

and similarly for a D-optimal policy π^*, where **AC-optimal** means that

$$J(\pi^*, x) = J^*(x) \quad \forall x. \qquad (10.1.16)$$

Furthermore, if a policy π is *not* AC-optimal, that is, if $J(\pi, x) > J^*(x)$ for some state x, then a straightforward argument shows that $D(\pi, x)$ *and* $OC(\pi, x)$ *are both infinite*, i.e.,

$$J(\pi, x) > J^*(x) \quad \Rightarrow \quad D(\pi, x) = +\infty \text{ and } OC(\pi, x) = +\infty. \qquad (10.1.17)$$

From the latter fact, together with (10.1.9), (10.1.14) and (10.1.15), we can see not only that there are several "natural" undiscounted cost criteria and that they all lead in an obvious manner to the AC criterion, but also that [by (10.1.17)] to find optimal policies with *finite* undiscounted costs it suffices to restrict ourselves to the class of AC-optimal policies.

In fact, one of the main objectives in this chapter is to show that *within the class* Π_{DS} *of deterministic stationary policies*, all of the following concepts

$$\text{weakly } O.O., \; OC\text{-optimal, } D\text{-optimal,} \qquad (10.1.18)$$
$$\text{and } \mathbf{bias\text{-}optimal} \; \text{are equivalent,}$$

where bias optimality is an AC-related criterion defined in §10.3.D.

C. Outline of the chapter

Section 10.2 introduces the Markov control model dealt with in this chapter. In §10.3 we present the main results, starting with AC-optimality and then going to special classes of AC-optimal policies (namely, canonical and bias-optimal polcies), and to the undiscounted criteria in subsection A, above.

For the sake of "continuity" in the exposition, §10.3 contains only the statements of the main results—the proofs are given in §10.4 to §10.8. The chapter closes in §10.9 with some examples.

10.1.1 Remark. Theorem 10.3.1 establishes the existence of solutions to the *Average Cost Optimality Inequality* (ACOI) by using the same approach already used to prove Theorem 5.4.3, namely, the "vanishing discount" approach. However, to obtain Theorem 10.3.1 we cannot just refer to Theorem 5.4.3 because the hypotheses of these theorems are *different*, an important difference being that the latter theorem assumes the cost-per-stage $c(x, a)$ to be *nonnegative*—this condition is not required in the present chapter. Moreover, the hypotheses here (Assumptions 10.2.1 and 10.2.2) are on the components of the control model itself, whereas in Chapter 5 the assumptions are based on the associated α-discounted cost problem.

10.1.2 Remark. The reader should be warned that not everyone uses the same terminology for the several optimality criteria in this chapter. For instance, what we call "strong overtaking optimality" [see (10.1.4)] is referred to as "overtaking optimality" by, say, Fernández-Gaucherand *et al.* [1], whereas our "weak O.O." [see (10.1.5)] is sometimes called "catching-up" in the economics literature, for example, in Dutta [1] and Gale [1].

10.2   Preliminaries

Let $\mathcal{M} := (X, A, \{A(x)|x \in X\}, Q, c)$ be the Markov control model introduced in §8.2. In this chapter we shall impose two sets of hypotheses on $\mathcal{M}$. The first one, Assumption 10.2.1 below, is in fact a combination of Assumptions 8.3.1, 8.3.2 and 8.3.3, with (8.3.5) being replaced by an inequality of the form (8.3.11) [see also (7.3.9) or (7.3.10)].

A. Assumptions

10.2.1 Assumption. For every state $x \in X$:

(a) The control-constraint set $A(x)$ is compact;

(b) The cost-per-stage $c(x, a)$ is l.s.c. in $a \in A(x)$; and

(c) The function $a \mapsto \int_X u(y)Q(dy|x, a)$ is continuous on $A(x)$ for every function u in $\mathbb{B}(X)$.

Moreover, there exists a weight function $w \geq 1$, a bounded measurable (possible constant) function $b \geq 0$, and nonnegative constants $\bar{c}$ and β, with $\beta < 1$, such that for every $x \in X$:

(d) $\sup_{A(x)} |c(x, a)| \leq \bar{c}w(x)$;

(e) $a \mapsto \int_X w(y)Q(dy|x, a)$ is continuous on $A(x)$; and

(f) $\sup_{A(x)} \int_X w(y)Q(dy|x, a) \leq \beta w(x) + b(x)$.   (10.2.1)

As usual, one of the main purposes of parts (a) through (e) in Assumption 10.2.1 is to assure the existence of suitable "measurable selectors" $f \in \mathbb{F}$, as in, for instance, Lemma 8.3.8 and Proposition 8.3.9(b). On the other hand, the Lyapunov-like inequality in part (f) yields an "expected growth" condition on the weight function w, as in (8.3.29) when $b = 0$; see also Remark 8.3.5(a), Example 9.6.7 or (10.4.2) below.

We will next introduce an additional assumption according to which the Markov chains associated to deterministic stationary policies are w-geometrically ergodic (Definition 7.3.9) uniformly on Π_{DS}. To state this assumption it is convenient to slightly modify the notation (8.2.6) and (8.2.11) for the stochastic kernel $Q(\cdot|x, f)$, which now will be written as $Q_f(\cdot|x)$; that is, for every $f \in \mathbb{F}$, $B \in \mathcal{B}(X)$, $x \in X$, and $t = 0, 1, \ldots,$

$$Q_f^t(B|x) := Q^t(B|x, f) = P_x^f(x_t \in B).$$ (10.2.2)

For $t = 0$, (10.2.2) reduces, of course, to

$$Q_f^0(B|x) = \delta_x(B) = I_B(x).$$ (10.2.3)

Observe that [by (10.2.23) below and the argument to obtain (8.3.44)] the inequality (10.2.1) *is equivalent to*

$$\int_X w(y)Q_f(dy|x) \leq \beta w(x) + b(x) \quad \forall f \in \mathbb{F}, \ x \in X.$$ (10.2.4)

Thus, multiplying by $w(x)^{-1}$ one can see that the stochastic kernels Q_f have a uniformly bounded w-norm. Actually, as in Remark 7.3.12, the w-norm of Q_f, i.e.,

$$\|Q_f\|_w := \sup_X w(x)^{-1} \int_X w(y) Q_f(dy|x) \qquad (10.2.5)$$

satisfies

$$\|Q_f\|_w \le \beta + \|b\|_w \quad \forall f \in \mathbb{F}$$

or, alternatively, by (7.2.2),

$$\|Q_f\|_w \le \beta + \|b\| \qquad \forall f \in \mathbb{F}, \qquad (10.2.6)$$

where $\|b\| := \sup_X |b(x)|$ is the *sup-norm* of b. (Recall that b is assumed to be *bounded*.)

We will now state our second main assumption, and in the remainder of this section we discuss some consequences of—as well as some sufficient conditions for—it.

10.2.2 Assumption. (w-Geometric ergodicity.) For every decision function $f \in \mathbb{F}$ there exists a p.m. μ_f on X such that

$$\|Q_f^t - \mu_f\|_w \le R\rho^t \quad \forall t = 0, 1, \dots, \qquad (10.2.7)$$

where $R > 0$ and $0 < \rho < 1$ are constants independent of f.

This assumption implies in particular that μ_f *is the unique i.p.m. of* Q_f in $\mathbb{M}_w(X)$. (See the paragraph after Definition 7.3.9.) Furthermore, integration of both sides of (10.2.4) with respect to μ_f yields [by (7.3.1)]

$$\int w d\mu_f \le \beta \int w d\mu_f + \|b\|,$$

so that

$$\|\mu_f\|_w := \int w d\mu_f \le \|b\|/(1 - \beta) \quad \forall f \in \mathbb{F}. \qquad (10.2.8)$$

In words, (10.2.8) means that the w-norm of μ_f is *uniformly bounded* in $f \in \mathbb{F}$.

On the other hand, in analogy with (7.3.8), we may rewrite (10.2.7) as

$$\left| \int_X u(y) Q_f^t(dy|x) - \mu_f(u) \right| \le \|u\|_w R\rho^t w(x), \quad t = 0, 1, \dots, \qquad (10.2.9)$$

for all u in $\mathbb{B}_w(X)$, where

$$\mu_f(u) := \int_X u d\mu_f. \qquad (10.2.10)$$

Observe that this integral is well-defined because every function in $\mathbb{B}_w(X)$ is μ_f-integrable, that is, with $L_1(\mu_f) := L_1(X, \mathcal{B}(X), \mu_f)$,

$$u \in L_1(\mu_f) \qquad \forall u \in \mathbb{B}_w(X), \ f \in \mathbb{F}. \tag{10.2.11}$$

In fact, by (7.2.1) and (10.2.8),

$$\int |u| d\mu_f \le \|u\|_w \int w d\mu_f \le \|u\|_w \|b\|/(1-\beta) \tag{10.2.12}$$

for all $u \in \mathbb{B}_w$ and $f \in \mathbb{F}$.

B. Corollaries

We should remark that (10.2.11) is just a restatement of Theorem 7.5.10(a) applied to the stochastic kernel $P := Q_f$ and the i.p.m. $\mu = \mu_f$. Similarly, from parts (b) and (d) of Theorem 7.5.10 we obtain

$$\lim_{n \to \infty} \|Q_f^n u\|_w/n = 0 \qquad \forall u \in \mathbb{B}_w(X), \ f \in \mathbb{F}, \tag{10.2.13}$$

and if $u \in \mathbb{B}_w(X)$ is *harmonic*—or *invariant*—with respect to Q_f (meaning: $Q_f u(x) = u(x)$ for all $x \in X$), then u is the constant $\mu_f(u)$, i.e.,

$$Q_f u = u \ \Rightarrow \ u(x) = \mu_f(u) \quad \forall x \in X. \tag{10.2.14}$$

We will now consider the strictly unichain *Poisson equation* (Definition 7.5.9) for the kernel Q_f and the "charge" $c(\cdot, f)$ which sometimes we shall write as $c_f(\cdot)$, i.e., [using (8.2.6)],

$$c_f(x) := c(x, f) = c(x, f(x)) \qquad \forall x \in X, \ f \in \mathbb{F}. \tag{10.2.15}$$

As in (10.1.2),

$$J_n(f^\infty, x) := E_x^{f^\infty} \left[\sum_{t=0}^{n-1} c_f(x_t) \right] = \sum_{t=0}^{n-1} \int_X c_f(y) Q_f^t(dy|x) \tag{10.2.16}$$

denotes the n-stage expected total cost when using the deterministic stationary policy $f^\infty \in \Pi_{DS}$. Similarly, by (10.1.11), the long-run *expected average cost* (AC) is

$$J(f^\infty, x) := \limsup_{n \to \infty} J_n(f^\infty, x)/n. \tag{10.2.17}$$

Let $J(f)$ be the constant

$$J(f) := \mu_f(c_f) = \int_X c_f(y) \mu_f(dy), \qquad f \in \mathbb{F}. \tag{10.2.18}$$

10.2.3 Proposition. (The Poisson equation.) *Let $f \in \mathbb{F}$ be an arbitrary decision function, and f^∞ the corresponding deterministic stationary policy. Then [with $\bar{c}$ as in Assumption 10.2.1(d), and R, ρ as in (10.2.7)]*

(a) $|J_r(f^\infty, x) - nJ(f)| \leq \bar{c}Rw(x)/(1-\rho)$ $\forall x \in X, \; n = 1, 2, \ldots,$

so that, in particular,

(b) $J(f^\infty, x) = \lim\limits_{n \to \infty} J_n(f^\infty, x)/n = J(f)$ $\forall x \in X$ [cf. (10.2.17)], and

(c) The function

$$h_f(x) := \lim_{n \to \infty} [J_n(f^\infty, x) - nJ(f)]$$
$$= \sum_{t=0}^{\infty} E_x^{f^\infty}[c_f(x_t) - J(f)] \tag{10.2.19}$$

belongs to $\mathbb{B}_w(X)$ since, by (a),

$$\|h_f\|_w \leq \bar{c}R/(1-\rho). \tag{10.2.20}$$

(d) The pair $(J(f), h_f)$ in $\mathbb{R} \times \mathbb{B}_w(X)$ is the unique solution of the strictly unichain Poisson equation

$$J(f) + h_f(x) = c_f(x) + \int_X h_f(y)Q_f(dy|x), \qquad x \in X, \tag{10.2.21}$$

that satisfies the condition

$$\mu_f(h_f) := \int_X h_f d\mu_f = 0. \tag{10.2.22}$$

Proof. Part (a) follows directly from (10.2.9) applied to $u = c_f$, together with the elementary fact

$$\sum_{t=0}^{n-1} \rho^t = (1 - \rho^n)/(1-\rho) \leq 1/(1-\rho),$$

which was already used in (7.5.32).

Part (d), on the other hand, is a consequence of Theorem 7.5.10(e). $\square$

Additional relevant information on the Poisson equation can be found in §7.5; see, for instance, Theorem 7.5.5 and Remark 7.5.11.

The function h_f in (10.2.19), that is, the unique solution of (10.2.21)–(10.2.22) in $\mathbb{B}_w(X)$ will be referred to as the **bias** of the decision function $f \in \mathbb{F}$ or the deterministic stationary policy $f^\infty \in \Pi_{DS}$.

C. Discussion

The most restrictive hypotheses are of course (10.2.4) [which is equivalent to (10.2.1)] and (10.2.7). The former can be reduced to checking (10.2.4) for a *single* decision function.

Indeed, by Assumption 10.2.1(e), we can apply Lemma 8.3.8(a) to the function

$$v(x, a) := - \int w(y)Q(dy|x, a)$$

to see that there exists a decision function $f_w \in \mathbb{F}$ such that

$$\int w(y)Q_{f_w}(dy|x) = \max_{A(x)} \int w(y)Q(dy|x, a) \quad \forall x \in X. \qquad (10.2.23)$$

Then

$$\int w(y)Q_{f_w}(dy|x) \geq \int w(y)Q_f(dy|x) \quad \forall f \in \mathbb{F}, \ x \in X,$$

and so if the inequality in (10.2.4) holds for f_w, i.e.,

$$\int w(y)Q_{f_w}(dy|x) \leq \beta w(x) + b(x), \quad x \in X, \qquad (10.2.24)$$

then the inequality holds for all $f \in \mathbb{F}$.

To verify Assumption 10.2.2, on the other hand, we may try to apply suitable "MCP-versions" of the results in §7.3D—see Theorems 7.3.10, 7.3.11, 7.3.14, and Remark 7.3.13. [If the cost-per-stage $c(x, a)$ is *bounded*, it suffices to verify conditions of the form (7.3.13) of (7.3.16) for the total variation norm.] For example, one can easily check that the same proof of Theorem 7.3.14 yields the following.

10.2.4 Proposition. *Suppose there exists a weight function $w \geq 1$, a number $0 < \rho < 1$, a state $x^* \in X$, and a control action $a^* \in A(x^*)$ that satisfy*

(a) $w_* := \int_X w(y)Q(dy|x^*, a^*) < \infty$, *and*

(b₁) $\|Q_f(\cdot|x) - Q_f(\cdot|x')\|_w \leq \rho[w(x) + w(x')] \quad \forall f \in \mathbb{F}, \ x, x' \in X,$

or, equivalently,

(b₂) $\|\theta Q_f\|_w \leq \rho\|\theta\|_w$ *for every signed measure $\theta \in \mathbb{M}_w(X)$ with $\theta(X) = 0.$*

Then, for each $f \in \mathbb{F}$, the stochastic kernel Q_f satisfies the conclusions of Theorem 7.3.14, that is, for each $f \in \mathbb{F}$:

(i) $Q_f w \leq \rho w + b$, *with* $b := \rho w(x^*) + w_*$;

(ii) $\|Q_f\|_w \leq \rho + b < \infty$;

(iii) *Assumption 10.2.2 holds for some constant $R \leq 1 + b/(1 - \rho)$, and ρ as in* (b₁), (b₂).

In other words, the hypotheses of Proposition 10.2.4 yield both (10.2.7) and (10.2.4) with $\beta = \rho$ and $b(\cdot)$ the constant function $b = \rho w(x^*) + w_*$. These hypotheses were introduced for MCPs by Gordienko, Montes-de-Oca and Minjárez-Sosa [1], extending ideas of Kartashov [2] for noncontrolled Markov chains.

Similarly, the MCP-version of Theorem 7.3.10 is the following proposition, for the proof of which the reader is referred to Gordienko and Hernández-Lerma [2, Lemmas 3.3, and 3.4].

10.2.5 Proposition. *Suppose that, for each $f \in \mathbb{F}$, the stochastic kernel Q_f has a unique i.p.m. μ_f, and, in addition, there is a weight function $w \geq 1$, a p.m. ν in $\mathbb{M}_w(X)$ and positive numbers γ and β, with $\beta < 1$, that satisfy the following. For each $f \in \mathbb{F}$ there exists a measurable function $0 \leq l_f(\cdot) \leq 1$ such that*

(i) $Q_f(B|x) \geq l_f(x)\nu(B)$ $\forall x \in X, B \in \mathcal{B}(X)$;

(ii) $\nu(l_f) := \int_X l_f d\nu \geq \gamma$;

(iii) $\nu(w) := \int_X w d\nu = \|\nu\|_w < \infty$; and

(iv) $\int_X w(y)Q_f(dy|x) \leq \beta w(x) + l_f(x)\nu(w)$ for all $x \in X$.   (10.2.25)

Then there exist constants $R \geq 0$ and $0 < \rho < 1$ independent of $f \in \mathbb{F}$ for which (10.2.7) holds.

Following ideas of Kartashov [3, Theor. 6], [5, Theor. 3.6], it is possible to get estimates of the constants R and ρ in the conclusion of Proposition 10.2.5—see Gordienko and Hernández-Lerma [2] for details.

10.3   From AC optimality to undiscounted criteria

As was already noted in §10.1, undiscounted criteria lead directly to the average cost (AC) criterion—see (10.1.15), and also (10.1.9), (10.1.14), and above all (10.1.17).

However, *a priori* it is not obvious how to go in the reverse direction, from AC optimality to the undiscounted criteria, because AC-optimal policies can have a "nasty" finite-horizon behavior. For instance, we can have two AC-optimal policies π and π' with very different n-steps costs $J_n(\pi, \cdot)$, $J_n(\pi', \cdot)$ for all n, e.g.,

$$J_n(\pi, \cdot) = J_n(\pi', \cdot) + n^\theta, \text{ with } 0 < \theta < 1,$$

so that

$$\lim_{n \to \infty} [J_n(\pi, \cdot) - J_n(\pi', \cdot)] = \infty. \tag{10.3.1}$$

Thus to go from AC optimality to a result such as, say, (10.1.5) seems to be virtually impossible.

By consequence, to relate the AC criterion to the undiscounted criteria in §10.1.A it is necessary to study subclasses of AC-optimal policies with special properties, which is the main objective in this section. For the sake of "continuity" in the exposition, here we only state the main results; they are proved in subsequent sections.

The program for this section is as follows. We begin by distinguishing a class of deterministic stationary policies which are obtained from the *Average Cost Optimality Inequality* (ACOI), and then they are used to obtain the *Average Cost Optimality Equation* (ACOE). (See Theorems 10.3.1 and 10.3.6, respectively.) From the latter equation we shall obtain the subclass of so-called *canonical* policies, and, finally, a further subclass of *bias-optimal* policies (Theorem 10.3.10). In short, we get a hierarchy

$$\mathbb{F} \supset \mathbb{F}_{AC} \supset \mathbb{F}_{ca} \supset \mathbb{F}_{bias} \qquad (10.3.2)$$

of subsets of decision functions, where $\mathbb{F}_{AC}$ denotes the family of decision functions $f \in \mathbb{F}$ for which the corresponding deterministic stationary policy $f^\infty \in \Pi_{DS}$ is AC-optimal, and similarly for the subfamilies $\mathbb{F}_{ca}$ and $\mathbb{F}_{bias}$ of canonical and bias-optimal decision functions, respectively.

A particularly interesting feature of canonical policies is that they exclude possibilities such as (10.3.1). Namely, it will be shown that a canonical policy f^∞ satisfies a relation of the form

$$J_n(f^\infty, x) = n\rho^* + h^*(x) - E_x^{f^\infty} h^*(x_n) \quad \forall x \in X,\ n = 0, 1, \ldots, \quad (10.3.3)$$

for some constant ρ^* and some function h^* in $\mathbb{B}_w(X)$. [See (10.3.21).] Thus, if g^∞ is any other such policy, then, by (10.2.9) and (10.2.10),

$$\lim[J_n(f^\infty, x) - J_n(g^\infty, x)] = -\int h^* d\mu_f + \int h^* d\mu_g \quad \forall x \in X, \quad (10.3.4)$$

and so the finite-horizon behavior of two canonical policies cannot differ "too much".

Moreover, it will be shown [in (10.3.30), (10.3.31)] that if f^∞ is *bias-optimal* (that is, f is in $\mathbb{F}_{bias}$) then

$$\int h^* d\mu_f \geq \int h^* d\mu_g \quad \forall g \in \mathbb{F}_{AC}, \qquad (10.3.5)$$

so that instead of (10.3.4) we will get

$$\limsup[J_n(f^\infty, \cdot) - J_n(g^\infty, \cdot)] \leq 0 \quad \forall f \in \mathbb{F}_{bias},\ g \in \mathbb{F}_{AC}, \qquad (10.3.6)$$

with equality if and only if g is in $\mathbb{F}_{bias}$ also. The inequality (10.3.6) will turn out to be related to the defining property (10.1.5) of weakly O.O. policies. Combining these remarks with (10.1.9) and (10.1.14) we will be on our way to prove statement (10.1.18), which is the result that closes this section (see Theorem 10.3.11 and Corollary 10.3.12).

A. The AC optimality inequality

The following theorem states the existence of a solution (ρ^*, h_0) to the ACOI, as well as the existence of a deterministic stationary policy f_0^∞ that is AC-optimal in Π_{DS}, i.e.,

$$J(f_0^\infty, x) = \inf_{\Pi_{DS}} J(f^\infty, x) \quad \forall x \in X, \tag{10.3.7}$$

or, equivalently [by (10.2.18) and Proposition 10.2.3(b)],

$$J(f_0^\infty, x) = J(f_0) = \inf_{\mathbb{F}} J(f) \quad \forall x \in X. \tag{10.3.8}$$

10.3.1 Theorem. (The ACOI.) *Suppose that Assumptions 10.2.1 and 10.2.2 are satisfied. Then there exists a constant ρ^*, a function h_0 in $\mathbb{B}_w(X)$, and a decision function $f_0 \in \mathbb{F}$ such that for each state $x \in X$ the Average Cost Optimality Inequality (ACOI) holds, i.e.,*

$$\rho^* + h_0(x) \geq \min_{A(x)} \left[c(x,a) + \int_X h_0(y) Q(dy|x,a) \right], \tag{10.3.9}$$

and, moreover, $f_0(x) \in A(x)$ attains the minimum in (10.3.9), so that [using the notation (10.2.2) and (10.2.15)]

$$\rho^* + h_0(x) \geq c_{f_0}(x) + \int_X h_0(y) Q_{f_0}(dy|x). \tag{10.3.10}$$

In addition, the deterministic stationary policy $f_0^\infty \in \Pi_{DS}$ corresponding to f_0 is optimal for (that is, minimizes) the AC criterion in Π_{DS}, with ρ^* being the optimal value, i.e., f_0^∞ satisfies (10.3.7)–(10.3.8) and

$$\rho^* = J(f_0) = \inf_{\mathbb{F}} J(f). \tag{10.3.11}$$

In fact, any decision function f_0 that satisfies (10.3.10) also satisfies (10.3.11).

Proof. See §10.4.

10.3.2 Remark. (AC-optimality of f_0.) From the proof of Theorem 10.3.1 (see Lemma 10.4.3 and Remark 10.4.4) it will be clear that if the one-stage cost function $c(x,a)$ is *nonnegative*, then f_0^∞ is *AC-optimal* and ρ^* is *he AC value function*; that is, with $J^*(\cdot)$ as in (10.1.10), we can rewrite (10.3.11) as

$$\rho^* = J(f_0) = J^*(x) \quad \forall x \in X. \tag{10.3.12}$$

Furthermore, any decision function f_0 that satisfies (10.3.10) also satisfies (10.3.12). $\square$

B. The AC optimality equation

A deterministic stationary policy f_*^∞ is said to be **canonical** (or AC-canonical) if, for some function h^* in $\mathbb{B}_w(X)$, *equality* holds in (10.3.9) and (10.3.10) when h_0 and f_0 are replaced by h^* and f_*, respectively; that is, for each state $x \in X$,

$$\rho^* + h^*(x) = \min_{A(x)} \left[c(x,a) + \int_X h^*(y)Q(dy|x,a) \right], \qquad (10.3.13)$$

and

$$\rho^* + h^*(x) = c_{f_*}(x) + \int_X h^*(y)Q_{f_*}(dy|x). \qquad (10.3.14)$$

As in Chapter 5, (10.3.13) will be referred to as the *Average Cost Optimality Equation* (ACOE). From (10.3.13) and (10.2.13) it is easy to see (as in §5.2) that ρ^* is the AC value function, and that any deterministic stationary policy f_*^∞ for which (10.3.14) holds is AC-optimal, i.e.,

$$\rho^* = J(f_*) = J^*(x) \quad \forall x \in X. \qquad (10.3.15)$$

The converse is not true; that is, an AC-optimal policy is not necessarily canonical. [See, for instance, Example 10.9.1.] Nevertheless, in Theorem 10.3.6(b) we give conditions under which an *AC-optimal policy is "almost everywhere (a.e.)" canonical.* Furthermore, the proof of Theorem 10.3.6(a) will show in fact that *the existence of a deterministic stationary policy that minimizes the AC criterion in* Π_{DS} [as f_0 in (10.3.11)] *implies the existence of a canonical policy.*

The term "canonical" comes from Definition 5.2.1 and Theorem 5.2.2, which are next briefly recalled.

Let $h : X \to \mathbb{R}$ be a given measurable function, and let $J_n(\pi, x, h)$ be the n-stage expected total cost with *terminal cost* function h; that is, for each policy π and initial state x,

$$J_0(\pi, x, h) := h(x),$$

and for $n = 1, 2, \ldots$

$$J_n(\pi, x, h) := E_x^\pi \left[\sum_{t=0}^{n-1} c(x_t, a_t) + h(x_n) \right]. \qquad (10.3.16)$$

Of course, we have

$$J_n(\pi, x, h) = J_n(\pi, x) + E_x^\pi h(x_n), \qquad (10.3.17)$$

where $J_n(\pi, x) = J_n(\pi, x, 0)$ is the n-stage cost in (10.1.2). The value function corresponding to $J_n(\pi, x, h)$ is

$$J_n^*(x, h) := \inf_\Pi J_n(\pi, x, h). \qquad (10.3.18)$$

10.3.3 Definition. Let (ρ, h, f) be a triplet consisting of two real-valued measurable functions ρ and h on X, and a decision function $f \in \mathbb{F}$. We call (ρ, h, f) a **canonical triplet** if

$$J_n(f^\infty, x, h) = n\rho(x) + h(x) = J_n^*(x, h) \quad \forall x \in X, \ n = 0, 1, \ldots . \quad (10.3.19)$$

A decision function $f \in \mathbb{F}$, or the corresponding policy f^∞ in Π_{DS}, is said to be **canonical** if it enters into some canonical triplet.

In the context of this chapter, $\rho(\cdot) \equiv \rho^*$ is a constant, and the connection between a canonical triplet and the ACOE (10.3.13)–(10.3.14) is provided by Theorem 5.2.2, which can be especialized as follows.

10.3.4 Theorem. (ACOE $\Leftrightarrow$ canonical triplet.) *A triplet (ρ^*, h^*, f_*) consisting of a number ρ^*, a function h^* in $\mathbb{B}_w(X)$, and a decision function $f_* \in \mathbb{F}$ satisfies the ACOE (10.3.13)–(10.3.14) if and only if (ρ^*, h^*, f_*) is a canonical triplet, i.e.,*

$$J_n(f_*^\infty, x, h^*) = n\rho^* + h^*(x) = J_n^*(x, h^*) \quad \forall x \in X, \ n = 0, 1, \ldots . \quad (10.3.20)$$

Let us rewrite the first equality in (10.3.20) as

$$J_n(f_*^\infty, x) + E_x^{f_*^\infty} h^*(x_n) = n\rho^* + h^*(x). \quad (10.3.21)$$

Then, by Theorem 7.5.5(a), (b), we can see that (10.3.21) is just another way of writing the *Poisson equation* (10.3.14). In a little more generality, the Poisson equation (10.2.21) is "equivalent" [in the sense of Theorem 7.5.5(a), (b)] to

$$J_n(f^\infty, x) + E_x^{f^\infty} h_f(x_n) = nJ(f) + h_f(x) \quad (10.3.22)$$

for all $x \in X$ and $n = 0, 1, \ldots .$

We will require the following assumption, which uses the concept of λ-irreducibility [Definition 7.3.1(a), (a1), (a2)] of the stochastic kernel Q_f in (10.2.2).

10.3.5 Assumption. (Irreducibility.) There exists a σ-finite measure λ on $\mathcal{B}(X)$ with respect to which Q_f is λ-irreducible for all $f \in \mathbb{F}$. [Of course, λ is non-trivial: $\lambda(X) > 0$.]

10.3.6 Theorem. (Existence of canonical policies.) *Suppose that Assumptions 10.2.1, 10.2.2, and also 10.3.5 are satisfied. Then*

(a) *There exists a canonical policy.*

(b) *If $f^\infty \in \Pi_{DS}$ is AC-optimal, then*

$$\rho^* + h_f(x) = \min_{A(x)} \left[c(x, a) + \int_X h_f(y) Q(dy|x, a) \right] \quad \mu_f\text{-a.e.,} \quad (10.3.23)$$

where h_f is the solution to the Poisson equation (10.2.21)–(10.2.22).

Proof. See §10.5.

C. Uniqueness of the ACOE

In view of Theorem 10.3.4, part (a) in Theorem 10.3.6 gives, in other words, the *existence* of solutions (ρ^*, h^*) to the ACOE (10.3.13). Thus we already have the two nonempty sets $\mathbb{F}_{AC}$ and $\mathbb{F}_{ca}$ satisfying (10.3.2). Before introducing the third set, $\mathbb{F}_{bias}$, we need to consider the question of *uniqueness* of solutions to the ACOE.

It is obvious that ρ^* is unique—there can be no two different values of ρ^* that satisfy (10.3.15)! It is just as obvious that *if $h^*(\cdot)$ satisfies (10.3.13), then so does $h^*(\cdot) + k$ for any constant k.*

What is not obvious at all is that the solutions of (10.3.13) are all precisely of the form $h^*(\cdot) + k$; that is, two solutions of the ACOE can differ at most by a constant. Accordingly, understanding "uniqueness" as "uniqueness modulo an additive constant," we have the following.

10.3.7 Theorem. (Uniqueness of solutions to the ACOE.) *Suppose that the hypotheses of Theorem 10.3.6(a) hold. If h_1^* and h_2^* are two functions in $\mathbb{B}_w(X)$ such that (ρ^*, h_1^*) and (ρ^*, h_2^*) both satisfy the ACOE (10.3.13), then there exists a constant $k = k(h_1^*, h_2^*)$ for which*

$$h_1^*(x) = h_2^*(x) + k \quad \forall x \in X. \tag{10.3.24}$$

Proof. See §10.6.

10.3.8 Remark. As (10.3.14) is just a Poisson equation of the form (10.2.21), we may expect to be able to "fix" a solution to the ACOE in the same way we did in Remark 7.5.11(a) for the "noncontrolled" Poisson equation, which is indeed the case. For instance, let h^* be a function that satisfies (10.3.13) and choose an arbitrary, fixed state $\bar{x}$. Then, by (10.3.24), the function

$$h^*(\cdot) - h^*(\bar{x}) \tag{10.3.25}$$

is the *unique* solution of (10.3.13) that vanishes at $\bar{x}$. Similarly [as in (10.2.22)], if $h^* \in \mathbb{B}_w(X)$ and $f_* \in \mathbb{F}$ satisfy (10.3.13) and (10.3.14), then

$$h^*(\cdot) - \int_X h^* d\mu_{f_*} \tag{10.3.26}$$

is the *unique* solution of the ACOE (10.3.13) whose integral with respect to μ_{f_*} is zero.

D. Bias-optimal policies

Let $h_f \in \mathbb{B}_w(X)$ and $\mathbb{F}_{AC}$ be as in (10.2.19) and (10.3.2), respectively; that is, h_f is the bias function corresponding to $f \in \mathbb{F}$, and

$$\mathbb{F}_{AC} := \{f \in \mathbb{F} | J(f) = \rho^*\}.$$

The infimum of h_f over f in $\mathbb{F}_{AC}$ is called the **optimal bias function** and we shall denote it by $\widehat{h}$, i.e.,

$$\widehat{h}(x) := \inf\{h_f(x) | f \in \mathbb{F}_{AC}\} \quad \forall x \in X. \qquad (10.3.27)$$

Observe that the inequality (10.2.20) ensures that

$$\|\widehat{h}\|_w \leq \bar{c}R/(1-\rho). \qquad (10.3.28)$$

Hence $\widehat{h}$ is in the space $\mathbb{B}_w(X)$ if $\widehat{h}$ is measurable, which is the case, for instance, if there is a decision function $\widehat{f}$ such that $\widehat{h} = h_{\widehat{f}}$. This is related to the following definition.

10.3.9 Definition. A decision function $\widehat{f}$ (or the corresponding deterministic stationary policy $\widehat{f}^\infty$) is said to be **bias-optimal** if it attains the minimum in (10.3.27), i.e.,

$$J(\widehat{f}) = \rho^* \text{ and } h_{\widehat{f}}(x) = \widehat{h}(x) \quad \forall x \in X. \qquad (10.3.29)$$

We denote by $\mathbb{F}_{bias}$ the class of bias-optimal decision functions.

The concept of bias optimality was introduced by Veinott [2].

To prove the existence of bias-optimal policies we will use the fact that, under the hypotheses of Theorem 10.3.6(a), to obtain $\widehat{h}$ in (10.3.27) we may replace $\mathbb{F}_{AC}$ by the smaller class $\mathbb{F}_{ca}$ of *canonical* decision functions, i.e.,

$$\widehat{h}(x) = \inf\{h_f(x) | f \in \mathbb{F}_{ca}\}, \quad x \in X. \qquad (10.3.30)$$

In fact. we can write $\widehat{h}$ more explicitly as

$$\begin{aligned} \widehat{h}(x) &= \inf_{\mathbb{F}_{ca}} \left[h^*(x) - \int_X h^* d\mu_f \right] \\ &= h^*(x) - \sup_{\mathbb{F}_{ca}} \int_X h^* d\mu_f \quad \forall x \in X. \end{aligned} \qquad (10.3.31)$$

This is easily seen from the second equality in (10.3.20) and the definition (10.3.18) of $J_n^*(x, h^*)$. They yield that for each decision function f in $\mathbb{F}_{AC}$

$$n\rho^* + h^*(x) \leq J_n(f^\infty, x, h^*) = J_n(f^\infty, x) + E_x^{f^\infty} h^*(x_n) \quad \forall n, x,$$

or, equivalently,

$$J_n(f^\infty, x) - n\rho^* \geq h^*(x) - E_x^{f^\infty} h^*(x_n) \quad \forall n, x, \qquad (10.3.32)$$

with equality if f is canonical [see (10.3.21)]. Consequently, as $J(f) = \rho^*$, letting $n \to \infty$ we obtain [by (10.2.19) and (10.2.9)]

$$h_f(x) \geq h^*(x) - \int_X h^* d\mu_f \quad \forall x \in X, \ f \in \mathbb{F}_{AC}, \tag{10.3.33}$$

with equality if f is canonical, that is, if f is in $\mathbb{F}_{ca}$. This fact, gives (10.3.30) and (10.3.31).

Observe that (10.3.31), (10.3.27) and (10.3.29) give an alternative characterization of a bias-optimal decision function: $\widehat{f} \in \mathbb{F}_{AC}$ is bias-optimal if and only if it maximizes the integral $\int h^* d\mu_f$ over all $f \in \mathbb{F}_{ca}$, that is,

$$\int_X h^* d\mu_{\widehat{f}} = \sup_{\mathbb{F}_{ca}} \int_X h^* d\mu_f. \tag{10.3.34}$$

In addition to (10.3.34), other characterizations of bias-optimal policies are given in Theorem 10.3.10, which uses the following *notation*: For each state $x \in X$, $A^*(x) \subset A(x)$ is the set of actions for which the minimum is attained in (10.3.13), i.e.,

$$A^*(x) := \{a \in A(x) | c(x,a) + \int_X h^*(y) Q(dy|x,a) = \rho^* + h^*(x)\}. \tag{10.3.35}$$

Thus, in particular, a decision function f is canonical if and only if $f(x)$ is in $A^*(x)$ for all $x \in X$, i.e.,

$$f \in \mathbb{F}_{ca} \Leftrightarrow f(x) \in A^*(x) \quad \forall x \in X. \tag{10.3.36}$$

Furthermore, in (10.3.13) we may replace $A(x)$ by $A^*(x)$, i.e.,

$$\rho^* + h^*(x) = \min_{A^*(x)} \left[c(x,a) + \int_X h^*(y) Q(dy|x,a) \right], \tag{10.3.37}$$

and, on the other hand, as $\widehat{h}$ differs from h^* only by a constant [see (10.3.31)], in (10.3.13) and (10.3.37) we may replace h^* by $\widehat{h}$, which gives

$$\rho^* + \widehat{h}(x) = \min_{A(x)} \left[c(x,a) + \int_X \widehat{h}(y) Q(dy|x,a) \right] \tag{10.3.38}$$

or, equivalently,

$$\rho^* + \widehat{h}(x) = \min_{A^*(x)} \left[c(x,a) + \int_X \widehat{h}(y) Q(dy|x,a) \right]. \tag{10.3.39}$$

In either case, a canonical decision function $f \in \mathbb{F}_{ca}$ satisfies [by (10.3.36)]

$$\rho^* + \widehat{h}(x) = c_f(x) + \int_X \widehat{h}(y) Q_f(dy|x) \quad \forall x \in X. \tag{10.3.40}$$

The following theorem shows that, among other things, (10.3.40) with an additional condition characterizes bias-optimal policies.

10.3.10 Theorem. (Existence and characterization of bias-optimal policies.) *Suppose that the hypotheses of Theorem 10.3.6(a) are satisfied. Then:*

(a) *There exists a bias-optimal decision function $\widehat{f} \in \mathbb{F}_{bias}$; moreover,*

(b) *$(\rho^*, \widehat{h}, \widehat{f})$ is a canonical triplet [that is, it satisfies (10.3.39)—or (10.3.38)—and (10.3.40)] and there exists a function h' in $\mathbb{B}_w(X)$ such that*

$$\widehat{h}(x) + h'(x) = \min_{A^*(x)} \int_X h'(y)Q(dy|x,a) \quad \forall x \in X. \qquad (10.3.41)$$

In addition, if $f' \in \mathbb{F}_{ca}$ is a canonical decision function that attains the minimum in (10.3.41), i.e.,

$$\widehat{h}(x) + h'(x) = \int_X h'(y)Q_{f'}(dy|x) \quad \forall x \in X, \qquad (10.3.42)$$

then also f' is bias-optimal.

(c) *Conversely, if (ρ^*, h, f) is a canonical triplet and if there is a function h' in $\mathbb{B}_w(X)$ that together with h and f satisfies (10.3.41) and (10.3.42) for all $x \in X$, i.e.,*

$$
\begin{aligned}
h(x) + h'(x) &= \int_X h'(y)Q_f(dy|x) \qquad\qquad (10.3.43) \\
&= \min_{A^*(x)} \int_X h'(y)Q(dy|x,a),
\end{aligned}
$$

then f is bias-optimal and h is the optimal bias function, i.e., $h = \widehat{h}$.

(d) *The following statements are equivalent:*

(d₁) *$f \in \mathbb{F}$ is bias-optimal.*

(d₂) *$f \in \mathbb{F}$ is a canonical decision function and*

$$\int_X \widehat{h}\,d\mu_f = 0, \qquad (10.3.44)$$

where μ_f is the i.p.m. in Assumption 10.2.2.

Proof. See §10.7.

It is worth noting that Theorem 10.3.10 provides two "optimality equations" for the bias-minimization problem, namely:

(i) From Theorem 10.3.10(c), the ACOE (10.3.13)–(10.3.14) [see also (10.3.36)–(10.3.40)] and (10.3.43) together form an "optimality equation" for bias minimization; and

(ii) From Theorem 10.3.10(d), the ACOE (10.3.13)–(10.3.14) [or (10.3.39)–(10.3.40)] and (10.3.44) form another "optimality equation".

Furthermore (as shown in the proof of the theorem, in §10.7), case (i) occurs when the bias-minimization problem is viewed as an "average cost" problem, and then (i) is a direct consequence of Theorem 10.3.6(a). Case (ii), on the other hand, appears when bias minimization is posed as an "expected total cost (ETC)" problem, which is done in Remark 10.7.1. The latter remark provides a *second* proof of Theorem 10.3.10(a) and it is based on the ETC results in §9.5. Interestingly enough, our (first) proof of Theorem 10.3.10(a)—following an "average cost" approach [see (10.7.3), (10.7.4)]—is basically the same as Nowak's [1] proof of the existence (in π_{DS}) of *weakly overtaking optimal* policies!!! It was precisely this observation that suggested the equivalence of the several optimality concepts in (10.1.18), which is the content of Theorem 10.3.11 below.

E. Undiscounted criteria

In §10.1 we saw that some undiscounted cost criteria naturally lead to the AC criterion. The following result, on the other hand, states that we can go backwards—via bias optimality—in the sense that (10.1.18) holds in the class Π_{DS} of deterministic stationary policies.

10.3.11 Theorem. (Equivalence of undiscounted criteria.) *Suppose that the hypotheses of Theorem 10.3.6(a) are satisfied. Then the following statements are equivalent:*

(a) $f^\infty \in \Pi_{DS}$ *is bias-optimal, that is, f is in* $\mathbb{F}_{bias}$.

(b) f^∞ *is OC-optimal in* Π_{DS}, *i.e.* [see (10.1.8)],

$$OC(f^\infty, x) = \inf_{\pi_{DS}} OC(g^\infty, x) \text{ and } OC(f^\infty, x) < \infty \quad \forall x \in X.$$

(c) f^∞ *is D-optimal in* Π_{DS}, *i.e.,* [see (10.1.13)],

$$D(f^\infty, x) = \inf_{\Pi_{DS}} D(g^\infty, x) \quad \text{and} \quad D(f^\infty, x) < \infty \quad \forall x \in X.$$

(d) f^∞ *is weakly O.O. in* Π_{DS}, *i.e.,* [see (10.1.5)],

$$\limsup_{n\to\infty}[J_n(f^\infty, x) - J_n(g^\infty, x)] \leq 0 \quad \forall g^\infty \in \Pi_{DS}, \ x \in X.$$

Proof. See §10.8.

Finally, as a direct consequence of Theorems 10.3.10 and 10.3.11 we have:

10.3.12   Corollary. (Existence of "undiscounted" optimal policies.) *Under the hypotheses of Theorem 10.3.6(a), there is a deterministic stationary policy f^∞ for which the statements (a) to (d) in Theorem 10.3.11 hold.*

Notes on §10.3

1. All of the results in this section are essentially from Vega-Amaya [2], and Hernández-Lerma and Vega-Amaya [1]. Theorems 10.3.1 and 10.3.6 are also obtained in Gordienko and Hernández-Lerma [2] but under additional assumptions. In particular, the latter reference requires the cost-per-stage $c(x, a)$ to be *nonnegative*, which allows a direct application of the Abelian theorem (10.4.13) to obtain the result mentioned in Remark 10.3.2. Moreover, the ACOE (10.3.13) is obtained via the Ascoli Theorem, which of course requires to impose suitable "equicontinuity" hypotheses on the control model. The proof presented here (in §10.5) of the ACOE uses a "policy iteration" argument instead of the Ascoli Theorem.

For additional comments (with references) on how to obtain the ACOE see the Notes on §5.5.

2. Concerning the relation (10.3.2), we may recall from §5.2 that there are intermediate optimality concepts between "canonical" and "AC-optimal". For instance, a policy $\pi^* \in \Pi$ is said to be *F-strong AC-optimal* (or strong AC-optimal in the sense of Flynn [1]) if

$$\lim_{n \to \infty} [J_n(\pi^*, x) - J_n^*(x)]/n = 0 \quad \forall x \in X. \tag{10.3.45}$$

Thus, denoting by $\mathbb{F}_{F-SAC}$ the class of decision functions $f \in \mathbb{F}$ for which f^∞ is F-strong AC-optimal, it is easy to see that $\mathbb{F}_{F-SAC}$ lies between $\mathbb{F}_{ca}$ and $\mathbb{F}_{AC}$, i.e.,

$$\mathbb{F}_{ca} \subset \mathbb{F}_{F-SAC} \subset \mathbb{F}_{AC}. \tag{10.3.46}$$

3. Examples by Brown [1] and Nowak and Vega-Amaya [1] show that, without additional assumptions, the results in Theorem 10.3.11 and Corollary 10.3.12 cannot be extended to class Π of *all* policies. (See Remark 10.9.2.)

4. Haviv and Puterman [1] use bias optimality to distinguish between two AC-optimal policies for a certain admission control queueing system. To discriminate AC-optimal policies one can use the minimum *average variance* (see §11.3) instead of the minimum bias.

10.4   Proof of Theorem 10.3.1

The proof of Theorem 10.3.1 requires several preliminary results that are presented in the following subsection.

A. Preliminary lemmas

As the function $b(\cdot)$ in (10.2.1) satisfies that $0 \leq b(x) \leq \|b\|$ for all $x \in X$, we will assume that $b(\cdot)$ *is a constant* to be denoted by b again, i.e., $b(\cdot) \equiv b$. Thus, instead of (10.2.1) we now have

$$\sup_{A(x)} \int_X w(y)Q(dy|x,a) \leq \beta w(x) + b \quad \forall x \in X, \tag{10.4.1}$$

and similarly for (10.2.4).

10.4.1 Lemma. *Let $\pi \in \Pi$ be an arbitrary policy. Then for each $x \in X$ and $t = 1, 2, \ldots$*

$$E_x^\pi w(x_t) \leq \beta^t w(x) + b \sum_{j=0}^{t-1} \beta^j \leq [1 + b/(1 - \beta)]w(x). \tag{10.4.2}$$

Hence [with $\bar{c}$ as in Assumption 10.2.1(d)] for each $t = 0, 1, \ldots,$ $x \in X$, and $u \in \mathbb{B}_w(X)$

$$E_x^\pi |c(x_t, a_t)| \leq \bar{c}[1 + b/(1 - \beta)]w(x), \tag{10.4.3}$$

and

$$E_x^\pi |u(x_t)| \leq \|u\|_w [1 + b/(1 - \beta)]w(x),$$

so that

$$\lim_{t \to \infty} \sup_\Pi t^{-1} E_x^\pi |u(x_t)| = 0.$$

Proof. As in (8.3.31),

$$
\begin{aligned}
E_x^\pi[w(x_t)|h_{t-1}, a_{t-1}] &= \int w(y)Q(dy|x_{t-1}, a_{t-1}) \\
&\leq \beta w(x_{t-1}) + b \quad \text{[by (10.4.1)]}.
\end{aligned}
$$

Hence, taking the expectation $E_x^\pi(\cdot)$,

$$E_x^\pi w(x_t) \leq \beta E_x^\pi w(x_{t-1}) + b,$$

which iterated gives the first inequality in (10.4.2). The second inequality in (10.4.2) is obvious (recall that $w \geq 1$).

To obtain (10.4.3) it suffices to note that, by Assumption 10.2.1(d),

$$E_x^\pi |c(x_t, a_t)| \leq \bar{c} E_x^\pi w(x) \tag{10.4.4}$$

because then (10.4.3) follows from (10.4.2). Finally, as

$$E_x^\pi |u(x_t)| \le \|u\|_w E_x^\pi w(x_t),$$

the proof of the lemma can be completed in the obvious manner. □

Let us now consider the α-discounted cost ($0 < \alpha < 1$) and the α-discount value function in (8.3.1) and (8.3.2), i.e.,

$$V_\alpha(\pi, x) := E_x^\pi \left[\sum_{t=0}^\infty \alpha^t c(x_t, a_t) \right] \tag{10.4.5}$$

and

$$V_\alpha^*(x) := \inf_\Pi V_\alpha(\pi, x). \tag{10.4.6}$$

From (10.4.3) it is evident that

$$|V_\alpha(\pi, x)| \le \widehat{b}w(x)/(1 - \alpha) \quad \forall \pi, x,$$

and

$$|V_\alpha^*(x)| \le \widehat{b}w(x)/(1 - \alpha), \quad \text{with } \widehat{b} := \bar{c}[1 + b/(1 - \beta)]. \tag{10.4.7}$$

Thus, for each fixed $0 < \alpha < 1$, both functions $V_\alpha(\pi, \cdot)$ and $V_\alpha^*(\cdot)$ belong to $\mathbb{B}_w(X)$. On the other hand, note that the inequality (10.4.1) is of the same form as (8.3.11). Therefore, in view of Remark 8.3.5(a), all the results of Chapter 8 are valid in our present context. This means in particular, that, by Theorem 8.3.6(b), we may rewrite V_α^* in (10.4.6) as an infimum over the class of deterministic stationary policies, i.e.,

$$V_\alpha^*(x) = \inf_{\Pi_{DS}} V_\alpha(f^\infty, x). \tag{10.4.8}$$

Now fix an arbitrary state z in X, and for every $0 < \alpha < 1$ consider the function

$$u_\alpha(x) := V_\alpha^*(x) - V_\alpha^*(z). \tag{10.4.9}$$

We will next show that u_α belongs to the space $\mathbb{B}_w(X)$ for *all* $0 < \alpha < 1$.

10.4.2 Lemma. *Let $z \in X$ be the (fixed) state in (10.4.9). Then for every f^∞ in Π_{DS}, $x \in X$, and $t = 0, 1, \dots$*

$$|E_x^{f^\infty} c_f(x_t) - E_z^{f^\infty} c_f(x_t)| \le \bar{c}R\rho^t[1 + w(z)]w(x), \tag{10.4.10}$$

with R and ρ as in (10.2.9). Hence

$$|V_\alpha(f^\infty, x) - V_\alpha(f^\infty, z)| \le \bar{c}R(1 - \rho)^{-1}[1 + w(z)]w(x) \tag{10.4.11}$$

and

$$|u_\alpha(x)| \le \bar{c}R(1 - \rho)^{-1}[1 + w(z)]w(x) \tag{10.4.12}$$

for all $0 < \alpha < 1$ and $x \in X$.

Proof. Inside the absolute value in the left-hand side of (10.4.10) add and subtract $\mu_f(c_f) := \int c_f d\mu_f$, where μ_f is the i.p.m. of Q_f. Then, by (10.2.9), the left-hand side of (10.4.10) turns out to be less than or equal to

$$|\int c_f(y)Q_f^t(dy|x) - \mu_f(c_f)| + |\int c_f(y)Q_f^t(dy|z) - \mu_f(c_f)|$$
$$\leq \bar{c}R\rho^t[w(x) + w(z)],$$

and (10.4.10) follows because $w(x) \geq 1$.

To obtain (10.4.11) note that (10.4.5) and (10.4.10) yield

$$|V_\alpha(f^\infty, x) - V_\alpha(f^\infty, z)| \leq \sum_{t=0}^{\infty} \alpha^t |E_x^{f^\infty} c_f(x_t) - E_z^{f^\infty} c_f(x_t)|$$

$$\leq \bar{c}R[1 + w(z)]w(x) \sum_{t=0}^{\infty} \alpha^t \rho^t.$$

This implies (10.4.11) because, as $0 < \alpha < 1$,

$$\sum_{t=0}^{\infty} \alpha^t \rho^t \leq \sum_{t=0}^{\infty} \rho^t = 1/(1 - \rho).$$

Finally, to get (10.4.12) observe that (10.4.9) and (10.4.8) give

$$|u_\alpha(x)| \leq \sup_{\Pi_{DS}} |V_\alpha(f^\infty, x) - V_\alpha(f^\infty, z)|,$$

so (10.4.12) follows from (10.4.11). □

The following Lemma 10.4.3 determines a candidate for the number ρ^* in Theorem 10.3.1. We should also note that the inequality (10.4.16) is *not* needed to prove Theorem 10.3.1; we introduce (10.4.16) simply to complement Remark 10.3.2.

In the proof of (10.4.15) and (10.4.16) we use the Abelian theorem in Lemma 5.3.1, which states the following:

If $\{b_t, t = 0, 1, \ldots\}$ is a sequence of nonnegative numbers, then

$$\liminf_{n \to \infty} \frac{1}{n} \sum_{t=0}^{n-1} b_t \leq \liminf_{\alpha \uparrow 1} (1 - \alpha) \sum_{t=0}^{\infty} \alpha^t b_t$$

$$\leq \limsup_{\alpha \uparrow 1} (1 - \alpha) \sum_{t=0}^{\infty} \alpha^t b_t \qquad (10.4.13)$$

$$\leq \limsup_{n \to \infty} \frac{1}{n} \sum_{t=0}^{n-1} b_t.$$

10.4.3 Lemma. *There exists a number ρ^* such that*

$$\limsup_{\alpha\uparrow 1}(1-\alpha)V_\alpha^*(x) = \rho^* \quad \forall x \in X, \tag{10.4.14}$$

and

$$\rho^* \le \inf_{\Pi_{DS}} J(f^\infty, x) = \inf_{\mathbb{F}} J(f) \quad \forall x. \tag{10.4.15}$$

If, in addition, $c(x,a)$ is **nonnegative,** *then*

$$\rho^* \le J^*(x) := \inf_{\Pi} J(\pi, x) \quad \forall x \in X. \tag{10.4.16}$$

Proof. Let $z \in X$ be the (fixed) state in (10.4.9), and for every $0 < \alpha < 1$ define

$$\rho(\alpha) := (1-\alpha)V_\alpha^*(z). \tag{10.4.17}$$

By (10.4.7), $\rho(\alpha)$ is bounded since [with $\hat{b}$ as in (10.4.7)]

$$|\rho(\alpha)| \le \hat{b}w(z) \quad \forall 0 < \alpha < 1.$$

Therefore, there is a number ρ^* such that

$$\limsup_{\alpha\uparrow 1} \rho(\alpha) = \rho^*. \tag{10.4.18}$$

To prove that ρ^* satisfies (10.4.14), observe that (10.4.9) and (10.4.17) yield

$$|(1-\alpha)V_\alpha^*(x) - \rho^*| \le (1-\alpha)|u_\alpha(x)| + |\rho(\alpha) - \rho^*| \quad \forall x \in X.$$

Thus, by (10.4.18) and (10.4.12), letting $\alpha \uparrow 1$ we obtain (10.4.14).

We will now prove (10.4.16), which assumes $c(x,a) \ge 0$. Choose an arbitrary policy π and an arbitrary initial state x, and in (10.4.13) write $E_x^\pi c(x_t, a_t)$ in lieu of b_t. Then the third inequality of (10.4.13) gives [by (10.4.5) and (10.1.11)]

$$\limsup_{\alpha\uparrow 1}(1-\alpha)V_\alpha(\pi, x) \le J(\pi, x), \tag{10.4.19}$$

which in turn, by (10.4.6), yields

$$\limsup_{\alpha\uparrow 1}(1-\alpha)V_\alpha^*(x) \le J(\pi, x).$$

Hence, as π and x were arbitrary, the latter inequality and (10.4.14) give (10.4.16).

To complete the proof of the lemma, let us consider (10.4.15). We cannot proceed as in (10.4.19) because now $E_x^\pi c(x_t, a_t)$ may take negative values and so (10.4.13) is not directly applicable. Hence, we will use (10.4.4) to replace $E_x^\pi c(x_t, a_t)$ by

$$E_x^\pi[c(x_t, a_t) + \bar{c}w(x_t)] \geq 0;$$

in other words, we will write V_α as

$$V_\alpha(\pi, x) = \sum_{t=0}^\infty \alpha^t E_x^\pi[c(x_t, a_t) + \bar{c}w(x_t)] - \bar{c}\sum_{t=0}^\infty \alpha^t E_x^\pi w(x_t).$$

Moreover, let

$$W^s(\pi, x) := \limsup_{n\to\infty} \frac{1}{n}\sum_{t=0}^{n-1} E_x^\pi w(x_t),$$

$$W^i(\pi, x) := \liminf_{n\to\infty} \frac{1}{n}\sum_{t=0}^{n-1} E_x^\pi w(x_t).$$

Then (10.4.13) gives

$$\limsup_{\alpha\uparrow 1}(1-\alpha)V_\alpha(\pi, x) \leq J(\pi, x) + \bar{c}[W^s(\pi, x) - W^i(\pi, x)],$$

so that, as $V_\alpha^*(\cdot) \leq V_\alpha(\pi, \cdot)$ [by (10.4.6)], we can use again (10.4.14) to obtain

$$\rho^* \leq J(\pi, x) + \bar{c}[W^s(\pi, x) - W^i(\pi, x)] \quad \forall \pi, x. \tag{10.4.20}$$

Finally, note that if π is a deterministic stationary, say $\pi = f^\infty$, then (10.4.20) reduces to

$$\rho^* \leq J(f^\infty, x) = J(f) \tag{10.4.21}$$

because, by (10.2.9) and Proposition 10.2.3(b),

$$W^s(f^\infty, \cdot) = W^i(f^\infty, \cdot) = \mu_f(w), \quad \text{and} \quad J(f^\infty, \cdot) = J(f).$$

As (10.4.21) holds for all f^∞ in Π_{DS}, (10.4.15) follows. $\square$

We are now ready to complete the proof of Theorem 10.3.1.

B. Completion of the proof

Consider the α-*discounted cost optimality equation* in (8.3.4), i.e.,

$$V_\alpha^*(x) = \min_{A(x)}\left[c(x, a) + \alpha\int_X V_\alpha^*(y)Q(dy|x, a)\right], \quad x \in X. \tag{10.4.22}$$

With $\rho(\alpha)$ and $u_\alpha(x)$ as in (10.4.17) and (10.4.9), respectively, we can rewrite (10.4.22) as

$$\rho(\alpha) + u_\alpha(x) = \min_{A(x)}\left[c(x, a) + \alpha\int_X u_\alpha(x)Q(dy|x, a)\right], \quad x \in X. \tag{10.4.23}$$

On the other hand, by (10.4.18), there is a sequence of "discount factors" $\alpha(n) \uparrow 1$ such that

$$\rho^* = \lim_{n \to \infty} \rho[\alpha(n)]. \tag{10.4.24}$$

Define

$$h_0(x) := \liminf_{n \to \infty} u_{\alpha(n)}(x), \quad x \in X. \tag{10.4.25}$$

Observe that, by (10.4.12), h_0 is a function in $\mathbb{B}_w(X)$ and, moreover, the sequence $\{u_{\alpha(n)}\}$ is bounded in $\mathbb{B}_w(X)$ because

$$\|u_{\alpha(n)}\|_w \leq \bar{c}R(1 - \rho)^{-1}[1 + w(z)] \quad \forall n.$$

Thus, if in (10.4.23) we replace α by $\alpha(n)$ and take $\liminf_n$, then (10.4.24), (10.4.25), and Fatou's Lemma 8.3.7(b) [more precisely, (8.3.18)] yield

$$\rho^* + h_0(x) \geq \min_{A(x)} \left[c(x, a) + \int h_0(y)Q(dy|x, a) \right] \quad \forall x \in X,$$

which is precisely the ACOI (10.3.9).

Also note that the existence of a decision function (or selector) $f_0 \in \mathbb{F}$ that satisfies (10.3.10) is ensured by Proposition 8.3.9(b).

Furthermore, iterating (10.3.10) we get

$$n\rho^* + h_0(x) \geq J_n(f_0^\infty, x) + \int h_0(y)Q_{f_0}^n(dy|x) \quad \forall n = 1, 2, \ldots. \tag{10.4.26}$$

Thus, multiplying by $1/n$ and letting $n \to \infty$, we obtain [from (10.2.13) and Proposition 10.2.3(b)]

$$\rho^* \geq J(f_0^\infty, x) = J(f_0) \quad \forall x \in X. \tag{10.4.27}$$

This implies (10.3.11) because, by (10.4.15),

$$\rho^* \leq \inf_{\mathbb{F}} J(f) \leq J(f_0) \leq \rho^*.$$

It is also clear that any decision function f_0 that satisfies (10.3.10) also satisfies (10.4.27), and, therefore, (10.3.11). This completes the proof of Theorem 10.3.1. □

10.4.4 Remark. If $c(x, a)$ is *nonnegative*, then (10.4.27) and (10.4.16) give

$$\rho^* \leq J^*(x) \leq J(f_0) \leq \rho^* \quad \forall x \in X,$$

and (10.3.12) follows. □

10.5 Proof of Theorem 10.3.6

We will use the following fact—for a proof see, for instance, Orey [1, Theorem 7.2] or Meyn and Tweedie [1, Proposition 10.1.2].

10.5.1 Remark. Under Assumption 10.3.5, the irreducibility measure λ is *absolutely continuous* with respect to the i.p.m. μ_f for each $f \in \mathbb{F}$ (in symbols: $\lambda \ll \mu_f \ \forall f \in \mathbb{F}$); that is, if a set $B \in \mathcal{B}(X)$ is such that $\mu_f(B) = 0$, then $\lambda(B) = 0$. $\square$

A. Proof of part (a)

We wish to show that there exists a canonical triplet or, equivalently (by Theorem 10.3.4), a solution (ρ^*, h^*, f_*) to the ACOE (10.3.13), (10.3.14), with h^* in $\mathbb{B}_w(X)$.

It will be convenient to use the dynamic programming operator T in (9.5.1), with "min" instead of "inf", to write (10.3.13) in the form

$$\rho^* + h^*(x) = Th^*(x), \quad x \in X. \tag{10.5.1}$$

Moreover, to simplify the notation, given a sequence $\{f_n\}$ in $\mathbb{F}$ we shall write

$$c_{f_n}, h_{f_n}, Q_{f_n}, \mu_{f_n}, \dots \quad \text{as} \quad c_n, h_n, Q_n, \mu_n, \dots, \tag{10.5.2}$$

respectively, where $h_n \in \mathbb{B}_w(X)$ is the solution to the Poisson equation (10.2.21), (10.2.22) for f_n.

Now, to begin the proof itself, let $\rho^* \in \mathbb{R}$, $h_0 \in \mathbb{B}_w(X)$, and $f_0 \in \mathbb{F}$ be as in Theorem 10.3.1. In particular, as in (10.3.11) we have

$$J(f_0) = \rho^* = \inf_{\mathbb{F}} J(f) \tag{10.5.3}$$

and so we can write the Poisson equation (10.2.21) for f_0 as

$$\rho^* + h_0(x) = c_0(x) + \int_X h_0(y)Q_0(dy|x), \quad x \in X. \tag{10.5.4}$$

From this inequality and Proposition 8.3.9(b), there is a decision function $f_1 \in \mathbb{F}$ such that, for all $x \in X$,

$$
\begin{aligned}
\rho^* + h_0(x) &\geq \min_{A(x)} \left[c(x, a) + \int_X h_0(y)Q(dy|x, a) \right] \\
&= Th_0(x) \\
&= c_{f_1}(x) + \int_X h_0(y)Q_{f_1}(dy|x);
\end{aligned}
$$

that is, using the notation (10.5.2),

$$\rho^* + h_0(x) \geq c_1(x) + \int_X h_0(y)Q_1(dy|x) \quad \forall x \in X. \tag{10.5.5}$$

Furthermore, by the last statement in Theorem 10.3.1, also f_1 satisfies (10.5.3), i.e., $J(f_1) = \rho^*$. This implies in particular that the Poisson equation for f_1 is again of the form (10.5.4), namely,

$$\rho^* + h_1(x) = c_1(x) + \int_X h_1(y)Q_1(dy|x). \qquad (10.5.6)$$

On the other hand, from (10.5.6) and (10.5.5),

$$h_0(x) - h_1(x) \geq \int_X [h_0(y) - h_1(y)]Q_1(dy|x) \quad \forall x \in X;$$

in other words, the function $u := h_0 - h_1 \in \mathbb{B}_w(X)$ is *subinvariant* (or subharmonic) for the stochastic kernel $Q_1(\cdot|x)$. Hence, by Lemma 7.5.12(a), it follows that $u = h_0 - h_1$ equals μ_1-a.e. the constant

$$\int_X [h_0(y) - h_1(y)]\mu_1(dy) = \inf_X [h_0(x) - h_1(x)] =: \Delta_1.$$

More precisely, there is a Borel set $N_1 \in \mathcal{B}(X)$ such that $\mu_1(N_1) = 1$ and

$$h_0(\cdot) = h_1(\cdot) + \Delta_1 \quad \text{on} \quad N_1,$$

and

$$h_0(\cdot) > h_1(\cdot) + \Delta_1 \quad \text{on} \quad N_1^c,$$

where $N_1^c := X - N_1$ denotes the complement of N_1.

Repeating this procedure we obtain sequences $\{f_n\}$ in $\mathbb{F}$, $\{h_n\}$ in $\mathbb{B}_w(X)$, and $\{N_n\}$ in $\mathcal{B}(X)$ for which the following holds: For every $x \in X$ and $n = 0, 1, \ldots$ [and using the notation (10.5.2)]

(i) $J(f_n) = \rho^*$;

(ii) (ρ^*, h_n) satisfies the Poisson equation

$$\rho^* + h_n(x) = c_n(x) + \int_X h_n(y)Q_n(dy|x); \qquad (10.5.7)$$

(iii) $Th_n(x) = c_{n+1}(x) + \int_X h_n(y)Q_{n+1}(dy|x);$ \qquad (10.5.8)

(iv) $\mu_{n+1}(N_{n+1}) = 1$ and, with $\Delta_{n+1} := \int (h_n - h_{n+1})d\mu_{n+1} = \inf_X [h_n(x) - h_{n+1}(x)]$,

$$\begin{aligned} h_n(\cdot) &= h_{n+1}(\cdot) + \Delta_{n+1} \quad \text{on} \quad N_{n+1}, \\ h_n(\cdot) &> h_{n+1}(\cdot) + \Delta_{n+1} \quad \text{on} \quad N_{n+1}^c. \end{aligned} \qquad (10.5.9)$$

In addition, we claim that the set

$$N_* := \bigcap_{n=1}^{\infty} N_n \qquad (10.5.10)$$

is *nonempty*. Indeed, if N_* were empty, then from Assumption 10.3.5 and Remark 10.5.1 we would have

$$\lambda(X) = \lambda(N_*^c) \leq \sum_{n=1}^{\infty} \lambda(N_n^c) = 0 \quad \text{as} \quad \mu_n(N_n^c) = 0 \quad \forall n \geq 1;$$

that is $\lambda(X) = 0$, which contradicts that λ is a nontrivial measure.

Now choose x_* in N_*. Then, by (10.5.9),

$$h_n(x) \geq h_{n+1}(x) + \Delta_{n+1} \quad \forall x \in X,$$

and

$$h_n(x_*) = h_{n+1}(x_*) + \Delta_{n+1},$$

which implies that the functions

$$h_n^*(\cdot) := h_n(\cdot) - h_n(x_*)$$

form a nonincreasing sequence, i.e.,

$$h_n^* \geq h_{n+1}^* \quad \forall n = 0, 1, \ldots . \tag{10.5.11}$$

Moreover [by (10.2.20)], the sequence $\{h_n\}$ is uniformly bounded in $\mathbb{B}_w(X)$ and, therefore, so is $\{h_n^*\}$. Hence, there is a function h^* in $\mathbb{B}_w(X)$ such that

$$h^*(x) = \lim_{n \to \infty} h_n^*(x) \quad \forall x \in X. \tag{10.5.12}$$

Finally, to obtain (10.5.1), first note that the Poisson equation (10.5.7) remains valid if we replace h_n by h_n^*, i.e.,

$$\rho^* + h_n^*(x) = c_n(x) + \int_X h_n^*(y)Q_n(dy|x). \tag{10.5.13}$$

This in turn gives $\rho^* + h_n^* \geq Th_n^*$, i.e.,

$$\rho^* + h_n^*(x) \geq \min_{A(x)} \left[c(x,a) + \int_X h_n^*(y)Q(dy|x,a) \right] \quad \forall x \in X.$$

Therefore, letting $n \to \infty$, (10.5.12) and the Fatou Lemma (8.3.8) yield

$$\rho^* + h^*(x) \geq Th^*(x) \quad \forall x \in X. \tag{10.5.14}$$

To get the reverse inequality, i.e.,

$$\rho^* + h^*(x) \leq Th^*(x) \quad \forall x \in X, \tag{10.5.15}$$

write (10.5.8) with h_n^* and $n-1$ in lieu of h_n and n, respectively, to obtain

$$c_n(x) = Th_{n-1}^*(x) - \int_X h_{n-1}^*(y)Q(dy|x).$$

Thus, from (10.5.13),

$$
\begin{aligned}
\rho^* + h_n^*(x) &= Th_{n-1}^*(x) - \int [h_{n-1}^*(y) - h_n^*(y)]Q_n(dy|x) \\
&\leq Th_{n-1}^*(x) \quad \text{[by (10.5.11)]} \\
&\leq c(x,a) + \int h_{n-1}^*(y)Q(dy|x,a) \quad \text{[by (9.5.1)]}
\end{aligned}
$$

for all $x \in X$ and $a \in A(x)$. It follows that [from (10.5.12) and the Fatou Lemma (8.3.19)]

$$
\rho^* + h^*(x) \leq c(x,a) + \int h^*(y)Q(dy|x,a) \quad \forall x \in X, \ a \in A(x),
$$

which implies (10.5.15).

This completes the proof of part (a) because (10.5.14)–(10.5.15) give the ACOE (10.5.1) and, as usual, the existence of a decision function $f_* \in \mathbb{F}$ that satisfies (10.3.14) is obtained from Proposition 8.3.9(b).

B. Proof of part (b)

Let $f^\infty \in \Pi_{DS}$ be an AC-optimal policy, i.e., $J(f) = \rho^*$. Then, as in (10.5.4), the Poisson equation for f is

$$
\rho^* + h_f(x) = c_f(x) + \int_X h_f(y)Q_f(dy|x) \quad \forall x.
$$

On the other hand, from the ACOE (10.5.1),

$$
\rho^* + h^*(x) \leq c_f(x) + \int_X h^*(y)Q_f(dy|x) \quad \forall x.
$$

Hence

$$
h_f(x) - h^*(x) \geq \int [h_f(y) - h^*(y)]Q_f(dy|x) \quad \forall x,
$$

which means that the function $h_f - h^* \in \mathbb{B}_w(X)$ is *subvariant* for the stochastic kernel Q_f. This implies that [by Lemma 7.5.12(a)]

$$
h_f(\cdot) = h^*(\cdot) + k \quad \mu_f\text{-a.e.,} \tag{10.5.16}
$$

with $k := \int (h_f - h^*)d\mu_f = -\int h^* d\mu_f$, by (10.2.22). Therefore, replacing h^* with $h_f - k$ in (10.5.1) we obtain (10.3.23), and part (b) follows.

This completes the proof of Theorem 10.3.6 $\square$

C. Policy iteration

The approach used to prove Theorem 10.3.6(a) is a special case of the **policy iteration algorithm** (PIA) for the AC problem, which in general can be described as follows. (See §9.6.D for the PIA associated to the expected total cost problem.)

Step 0. Initialization: Set $n = 0$ and choose an arbitrary decision function f_n in $\mathbb{F}$.

Step 1. Policy evaluation: Find $J(f_n) \in \mathbb{R}$ and $h_n \in \mathbb{B}_w(X)$ that satisfy the Poisson equation for f_n; that is, using the notation (10.5.2),

$$J(f_n) + h_n(x) = c_n(x) + \int_X h_n(y)Q_n(dy|x) \quad \forall x \in X. \quad (10.5.17)$$

Step 3. Policy improvement: Determine a decision function $f_{n+1} \in \mathbb{F}$ such that

$$Th_n(x) = c_{n+1}(x) + \int_X h_n(y)Q_{n+1}(dy|x) \quad \forall x \in X. \quad (10.5.18)$$

Comparing (10.5.17)–(10.5.18) with (10.5.7)–(10.5.8) we see that the proof of Theorem 10.3.6(a) consisted precisely of the PIA when the initial decision function f_0 satisfies (10.3.10)–(10.3.11), which gave us $J(f_n) = \rho^*$ for all n, as well as (10.5.11).

In general, the objective of the PIA is to find a solution (ρ^*, h^*) to the ACOE (10.3.13). The idea is that combining (10.5.17) and (10.5.18) we obtain

$$J(f_n) + h_n(x) \geq c_{n+1}(x) + \int_X h_n(y)Q_{n+1}(dy|x), \quad (10.5.19)$$

so that integration with respect to the i.p.m. μ_{m+1} yields [by (10.2.18) or Proposition 10.2.3(b)]

$$J(f_n) \geq J(f_{n+1}); \quad (10.5.20)$$

that is, the sequence of average costs $J(f_n)$ is *nonincreasing*. Moreover, it is obviously *bounded* since [by Assumption 10.2.1(d), (10.2.12), and (10.2.18)]

$$|J(f)| \leq \int |c_f| d\mu_f \leq \bar{c}\|b\|/(1 - \rho) \quad \forall f \in \mathbb{F}. \quad (10.5.21)$$

Therefore, there exists a constant $\hat{\rho}$ such that

$$J(f_n) \downarrow \hat{\rho}. \quad (10.5.22)$$

Of course, we necessarily have that $\hat{\rho} \geq \rho^*$ because ρ^* is the AC value function [see (10.3.15)].

Thus, the PIA's objective is to show that $\hat{\rho} = \rho^*$ and that $\{h_n\}$ or a subsequence thereof, or even a modified sequence [such as h_n^* in (10.5.11)], converges to a function that satisfies the ACOE. It seems to be an *open problem* to determine whether the latter fact is possible under the hypotheses of Theorem 10.3.6. However, as stated in the following result, under some extra condition, the PIA does converge. We shall omit the proof of Theorem 10.5.2 because it is very similar to the proof of Theorem 10.3.6 a)—the reader may refer to Hernández-Lerma and Lasserre [11] for details.

10.5.2 Theorem. (Convergence of the PIA.) *Suppose that the hypotheses of Theorem 10.3.6(a) are satisfied, and, in addition, the sequence $\{h_n\} \subset \mathbb{B}_w(X)$ in (10.5.17) has a convergent subsequence $\{h_m\}$, i.e.,*

$$\lim_{m \to \infty} h_m(x) = h(x) \quad \forall x \in X, \tag{10.5.23}$$

for some function h on X. Then the PIA converges; in fact, the pair

$$(\rho^*, h^*) := (\widehat{\rho}, h) \quad [\text{with } \widehat{\rho} \text{ as in (10.5.22)}]$$

satisfies the ACOE (10.3.13).

10.5.3 Remark. (a) By (10.2.20), the function h in (10.5.23) is necessarily in $\mathbb{B}_w(X)$. Moreover, (10.2.20) gives that the sequence $\{h_n\}$ in (10.5.17) is *locally bounded*. Hence, for instance, if it can be shown that $\{h_n\}$ is *equicontinuous*, then the existence of a subsequence $\{h_m\}$ that satisfies (10.5.23) is ensured by the well-known Ascoli Theorem. (See, for instance, Remark 5.5.2 or Royden [1] for the statement of the Ascoli Theorem.) Remark 5.5.3 and Assumption 2.7 in Gordienko and Hernández-Lerma [2] give conditions for $\{h_n\}$ to be equicontinuous.

On the other hand, if X is a *denumerable* set (with the discrete topology), then any collection of functions on X, in particular $\{h_n\}$, is equicontinuous. Thus, in the denumerable-state case, Theorem 10.5.2 gives that the PIA converges.

Finally, it is important to keep in mind that in many cases the sequence $\{h_n\}$ [or a modified sequence—as h_n^* in (10.5.11)] can be shown to be monotone or "nearly monotone", so that $\{h_n\}$ itself satisfies (10.5.23); see, for instance, Meyn [1] of Puterman [1, Propos. 8.6.5].

(b) The difference between the left-hand side and the right-hand side of (10.5.19), i.e.,

$$D_n(x) := J(f_n) + h_n(x) - Th_n(x),$$

is called the PIA's *discrepancy function* at the nth iteration. Similarly, from (10.5.20) we get the *cost decrease* $C_n := J(f_n) - J(f_{n+1})$, which can also be written as

$$C_n = \int_X D_n(x)\mu_{n+1}(dx), \quad n = 0, 1, \dots.$$

If we now define $\widehat{h}_n := h_n - h_{n+1}$, we obtain

$$C_n + \widehat{h}_n(x) = D_n(x) + \int_X \widehat{h}_n(y)Q_{n+1}(dy|x) \quad \forall x \in X, \tag{10.5.24}$$

which means that the pair $(C_n, \widehat{h}_n)$ is a solution to the Poisson equation for the transition kernel $Q_{n+1}(\cdot|x) = Q(\cdot|x, f_{n+1})$ with "cost" (or charge) function D_n. This fact can be used to prove the convergence of the PIA, at least when the state space X is a *finite* set; see, for instance, Puterman [1, §8.6]. Alternatively, one could try to show that $D_n(x) \to 0$ for all $x \in X$, as $n \to \infty$.

10.6 Proof of Theorem 10.3.7

Let h_1^* and h_2^* be two functions in $\mathbb{B}_w(X)$ such that (ρ^*, h_1^*) and (ρ^*, h_2^*) both satisfy the ACOE (10.3.13); that is, for each state $x \in X$,

$$\rho^* + h_1^*(x) = \min_{A(x)} \left[c(x,a) + \int_X h_1^*(y)Q(dy|x,a) \right], \qquad (10.6.1)$$

and

$$\rho^* + h_2^*(x) = \min_{A(x)} \left[c(x,a) + \int_X h_2^*(y)Q(dy|x,a) \right]. \qquad (10.6.2)$$

Let $f_1 \in \mathbb{F}$ be a decision function such that $f_1(x) \in A(x)$ attains the minimum in (10.6.1) for each $x \in X$. Then, using the notation (10.5.2),

$$\rho^* + h_1^*(x) = c_1(x) + \int h_1^*(y)Q_1(dy|x) \quad \forall x,$$

whereas using f_1 in (10.6.2) we get

$$\rho^* + h_2^*(x) \le c_1(x) + \int h_2^*(x)Q_1(dy|x) \quad \forall x.$$

Hence

$$h_1^*(x) - h_2^*(x) \ge \int [h_1^*(y) - h_2^*(y)]Q_1(dy|x) \quad \forall x,$$

so that [as in the argument used after (10.5.6) or to obtain (10.5.16)] Lemma 7.5.12(a) yields the existence of a set $N_1 \in \mathcal{B}(X)$ and a constant k_1 such that $\mu_1(N_1) = 1$ and

$$h_1^*(x) \ge h_2^*(x) + k_1 \quad \forall x \in X, \text{ with } equality \text{ on } N_1. \qquad (10.6.3)$$

We now repeat the above argument but interchanging the roles of (10.6.1) and (10.6.2), and using part (b) of Lemma 7.5.12 instead of part (a). That is, we take a decision function $f_2 \in \mathbb{F}$ that attains the minimum in (10.6.2) and we get a set $N_2 \in \mathcal{B}(X)$ and a constant k_2 such that $\mu_2(N_2) = 1$ and

$$h_1^*(x) \le h_2^*(x) + k_2 \quad \forall x \in X, \text{ with } equality \text{ on } N_2. \qquad (10.6.4)$$

Then, as in the proof of Theorem 10.3.6(a) [see (10.5.10)], we can use Assumption 10.3.5 and Remark 10.5.1 to show that the set

$$N := N_1 \cap N_2$$

is *nonempty*; otherwise we would get $\lambda(X) = \lambda(N^c) = 0$, a contradiction.
Now let $\widehat{x}$ be a point in N and define the functions

$$\widehat{h}_1(\cdot) := h_1^*(\cdot) - h_1^*(\widehat{x}), \quad \text{and} \quad \widehat{h}_2(\cdot) := h_2^*(\cdot) - h_2^*(\widehat{x}).$$

Thus as in (10.5.11)], (10.6.3) and (10.6.4) yield

$$\widehat{h}_1(\cdot) \geq \widehat{h}_2(\cdot) \quad \text{and} \quad \widehat{h}_1(\cdot) \leq \widehat{h}_2(\cdot),$$

respectively. Hence $\widehat{h}_1(\cdot) = \widehat{h}_2$, which gives (10.3.24) with $k := h_1^*(\widehat{x}) - h_2^*(\widehat{x})$ $\square$

10.7   Proof of Theorem 10.3.10

We will present *two proofs* of part (a). The first one is a direct application of Theorem 10.3.6(a), the reasoning being that the first equality in (10.3.31) can be written as

$$\widehat{h}(x) = h^*(x) + \inf_{\mathbb{F}_{ca}} \int_X (-h^*) d\mu_f, \tag{10.7.1}$$

and, therefore, the problem of finding a bias-optimal policy reduces to an AC problem with cost-per-stage function

$$c'(x, a) := -h^*(x). \tag{10.7.2}$$

Moreover, since the minimization in (10.7.1) is over the class $\mathbb{F}_{ca}$ of *canonical* decision functions, it suffices to consider "canonical control actions" $a \in A^*(x)$, where $A^*(x) \subset A(x)$ is the set in (10.3.35) [see also (10.3.36)]. The second proof of part (a) is given in Remark 10.7.1.

Proof of (a). Consider the Markov control model

$$\mathcal{M}_{bias} := (X, A, \{A^*(x) | x \in X\}, Q, c'), \tag{10.7.3}$$

which is the same as the original control model $\mathcal{M} = (X, A, \{A(x) | x \in X\}, Q, c)$ except that $A(x)$ and $c(x, a)$ have been replaced by $A^*(x)$ [in (10.3.35)] and $c'(x, a)$ [in (10.7.2)], respectively. Hence, as $\mathcal{M}$ satisfies the hypotheses of Theorem 10.3.6(a), it is clear that so does $\mathcal{M}_{bias}$ and, by consequence, there is a canonical policy $\widehat{f} \in \mathbb{F}_{ca}$ [see (10.3.36)] for the new model $\mathcal{M}_{bias}$, i.e.,

$$\int_X (-h^*) d\mu_{\widehat{f}} = \inf_{\mathbb{F}_{ca}} \int_X (-h^*) d\mu_f =: \widehat{\rho}. \tag{10.7.4}$$

This fact and (10.7.1) yield that $\widehat{f}$ is bias-optimal.

Proof of (b). The bias-optimal decision function $\widehat{f}$ in (a) is canonical, and so it satisfies (10.3.39) [or (10.3.38)] and (10.3.40); that is, $(\rho^*, \widehat{h}, \widehat{f})$ is a canonical triplet (see Theorem 10.3.4). Moreover, as was already mentioned in the proof of (a), $\mathcal{M}_{bias}$ satisfies the hypotheses of Theorem 10.3.6(a). Therefore, there exists a function h' in $\mathbb{B}_w(X)$ and a canonical decision function f' in $\mathbb{F}_{ca}$ such that $(\widehat{\rho}, h', f')$ *is a canonical triplet for* $\mathcal{M}_{bias}$, i.e. (by Theorem 10.3.4),

$$\widehat{\rho} + h'(x) = \min_{A^*(x)} \left[c'(x, a) + \int_X h'(y) Q(dy|x, a) \right]$$

or [by (10.7.2)]

$$\widehat{\rho} + h'(x) = -h^*(x) + \min_{A^*(x)} \int_X h'(y) Q(dy|x, a),$$

and

$$\widehat{\rho} + h'(x) = -h^*(x) + \int_X h'(y) Q_{f'}(dy|x) \quad \forall x \in X, \tag{10.7.5}$$

which [by (10.7.4) and (10.7.1)] yield (10.3.41) and (10.3.42), respectively. Finally, integrating both sides of (10.7.5) with respect to the i.p.m. $\mu_{f'}$ we get

$$\widehat{\rho} = \int_X (-h^*) d\mu_{f'},$$

that is, f' is bias-optimal.

Proof of (c). The first equality in (10.3.43) implies

$$\int_X h d\mu_f = 0. \tag{10.7.6}$$

On the other hand, as (ρ^*, h, f) is a canonical triplet, the uniqueness Theorem 10.3.7 yields that $h = h^* + k$ for some constant k, which combined with (10.7.6) implies

$$k = -\int_X h^* d\mu_f.$$

Hence, by (10.3.33), h coincides with the bias function h_f since

$$h = h^* - \int_X h^* d\mu_f = h_f. \tag{10.7.7}$$

From this fact and the second equality in (10.3.43) it follows that

$$h^*(x) - \int h^* d\mu_f + h'(x) \leq \int h'(y) Q_g(dy|x) \quad \forall g \in \mathbb{F}_{ca},$$

and integration with respect to the i.p.m. μ_g gives

$$\int h^* d\mu_g \le \int h^* d\mu_f \quad \forall g \in \mathbb{F}_{ca},$$

Thus

$$\int h^* d\mu_f = \sup_{\mathbb{F}_{ca}} \int h^* d\mu_g,$$

which together with (10.3.34) and (10.7.7) yields that f is bias-optimal and that $h = h_f = \hat{h}$.

Proof of (d). $(d_1) \Rightarrow (d_2)$. If f is bias-optimal, then [by (10.3.31)] f is in $\mathbb{F}_{ca}$ and

$$\hat{h}(\cdot) = h^*(\cdot) - \int_X h^* d\mu_f,$$

so that integrating with respect to μ_f we obtain (10.3.44).

$(d_2) \Rightarrow (d_1)$. If (d_2) holds, then $f \in \mathbb{F}_{ca}$ satisfies (10.3.40) and, on the other hand, the Poisson equation for f is

$$\rho^* + h_f(x) = c_f(x) + \int_X h_f(y) Q_f(dy|x).$$

Subtracting the latter equation from (10.3.40) we see that the function $u(\cdot) := \hat{h}(\cdot) - h_f(\cdot)$ is invariant with respect to Q_f, i.e.,

$$u(x) = \int_X u(y) Q_f(dy|x) \quad \forall x \in X.$$

Consequently [by Theorem 7.5.10(d)] $u(\cdot) = \mu_f(u)$, i.e.,

$$\hat{h}(\cdot) - h_f(\cdot) = \int_X (\hat{h} - h_f) d\mu_f = 0,$$

where the second equality is due to (10.3.44) and (10.2.22). Therefore $\hat{h} = h_f$ and (d_1) follows. This completes the proof of Theorem 10.3.10. □

10.7.1 Remark: Another proof of Theorem 10.3.10. In this proof the idea is to use (10.3.30) and the definition (10.2.19) of h_f [with $J(f) = \rho^*$] to pose the question of existence of bias-optimal policies as an *expected total cost* (ETC) problem.

Instead of the Markov control model $\mathcal{M}_{bias}$ in (10.7.3) consider the control model

$$\mathcal{M}^* := (X, A, \{A^*(x) | x \in X\}, Q, c^*)$$

with cost-per-stage c^* given by

$$c^*(x, a) := c(x, a) - \rho^* \quad \forall x \in X, \ a \in A^*(x),$$

and $A^*(x)$ as in (10.3.35).

Note that for a canonical decision function $f \in \mathbb{F}_{ca}$, the bias h_f in (10.2.19) becomes

$$h_f(x) = \sum_{t=0}^{\infty} E_x^{f^{\infty}}[c_f(x_t) - \rho^*] = \sum_{t=0}^{\infty} E_x^{f^{\infty}} c_f^*(x_t), \tag{10.7.8}$$

which is the same as the ETC $V_1(f^{\infty}, x)$ in (9.1.1) with $c^*(x,a)$ in lieu of $c(x,a)$. Let us write $V_1(f^{\infty}, x)^*$ for the ETC thus obtained so that

$$h_f(x) = V_1(f^{\infty}, x)^* = \sum_{t=0}^{\infty} E_x^{f^{\infty}} c_f^*(x_t), \quad f \in \mathbb{F}_{ca}.$$

Furthermore, if in (9.1.2) we replace Π and $V_1(\cdot, x)$ by $\mathbb{F}_{ca}$ and $V_1(\cdot, x)^*$, respectively, we get that the value function for the new ETC problem is precisely the optimal bias function $\widehat{h}$ in (10.3.30), and, on the other hand, the ETC optimality equation (9.5.3) becomes

$$\widehat{h}(x) = \min_{A^*(x)} \left[c^*(x,a) + \int_X \widehat{h}(y)Q(dy|x,a) \right], \tag{10.7.9}$$

which can also be written as

$$\rho^* + \widehat{h}(x) = \min_{A^*(x)} \left[c(x,a) + \int_X \widehat{h}(y)Q(dy|x,a) \right], \quad x \in X. \tag{10.7.10}$$

In addition, the control model $\mathcal{M}^*$ satisfies the hypotheses of Proposition 8.3.9(b), and so there exists a decision function $\widehat{f}$ in $\mathbb{F}_{ca}$ such that $\widehat{f}(x) \in A^*(x)$ attains the minimum in (10.7.10) for each $x \in X$, i.e.,

$$\rho^* + \widehat{h}(x) = c_{\widehat{f}}(x) + \int_X \widehat{h}(y)Q_{\widehat{f}}(dy|x) \quad \forall x \in X. \tag{10.7.11}$$

Therefore, by Theorem 9.5.12, to conclude that $\widehat{f} \in \mathbb{F}_{ac}$ is bias-optimal it only remains to verify that Assumption 9.5.2 holds in the present context, and also that $\widehat{f}$ and $\widehat{h}$ satisfy (9.5.38), i.e.,

$$\limsup_{n \to \infty} E_x^{\widehat{f}^{\infty}} \widehat{h}(x_n) = 0 \quad \forall x \in X. \tag{10.7.12}$$

To verify Assumption 9.5.2, replace Π and $V_1(\pi, x)$ in (9.3.8) and (9.3.16) by $\mathbb{F}_{ca}$ and $V_1(f^{\infty}, x)^* = h_f(x)$, respectively. Then Assumptions 9.3.2 and 9.3.4 both follow from (10.2.20), so that part (a) in Assumption 9.5.2 is satisfied. Similarly, replacing $\mathcal{U}$, T and W by $\mathbb{B}_w(X)$,

$$T^*u(x) := \min_{A^*(x)} \left[c^*(x,a) + \int u(y)Q(dy|x,a) \right] \quad \text{[see (10.7.9)]},$$

and

$$W(x) := \bar{c}Rw(x)/(1 - \rho) \quad \text{[see (10.2.20)]},$$

respectively, we obtain parts (b) and (c) in Assumption 9.5.2.

Finally, (10.7.12) follows from (10.7.11), which gives [as in (10.3.21) or (7.5.4)]

$$
\begin{aligned}
E_x^{\widehat{f}^\infty}\widehat{h}(x_n) &= \widehat{h}(x) - [J_n(\widehat{f}, x) - n\rho^*] \\
&= \sum_{t=n}^{\infty} E_x^{\widehat{f}^\infty}[c_{\widehat{f}}(x_t) - \rho^*] \\
&\to \quad 0 \quad \text{as} \quad n \to \infty \quad \text{[by (10.7.8)]}.
\end{aligned}
$$

Therefore, the conditions of Theorem 9.5.12 hold, and so we conclude that the decision function $\widehat{f} \in \mathbb{F}_{ca}$ is bias-optimal. $\square$

10.8   Proof of Theorem 10.3.11

We already have several implications between the concepts in (a), (b), (c), and (d), and AC optimality. For instance by (10.3.30),

$$\mathbb{F}_{bias} \subset \mathbb{F}_{ca} \subset \mathbb{F}_{AC}. \tag{10.8.1}$$

Similarly, by (10.1.5), it is clear that a weakly O.O. policy is AC-optimal; in fact, by (10.1.9) and (10.1.15),

$$\pi^* \text{ weakly O.O.} \Rightarrow \pi^* \text{ OC-optimal} \Rightarrow \pi^* \text{ AC-optimal}, \tag{10.8.2}$$

where the second implication holds if π^* has a finite opportunity cost.

To obtain further relations between the above concepts, let $J_n^*(x)$ and $J^*(x)$ be as in (10.1.16) and (10.1.10), respectively, and define the upper and lower limit functions

$$L^u(x) \quad := \quad \limsup_{n\to\infty}[J_n^*(x) - nJ^*(x)], \tag{10.8.3}$$

$$L^l(x) \quad := \quad \liminf_{n\to\infty}[J_n^*(x) - nJ^*(x)]. \tag{10.8.4}$$

Then, adding $\pm nJ^*(x)$ to $J_n(\pi, x) - J_n^*(x)$, from (10.1.7) and (10.1.12) we get

$$D(\pi, x) - L^u(x) \le OC(\pi, x) \le D(\pi, x) - L^l(x) \quad \forall \pi \in \Pi, \ x \in X, \tag{10.8.5}$$

which, of course, holds in particular when $\pi = f^\infty$ is a deterministic stationary policy, i.e.,

$$D(f^\infty, x) - L^u(x) \le OC(f^\infty, x) \le D(f^\infty, x) - L^l(x) \tag{10.8.6}$$

for each f^∞ in Π_{DS} and $x \in X$. Moreover, as a special case of (10.1.17), if f^∞ is *not* AC-optimal, then $D(f^\infty, \cdot)$ and $OC(f^\infty, \cdot)$ are both *infinite*.

In fact, in the context of Theorem 10.3.11, we can get a relation more explicit than (10.8.6) if f^∞ is AC-optimal [i.e., $J(f) = \rho^*$, or $f \in \mathbb{F}_{AC}$]; namely,

$$h_f(\cdot) = D(f^\infty, \cdot) = OC(f^\infty, \cdot) + L^l(\cdot) \quad \forall f \in \mathbb{F}_{AC}. \tag{10.8.7}$$

Indeed, as $J^*(\cdot) = \rho^*$, from (10.2.19) and (10.1.12) we obtain

$$h_f(x) = \lim_{n\to\infty}[J_n(f^\infty, x) - n\rho^*] = D(f^\infty, x) \quad \forall f \in \mathbb{F}_{AC}, \ x \in X, \tag{10.8.8}$$

which yields the first equality in (10.8.7). Similarly, by definition (10.1.7) of opportunity cost,

$$\begin{aligned}
OC(f^\infty, x) &= \limsup_{n\to\infty}\{[J_n(f^\infty, x) - n\rho^*] - [J_n^*(x) - n\rho^*]\} \\
&= h_f(x) - L^l(x) \quad \text{[by (10.8.8) and (10.8.4)]},
\end{aligned}$$

and (10.8.7) follows.

Thus, in view of (10.1.17) and (10.8.7), we have the equivalence of (a), (b) and (c) in Theorem 10.3.11, i.e.,

$$(a) \Leftrightarrow (b) \Leftrightarrow (c). \tag{10.8.9}$$

In particular, note that (10.8.7) and the definition (10.3.27) of the optimal bias function yield

$$\hat{h}(x) = \inf_{\mathbb{F}_{AC}} D(f^\infty, x) = \inf_{\mathbb{F}_{AC}} OC(f^\infty, x) + L^l(x) \quad \forall x \in X. \tag{10.8.10}$$

Finally, to prove the equivalence between (d) and, say, Dutta's optimality (c), we already have (d) $\Rightarrow$ (c), by (10.1.14). To prove the converse first note that (10.1.5) is *equivalent* to

$$\limsup_{n\to\infty}[J_n(\pi^*, x) - J_n(\pi, x)] \le 0 \quad \forall \pi \in \Pi, \ x \in X. \tag{10.8.11}$$

Thus to prove that (c) $\Rightarrow$ (d) we need to show that, with f^∞ as in (c),

$$\limsup_{n\to\infty}[J_n(f^\infty, x) - J_n(g^\infty, x)] \le 0 \quad \forall g^\infty \in \Pi_{DS}, \ x \in X. \tag{10.8.12}$$

In turn, to get (10.8.12) we consider two cases.

Case 1: g^∞ is not AC-optimal, i.e., $J(g) > \rho^*$. Then, by (10.1.12) and (10.2.19),

$$\begin{aligned}
D(g^\infty, x) &\ge \liminf_{n\to\infty}[J_n(g^\infty, x) - n\rho^*] \\
&\ge h_g(x) + \liminf_{n\to\infty} n[J(g) - \rho^*] \\
&= \infty \quad \forall x \in X.
\end{aligned}$$

Therefore,

$$\limsup_{n\to\infty}[J_n(f^\infty, x) - J_n(g^\infty, x)] \le D(f^\infty, x) - \liminf_{n\to\infty}[J_n(g^\infty, x) - n\rho^*]$$

$$\le 0 \quad \forall x \in X,$$

and (10.8.12) follows for $g \notin \mathbb{F}_{AC}$.

Case 2· g^∞ is AC-optimal, i.e., $J(g) = \rho^*$. Then, as in (10.8.8),

$$D(g^\infty, x) = \lim_{n\to\infty}[J_n(g^\infty, x) - n\rho^*],$$

so that for each $x \in X$,

$$\limsup_{n\to\infty}[J_n(f^\infty, x) - J_n(g^\infty, x)] \le D(f^\infty, x) - D(g^\infty, x) \le 0 \quad (10.8.13)$$

where the last inequality follows by the assumption on f^∞.

This completes the proof of (10.8.12), and, therefore, the proof of Theorem 10.3.11. $\square$

Observe that, by (10.8.10), the inequality in (10.8.13) is equivalent to the inequality informally introduced in (10.3.6).

10.9 Examples

The calculations in Example 10.9.1 can be done using the value iteration equation $J_n^* = T J_{n-1}^*$ in (9.5.4) for the optimal n-stage cost, i.e., for each $x \in X$ and $n = 1, 2, \dots$,

$$J_n^*(x) = \min_{A(x)}\left[c(x, a) + \int_X J_{n-1}^*(y)Q(dy|x, a)\right], \qquad (10.9.1)$$

with $J_0^*(\cdot) := 0$. We will also use the fact that for a deterministic stationary policy j^∞ the n-stage expected total cost

$$J_n(f^\infty, x) := E_x^{f^\infty}\left[\sum_{t=0}^{n-1} c_f(x_t)\right]$$

can be recursively computed as

$$J_n(f^\infty, x) = c_f(x) + \int_X J_{n-1}(f^\infty, y)Q_f(dy|x) \qquad (10.9.2)$$

for all $x \in X$, $n = 1, 2, \dots$, with $J_0(f^\infty, \cdot) := 0$. More generally, for an arbitrary policy π we have

$$J_n(\pi, x) = \int_X\left[c(x, a) + \int_X J_{n-1}(\pi, y)Q(dy|x, a)\right]\pi_0(da|x) \qquad (10.9.3)$$

for all $x \in X$, $n = 1, 2, \ldots$, with $J_0(\pi, \cdot) := 0$.

The following example shows that without the appropriate hypotheses, some of the results in the previous sections may fail. Specifically, the example shows a deterministic stationary policy f_*^∞ that is canonical but—in contrast to, say, Theorem 10.3.11—is not weakly O.O., nor OC-optimal, nor D-optimal. In addition, there is a deterministic stationary policy f^∞ that is AC-optimal and F-strong AC-optimal [see (10.3.45), (10.3.46)] but is not canonical.

10.9.1 Example. Consider a MCP with state space $X = \{0, 1, \ldots\}$, the set of nonnegative integers, with the discrete topology. The control (or action) sets are $A(x) = A := \{1, 2\}$ for all state x. The state $x = 0$ is absorbing with zero cost; that is, writing $Q(\{y\}|x, a)$ as $Q(y|x, a)$, we have

$$c(0, a) = 0 \quad \text{and} \quad Q(0|0, a) = 1 \quad \forall a \in A. \tag{10.9.4}$$

On the other hand, for $x \geq 1$ we have

$$c(x, 1) = 1/x - 1, \quad \text{and} \quad Q(0|x, 1) = 1; \tag{10.9.5}$$

and

$$c(x, 2) = 0, \quad \text{and} \quad Q(x + 1|x, 2) = 1. \tag{10.9.6}$$

From (10.9.4) and (10.9.3), $J_n(\pi, 0) = 0$ for each policy π and $n = 0, 1, \ldots$, so that $J_n^*(0) = 0$ for all n. Moreover, as $J_0^*(\cdot) := 0$, we can use (10.9.1) to obtain

$$J_n^*(x) = (x + n - 1)^{-1} - 1 \quad \forall x \geq 1, \ n \geq 1.$$

Let us now consider the decision functions $f_*(x) := 2$ and $f(x) := 1$ for all $x \in X$. Then, by (10.9.2),

$$J_n(f_*^\infty, x) = 0 \quad \forall n \geq 0, \ x \in X,$$

and

$$J_n(f^\infty, 0) = 0 \quad \forall n \geq 0; \quad J_n(f^\infty, x) = 1/x - 1 \quad \forall n \geq 1, \ x \geq 1.$$

We can also see that f_*^∞ is a *canonical* policy (i.e., f_* is in $\mathbb{F}_{ca}$) since (ρ^*, h^*, f_*) with

$$\rho^* = 0, \ h^*(0) = 0, \text{ and } h^*(x) = -1 \text{ for } x \geq 1$$

is a *canonical triplet*. However, f_*^∞ *is not weakly O.O.* since it does not satisfy (10.8.11) [which is equivalent to (10.1.5)]; namely,

$$\limsup_{n \to \infty} [J_n(f_*^\infty, x) - J_n(f^\infty, x)] = -(1/x - 1) > 0 \quad \forall x > 1.$$

Furthermore, f_*^∞ *is neither OC-optimal nor D-optimal* since (10.1.7) and (10.1.12) yield

$$OC(f_*^\infty, x) = 1 \quad \forall x \geq 1, \quad \text{and} \quad D(f_*^\infty, x) = 0 \quad \forall x \in X,$$

whereas

$$OC(f^\infty, x) = 1/x \quad \forall x \geq 1, \quad \text{and} \quad D(f^\infty, x) = 1/x - 1 \quad \forall x \geq 1;$$

hence

$$\inf_{\Pi} OC(\pi, x) \leq OC(f^\infty, x) < OC(f_*^\infty, x) \quad \forall x > 1,$$

and

$$\inf_{\Pi} D(\pi, x) \leq D(f^\infty, x) < D(f_*^\infty, x) \quad \forall x > 1.$$

Finally, it is evident that f^∞ *is AC-optimal and F-strong AC-optimal* [see (10.3.45)], but *it is not canonical*. $\square$

Example 10.9.1, which comes from Hernández-Lerma and Vega-Amaya [1], is a slight modification of an example by Nowak [1].

10.9.2 Remark. It is clear that, because of the condition on Q in (10.9.6), the MCP in Example 10.9.1 does not satisfy the w-geometric ergodicity requirement in Assumption 10.2.2. Thus one might be willing to conjecture that the conclusions in the example do not coincide with the results in §10.3 precisely because Assumption 10.2.2 fails. However, Brown [1] shows a simple (two states, two actions) MCP in which the *uniform geometric ergodicity* condition (7.3.16) holds but there is no deterministic stationary policy which is weakly O.O. in the class Π of all policies. What is even more interesting in Brown's MCP is that there exists a deterministic stationary policy which is *strongly O.O.* [see (10.1.4)] in the subclass Π_{DS} of deterministic stationary policies.

On the other hand, Nowak and Vega-Amaya [1] give an example in which again, as in Brown's MCP, (7.3.16) holds and yet there is no deterministic stationary policy which is strongly O.O., *not even in the subclass Π_{DS}!* The examples by Brown and by Nowak and Vega-Amaya both show in particular that Theorem 6.2 in Fernández-Gaucherand et al. [1] is false. (The latter theorem supposedly gives conditions for a deterministic stationary policy to be strongly O.O. in all of Π.)

Finally, in connection with (10.1.9), it is worth noting that Flynn [1, Example 1] shows a weakly O.O. policy—hence OC-optimal—but its opportunity cost is *infinite*, and so *every policy is OC-optimal!* In cases like this, (10.1.9) remains valid, of course, but it does not necessarily give "relevant" information. A similar comment is valid for (10.1.14). $\square$

10.9.3 Example. (Example 8.6.2 continued.) Consider the inventory-production system in Example 8.6.2, with system equation (8.6.5), state space $X = [0, \infty)$, and control sets

$$A(x) = A := [0, \theta] \quad \forall x \in X. \tag{10.9.7}$$

We again suppose that the demand process $\{z_t\}$ satisfies Assumption 8.6.1 and the condition 8.6.3, but in addition we will now suppose that, with $\bar{z} := E(z_0)$,

$$\theta < \bar{z}. \tag{10.9.8}$$

[Note that (10.9.8) states that the average demand $\bar{z}$ should exceed the maximum allowed production. Hence it excludes some frequently encountered cases that require the opposite, $\theta \geq \bar{z}$.] By the results in Example 8.6.2 we see all of the conditions (a) to (f) in Assumption 10.2.1 are satisfied, except that the constant β in (10.2.1) is greater than 1; in fact, after (8.6.14) we obtained $\beta = 1 + \varepsilon$. We will next use the new assumption (10.9.8) to see that β can chosen to satisfy $\beta < 1$, as required in (10.2.1).

Let $\psi(r) := E \exp[r(\theta - z_0)]$, $r \geq 0$, be the moment generating function of $\theta - z_0$. Then, as $\psi(0) = 1$ and $\psi'(0) = E(\theta - z_0) = \theta - \bar{z} < 0$ [by (10.9.8)], there is a positive number r_* such that

$$\psi(r_*) < 1.$$

[Compare the latter inequality with (8.6.11).] Therefore, defining the new weight function

$$w(x) := \exp[r_*(x + 2\bar{z})], \quad x \in X, \tag{10.9.9}$$

we see that $w(\cdot)$ satisfies (8.6.13) and (8.6.14) when $\hat{r}$ is replaced by r_*. In particular, (8.6.14) becomes

$$w'(x, a) \leq \beta w(x) + b \quad \forall x \in X, \ a \in A, \tag{10.9.10}$$

with

$$\beta := \psi(r_*) < 1, \quad \text{and} \quad b := w(0). \tag{10.9.11}$$

Thus, as (10.9.10)–(10.9.11) yield (10.2.1), we have that Assumption 10.2.1 holds in the present case. [Alternatively, to verify (10.2.1) we could use (10.9.20) below because $l_f(\cdot) \leq 1$.]

We will next verify Assumption 10.2.2 (using Proposition 10.2.5), and Assumption 10.3.5. We begin with the following lemma.

10.9.4 Lemma. *For each decision function $f \in \mathbb{F}$, let $\{x_t^f, t = 0, 1, \ldots\}$ be the Markov chain defined by (8.6.5) when $a_t := f(x_t)$ for all t, i.e.,*

$$x_{t+1} = [x_t + f(x_t) - z_t]^+, \quad t = 0, 1, \ldots. \tag{10.9.12}$$

Then, for each $f \in \mathbb{F}$, $\{x_t^f\}$ is positive recurrent, and so it has a unique i.p.m. μ_f.

Proof. Let $\{x_t^\theta\}$ be the Markov chain given by (10.9.12) when $f(x) := \theta$ for all state x, i.e.,

$$x_{t+1}^\theta = (x_t^\theta + \theta - z_t)^+, \quad \forall t = 0, 1, \ldots. \tag{10.9.13}$$

Observe that defining $y_t := \theta - z_t$, we can rewrite (10.9.13) as a random walk of the form (7.4.4), i.e.,

$$x_{t+1}^\theta = (x_t^\theta + y_t)^+. \tag{10.9.14}$$

Hence, as $E|y_0| \le \theta + \bar{z} < \infty$ and $E(y_0) = \theta - \bar{z} < 0$ [by (10.9.8)], the Markov chain $\{x_t^\theta\}$ is positive recurrent (see Example 7.4.2). This implies in particular that $E_0(\tau^\theta) < \infty$ where τ^θ denotes the time of first return to $x = 0$ given the initial state $x_0 = 0$.

Now choose an arbitrary decision function $f \in \mathbb{F}$, and let τ^f be the time of first return to $x = 0$ given $x_0^f = 0$. By (10.9.7), $f(x) \le \theta$ for all $x \in X$, and, therefore, $x_t^f \le x_t^\theta$ for all $t = 0, 1, \ldots$. This implies that

$$E_0(\tau^f) \le E_0(\tau^\theta) < \infty,$$

which yields that $\{x_t^f\}$ is positive (in fact, positive Harris) recurrent; see, for instance, Corollary 5.13 in Nummelin [1]. Thus, as $f \in \mathbb{F}$ was arbitrary, the lemma follows. $\square$

Lemma 10.9.4 gives the existence of the invariant probability measures μ_f required in Proposition 10.2.5. We will next verify the hypotheses (i)–(iv).

Let δ_0 be the Dirac measure at $x = 0$, and define

$$\nu(\cdot) := \delta_0(\cdot), \quad \text{and} \quad l_\theta(x) := 1 - G(x + \theta), \quad x \in X \tag{10.9.15}$$

where G denotes the common probability distribution of the demand variables z_t. Recall that G is supposed to satisfy the condition 8.6.3, the Assumption 8.6.1, and also (10.9.8). Moreover, for each decision function f, let

$$l_f(x) := 1 - G(x + f(x)).$$

By (10.9.7), $x + f(x) \le x + \theta$ for all $x \in X$ and $f \in \mathbb{F}$, so that

$$l_f(x) \ge l_\theta(x) \quad \forall x \in X, \ f \in \mathbb{F}. \tag{10.9.16}$$

Hence, if in (8.6.10) we replace the function $u(\cdot)$ by the indicator function $I_B(\cdot)$ of a Borel set $B \subset X$, we see that (8.6.10) [and (8.6.2)] yield

$$Q_f(B|x) \ge I_B(x)[1 - G(x + f(x))] = \nu(B)l_f(x) \tag{10.9.17}$$

for each $f \in \mathbb{F}$, $x \in X$, and $B \in \mathcal{B}(X)$; that is the hypotheses (i) in Proposition 10.2.5 is satisfied.

On the other hand, from (10.9.15), defining

$$\gamma := \nu(l_\theta) = l_\theta(0) = 1 - G(\theta) > 0 \tag{10.9.18}$$

we get the hypothesis (ii) since [by (10.9.16)]

$$\nu(l_f) = l_f(0) \ge l_\theta(0) = \gamma \quad \forall f \in \mathbb{F}. \tag{10.9.19}$$

Furthermore, (iii) is obvious since

$$\nu(w) = w(0) = \exp(2r_*\bar{z}) < \infty \quad [\text{see } (10.9.9)].$$

Finally, the hypothesis (iv) follows from (8.6.13) with r_* in lieu of $\hat{r}$, which gives

$$
\begin{aligned}
\int_X w(y)Q_f(dy|x) &\leq w(0)l_f(x) + \psi(r_*)w(x) \\
&= \nu(w)l_f(x) + \beta w(x) \quad [\text{by } (10.9.11)].
\end{aligned}
\tag{10.9.20}
$$

Thus, all of the hypotheses of Proposition 10.2.5 are satisfied, and so Assumption 10.2.2 holds.

Finally, to verify Assumption 10.3.5, simply note that (10.9.16) and (10.9.17) yield

$$Q_f(\cdot|x) \geq \nu(\cdot)l_\theta(x) \quad \forall f \in \mathbb{F}, \ x \in X, \tag{10.9.21}$$

which implies Assumption 10.3.5 with $\lambda(\cdot) := \nu(\cdot)$.

To conclude, the inventory-production system in Example 8.6.2, with the additional condition (10.9.8), satisfies Assumptions 10.2.1, 10.2.2, and 10.3.5, and, by consequence, all of the results in §10.3 hold in this case. $\square$

Example 10.9.3 is due to Vega-Amaya [1], [2]; see also Hernández-Lerma and Vega-Amaya [1].

10.9.5 Example. (Example 8.6.4 continued.) Let us again consider the queueing system (8.6.16) in Example 8.6.4. From that example we can see that all of the conditions in Assumption 10.2.1 are satisfied, except perhaps for part (f). On the other hand, by (10.2.4), to verify (10.2.1) it suffices to check that (10.2.25) holds because then, since $l_f(\cdot) \leq 1$, we may take $b := \nu(w)$ to obtain (10.2.1).

Hence we will proceed directly to verify Assumption 10.2.2 (via Proposition 10.2.5), and Assumption 10.3.5.

Let $z := \theta\eta_0 - \xi_0$ be as in Assumption 8.6.5(d), and let $\{x_t^\theta\}$ be the Markov chain obtained from (8.6.16) when $a_t := \theta$ for all t. Moreover, let $\{z_t\}$ be a sequence of i.i.d. random variables with the same distribution as z. Then we can write $\{x_t^\theta\}$ as

$$x_{t+1}^\theta = (x_t^\theta + z_t)^+, \tag{10.9.22}$$

which is a random walk of the form (10.9.14) [or (7.4.4)]. Hence, exactly as in Lemma 10.9.4, we can verify that the Markov chain $\{x_t^f\}$, $f \in \mathbb{F}$; obtained from (8.6.16) with $a_t := f(x_t)$ for all $t = 0, 1, \ldots$, i.e.,

$$x_{t+1}^f := [x_t^f + f(x_t)\eta_t - \xi_t]^+, \quad t = 0, 1, \ldots$$

is positive recurrent. For that reason, $\{x_t^f\}$ has a unique i.p.m. μ_f, for each $f \in \mathbb{F}$.

We will now verify the hypotheses (i) to (iv) in Proposition 10.2.5.

Let $\psi(r) := E\exp(rz)$, $r \geq 0$, be the moment generating function of $z := \theta\eta_0 - \xi_0$, and let w be the weight function in (8.6.18), i.e.,

$$w(x) := e^{rx}, \quad x \in X := [0, \infty), \tag{10.9.23}$$

where $r > 0$ is such that $\psi(r) < 1$. Furthermore, let $\nu(\cdot) := \delta_0(\cdot)$ be the Dirac measure at $x = 0$, and let l_θ, and l_f, for $f \in \mathbb{F}$, be defined as

$$
\begin{aligned}
l_\theta(x) &:= & P(x + \theta\eta_0 - \xi_0 \leq 0) = P(x + z \leq 0), \\
l_f(x) &:= & P(x + f(x)\eta_0 - \xi_0 \leq 0).
\end{aligned}
\tag{10.9.24}
$$

Now in (8.6.22) replace $u(\cdot)$ and $a \in A$ by $I_B(\cdot)$ and $f(x)$, respectively, where $B \subset X$ is a Borel set. This yields

$$Q_f(B|x) \geq l_f(x)\nu(B) \quad \forall f \in \mathbb{F}, \ B \in \mathcal{B}(X), \ x \in X, \tag{10.9.25}$$

which is precisely the hypothesis (i) in Proposition 10.2.5.

On the other hand, by Assumption 8.6.5(a), we have $a \leq \theta$ for all $a \in A$; hence,

$$l_f(x) \geq l_\theta(x) \quad \forall f \in \mathbb{F}, \ x \in X, \tag{10.9.26}$$

so that

$$\nu(l_f) := l_f(0) \geq l_\theta(0) \quad \forall f \in \mathbb{F},$$

which implies the hypothesis (ii) with $\gamma := l_\theta(0)$.

The hypothesis (iii) trivially holds since [by (10.9.23)] $\nu(w) = w(0) = 1$, and so does the hypothesis (iv) because taking $a = f(x)$ in (8.6.23) we get (10.2.25) with $\beta := \psi(r) < 1$ [see (8.6.24)], $l_f(\cdot)$ as in (10.9.24), and $\nu(w) = 1$.

Thus, Assumption 10.2.2 follows from Proposition 10.2.5.

Finally, by (10.9.25) and (10.9.26),

$$Q_f(\cdot|x) \geq l_\theta(x)\nu(\cdot) \quad \forall f \in \mathbb{F}, \ x \in X,$$

which yields Assumption 10.3.5 with $\lambda(\cdot) := \nu(\cdot)$.

Therefore, all of the results in §10.3 hold for the queueing system in Example 8.6.4. □

Example 10.9.5 (as well as Example 8.6.4) is due to Gordienko and Hernández-Lerma [1]. Other examples can be found, for instance, in Dynkin and Yushkevich [1], and Yushkevich [1]. For MCPs with a *countable* state space see, e.g., Bertsekas [1], Haviv and Puterman [1], Puterman [1], Ross [1].

One final comment: some of the optimality criteria discussed in §10.1 and §10.3 can be called "horizon sensitive" because they compare control policies on the basis of their performance on finite n-stage horizons. An alternative is to compare policies according to their performance with respect to the infinite-horizon α-discounted cost ($0 < \alpha < 1$) as α tends to 1. These alternative criteria are called *discount-sensitive* criteria. (See, for instance, Puterman [1], Yushkevich [1].)

11

Sample Path Average Cost

11.1 Introduction

In this chapter we study AC-related criteria, some of which have already been studied in previous chapters from a different viewpoint. We begin by introducing some notation and definitions, and then we outline the contents of this chapter.

A. Definitions

Let $\mathcal{M} = (X, A, \{A(x)|x \in X\}, Q, c)$ be a general Markov control model, and let

$$J_n(\pi, \nu) := E_\nu^\pi \left[\sum_{t=0}^{n-1} c(x_t, a_t) \right] \tag{11.1.1}$$

be the *expected* n-stage total cost when using the control policy π, given the initial distribution $\nu \in \mathcal{P}(X)$. We can also write (11.1.1) as

$$J_n(\pi, \nu) = \int_X J_n(\pi, x) \nu(dx), \tag{11.1.2}$$

where $J_n(\pi, x)$ is given by (11.1.1) when $\nu = \delta_x$ (the Dirac measure contrated at $x_0 = x$) or as

$$J_n(\pi, \nu) = E_\nu^\pi [J_n^0(\pi, \nu)], \tag{11.1.3}$$

where

$$J_n^0(\pi, \nu) := \sum_{t=0}^{n-1} c(x_t, a_t) \qquad (11.1.4)$$

in the *pathwise* (or *sample path*) n-stage total cost.

In addition to the usual "limit supremum" *expected* average cost (AC)

$$J(\pi, \nu) := \limsup_{n \to \infty} J_n(\pi, \nu)/n, \qquad (11.1.5)$$

in this chapter we can also consider

- the **sample path AC**

$$J^0(\pi, \nu) := \limsup_{n \to \infty} J_n^0(\pi, \nu)/n, \qquad (11.1.6)$$

- and the **"limit infimum" expected AC**

$$J^I(\pi, \nu) := \liminf_{n \to \infty} J_n(\pi, \nu)/n. \qquad (11.1.7)$$

In §11.3 we also study a particular case of the **limiting average variance**

$$\mathrm{Var}(\pi, \nu) := \limsup_{n \to \infty} \frac{1}{n} \mathrm{var}[J_n^0(\pi, \nu)], \qquad (11.1.8)$$

where (by definition of *variance* of a random variable)

$$\mathrm{var}[J_n^0(\pi, \nu)] := E_\nu^\pi [J_n^0(\pi, \nu) - J_n(\pi, \nu)]^2. \qquad (11.1.9)$$

In (10.1.16) we defined a policy π^* to be **AC-optimal** if

$$J(\pi^*, x) = J^*(x) \quad \forall x \in X, \qquad (11.1.10)$$

where

$$J^*(x) := \inf_\Pi J(\pi, x) \qquad (11.1.11)$$

is the *optimal expected AC function* (also known as the *AC value function*). We now wish to relate AC optimality with the performance criteria in the following definition.

11.1.1 Definition. Let π^* be a control policy and ν^* an initial distribution. Then

(a) (π^*, ν^*) is a **minimum pair** if

$$J(\pi^*, \nu^*) = \rho_{\min}$$

where

$$\rho_{\min} := \inf_{P(X)} J^*(\nu) = \inf_{P(X)} \inf_\Pi J(\pi, \nu) \qquad (11.1.12)$$

is the *minimum average cost*, and [as in (11.1.11)] $J^*(\nu) := \inf_\Pi J(\pi, \nu)$.

(b) π^* is **sample path AC-optimal** if [with J^0 as in (11.1.6)]

$$J^0(\pi^*, \nu) = \rho_{\min} \quad P_\nu^{\pi^*}\text{-a.s.} \quad \forall \nu \in \mathcal{P}(X), \qquad (11.1.13)$$

and, furthermore,

$$J^0(\pi, \nu) \geq \rho_{\min} \quad P_\nu^\pi\text{-a.s.} \quad \forall \pi \in \Pi, \ \nu \in \mathcal{P}(X). \qquad (11.1.14)$$

(See Remark 11.1.2 and Definition 11.1.3.)

(c) π^* is **strong AC-optimal** if it is AC-optimal and, in addition,

$$J(\pi^*, x) \leq J^I(\pi, x) \quad \forall \pi \in \Pi, \ x \in X,$$

with J^I as in (11.1.7).

11.1.2 Remark. The optimality concepts in Definition 11.1.1(a), (c) were already introduced in Chapter 5. On the other hand, Definition 11.1.1(b) is related to *pathwise AC optimality* in Definition 5.7.6(b), which only requires (11.1.13).

In §11.4 we study a class of Markov control problems for which there exists a sample path AC optimal policy. In general, however, Definition 11.1.1(b) turns out to be extremely demanding in the sense that requiring (11.1.13) and (11.1.14) to hold for *all $\pi \in \Pi$ and all $\nu \in \mathcal{P}(X)$* is a very strong condition. It is thus convenient to consider a weaker form of sample path AC optimality as follows.

11.1.3 Definition. Let $\widehat{\Pi} \subset \Pi$ be a subclass of control policies, and $\widehat{\mathcal{P}}(X) \subset \mathcal{P}(X)$ a subclass of probability measures ("initial distributions") on X. Let

$$\widehat{\rho} := \inf_{\widehat{\mathcal{P}}(X)} \inf_{\widehat{\Pi}} J(\pi, \nu). \qquad (11.1.15)$$

A policy $\widehat{\pi} \in \widehat{\Pi}$ is said to be **sample path AC-optimal with respect to $\widehat{\Pi}$ and $\widehat{\mathcal{P}}(X)$** if

$$J^0(\widehat{\pi}, \nu) = \widehat{\rho} \quad P_\nu^{\widehat{\pi}}\text{-a.s.} \quad \forall \nu \in \widehat{\mathcal{P}}(X), \qquad (11.1.16)$$

and

$$J^0(\pi, \nu) \geq \widehat{\rho} \quad P_\nu^\pi\text{-a.s.} \quad \forall \pi \in \widehat{\Pi}, \ \nu \in \widehat{\mathcal{P}}(X). \qquad (11.1.17)$$

If $\widehat{\Pi} = \Pi$ and $\widehat{\mathcal{P}}(X) = \mathcal{P}(X)$, then—as in Definition 11.1.1(b)—we simply say that $\widehat{\pi}$ is **sample path AC-optimal**.

For example, in §11.3 we consider a class of Markov control problems in which there exists a sample path AC-optimal policy with respect to

$$\widehat{\Pi} = \Pi_{DS} \quad \text{and} \quad \widehat{\mathcal{P}}(X) = \mathcal{P}(X), \qquad (11.1.18)$$

and another class of problems for which

$$\widehat{\Pi} = \Pi \quad \text{and} \quad \widehat{\mathcal{P}}(X) = \{\delta_x | x \in X\} =: \mathcal{P}_\delta(X), \tag{11.1.19}$$

where δ_x is the Dirac measure at the initial state $x_0 = x$. More explicitly, in the former case (11.1.18), there is a deterministic stationary policy f_0^∞ such that

$$J^0(f_0^\infty, \nu) = \rho_0 \quad P_\nu^{f_0^\infty}\text{-a.s.} \quad \forall \nu \in \mathcal{P}(X) \tag{11.1.20}$$

and

$$J^0(f^\infty, \nu) \geq \rho_0 \quad P_\nu^{f_0^\infty}\text{-a.s.} \quad \forall f^\infty \in \Pi_{DS}, \ \nu \in \mathcal{P}(X), \tag{11.1.21}$$

where

$$\rho_0 := \inf_{\mathcal{P}(X)} \inf_{\Pi_{DS}} J(f^\infty, \nu). \tag{11.1.22}$$

Similarly, in case (11.1.19) there is policy $\pi^* \in \Pi$ such that, with

$$\rho^* := \inf_X J^*(x) = \inf_X \inf_\Pi J(\pi, x), \tag{11.1.23}$$

we have

$$J^0(\pi^*, x) = \rho^* \quad P_x^{\pi^*}\text{-a.s.} \quad \forall x \in X \tag{11.1.24}$$

and

$$J^0(\pi, x) \geq \rho^* \quad P_x^\pi\text{-a.s.} \quad \forall \pi \in \Pi, \ x \in X. \tag{11.1.25}$$

Note that the condition "for all $x \in X$" in (11.1.24) and (11.1.25) can also be expressed as "for all ν in $\mathcal{P}_\delta(X)$", with $\mathcal{P}_\delta(X)$ as in (11.1.19).

B. Outline of the chapter

The rest of the chapter consists of four sections. Section 11.2 presents background material on positive Harris recurrence and the limiting average variance (11.1.8). The reader may go directly to §11.3 and refer to the concepts and results in §11.2 as they are needed. In §11.3 we consider a Markov control model which is "w-geometrically ergodic" (Definition 11.3.1) with respect to some weight function w. In this case we show the existence of deterministic stationary policies that satisfy (11.1.20) and (11.1.21), whereas under an additional condition (Assumption 11.3.4) they satisfy (11.1.24) and (11.1.25), and, moreover, the concepts in Definition 11.1.1 turn out to be "essentially" equivalent (see Theorem 11.3.5 for a precise statement). Also in §11.3 we prove the existence of a policy that minimizes the limiting average variance within the class of canonical policies (Theorem 11.3.8).

In §11.4 we turn our attention to Markov control models with a *strictly unbounded* cost-per-stage function $c(x, a)$ [see Assumption 11.4.1(c) and Remark 11.4.2(a)]. The main result in that section, Theorem 11.4.6, in particular gives conditions ensuring the existence of a sample path AC-optimal policy.

The chapter concludes in §11.5 with some examples that illustrate the results of §11.3 and §11.4.

11.2 Preliminaries

This section reviews background material that can be omitted on a first reading; the reader may refer to it as needed.

A. Positive Harris recurrence

Let $\{x_t, t = 0, 1, \ldots\}$ be a time-homogeneous X-valued Markov chain with transition probability function $P(B|x)$. The chain is said to be **positive Harris recurrent** if it satisfies that

(i) $\{x_t\}$ is Harris recurrent [Definition 7.3.1(b)], and

(ii) it has an i.p.m., which [by Theorem 7.3.4(a)] is necessarily the unique i.p.m. of $\{x_t\}$.

The next theorem presents, in particular, in parts (a) and (b), two characterizations of positive Harris recurrence.

11.2.1 Theorem. (Characterization and properties of positive Harris recurrence.)

(a) *Suppose that $\{x_t\}$ has an i.p.m. μ. Then the chain is positive Harris recurrent if and only if the* **strong Law of Large Numbers** *(LLN) holds for each function in $L_1(\mu) := L_1(X, \mathcal{B}(X), \mu)$; that is, for each function g in $L_1(\mu)$ and each initial distribution ν in $\mathcal{P}(X)$,*

$$\lim_{n \to \infty} \frac{1}{n} \sum_{t=0}^{n-1} g(x_t) = \mu(g) \quad P_\nu\text{-a.s.}, \tag{11.2.1}$$

where

$$\mu(g) := \int_X g \, d\mu. \tag{11.2.2}$$

(b) *The chain $\{x_t\}$ is positive Harris recurrent if and only if for each Borel set B in $\mathcal{B}(X)$ there is a nonnegative number α_B such that*

$$\lim_{n \to \infty} P^{(n)}(B|x) = \alpha_B \quad \forall x \in X, \tag{11.2.3}$$

where

$$P^{(n)}(B|x) := \frac{1}{n} \sum_{t=0}^{n-1} P^t(B|x), \quad n = 1, 2, \ldots \tag{11.2.4}$$

denotes the **expected average occupation measures.**

(c) *If $\{x_t\}$ is positive Harris recurrent with i.p.m. μ and g is in $L_1(\mu)$, then*

$$\lim_{n \to \infty} \frac{1}{n} E_x \left[\sum_{t=0}^{n-1} g(x_t) \right] = \mu(g) \quad \text{for} \quad \mu\text{-a.a. } x \in X,$$

with $\mu(g)$ as (11.2.2).

Proof. For parts (a) and (c) see, for instance, Revuz [1, pp. 139, 140]; for (b) see Glynn [1] or Hernández-Lerma and Lasserre [12]. (Glynn [2] proves (b) in the *continuous-time* case.) Part (a) is given also in Meyn and Tweedie [1], Theorem 17.1.7 and Proposition 17.1.6. (These references provide other characterizations of positive Harris recurrence. For additional comments see Note 1 at the end of this section.)

As an application of Theorem 11.2.1(b), let $w \geq 1$ be a weight function, and suppose that the Markov chain $\{x_t\}$ is w-geometrically ergodic (Definition 7.3.9). If in (7.3.8) we replace the function u by an indicator function I_B, then we get

$$P^t(B|x) \to \mu(B) \quad \text{as} \quad t \to \infty,$$

which of course implies (11.2.3) with $\alpha_B = \mu(B)$. Thus we have:

11.2.2 Corollary. *A w-geometrically ergodic Markov chain is positive Harris recurrent.*

B. Limiting average variance

In this subsection we suppose the following.

11.2.3 Assumption. The Markov chain $\{x_t\}$ is w-geometrically ergodic with limiting i.p.m. μ. Moreover, $c(\cdot)$ and $h(\cdot)$ are two functions in $\mathbb{B}_w(X)$, with h as in (7.5.30).

Under this assumption, Theorem 7.5.10 states that the pair (g, h) is the unique solution of the strictly unichain Poisson equation

$$c - \mu(c) = h - Ph \tag{11.2.5}$$

that satisfies

$$\mu(h) = 0 \quad \text{and} \quad g(x) = \mu(c) \quad \forall x \in X. \tag{11.2.6}$$

Equivalently, by Theorem 7.5.5(a), (b) [and (7.5.7) or (7.5.6)]

$$E_x\left[\sum_{t=0}^{n-1} c(x_t)\right] = n\mu(c) + h(x) - E_x h(x_n) \quad \forall x \in X, \, n \geq 1, \tag{11.2.7}$$

which gives the mean value $E_x[s_n(x)]$ of the random sum

$$s_n(x) := \sum_{t=0}^{n-1} c(x_t), \quad n = 1, 2, \ldots \tag{11.2.8}$$

for each initial state $x_0 = x$.

We now with to compute the **limiting average variance**

$$\sigma^2(c, x) := \lim_{n \to \infty} \frac{1}{n} \text{var}[s_n(x)], \tag{11.2.9}$$

where [by definition of variance, $\text{var}(\xi) := E(\xi - E\xi)^2$]

$$\text{var}[s_n(x)] := E_x[s_n(x) - E_x s_n(x)]^2 \qquad (11.2.10)$$

The next theorem gives conditions under which $\sigma^2(c, \cdot)$ is the (finite) *constant*

$$\sigma_c^2 := \mu[h^2 - (Ph)^2] = \int_X [h^2(x) - (Ph(x))^2]\mu(dx). \qquad (11.2.11)$$

This result is well known (see Doob [1], Duflo [1], Meyn and Tweedie [1], etc.) but we will give a proof of it because some of the arguments are also needed in later sections.

11.2.4 Theorem. *Suppose that Assumption 11.2.3 holds and, furthermore, $c^2(\cdot)$ is in $\mathbb{B}_w(X)$. Let ψ be the function on X defined by*

$$\psi(x) := \int_X h^2(y)P(dy|x) - \left[\int_X h(z)P(dz|x)\right]^2, \qquad (11.2.12)$$

which in short can be written as

$$\psi = P(h^2) - (Ph)^2. \qquad (11.2.13)$$

Then:

(a) *ψ is in $\mathbb{B}_w(X)$, and*

(b) *the limiting average variance satisfies*

$$\sigma^2(c, x) = \lim_{n \to \infty} \frac{1}{n} E_x \left[\sum_{t=0}^{n-1} \psi(x_t)\right] = \sigma_c^2 \quad \forall x \in X. \qquad (11.2.14)$$

where σ_c^2 is the constant in (11.2.11).

Of course, if part (a) in Theorem 11.2.4 is valid, then the second equality in (11.2.14) is evident because

$$\mu(\psi) := \int_X \psi d\mu = \sigma_c^2, \qquad (11.2.15)$$

and (7.3.8) gives

$$E_x[\psi(x_t)] \to \mu(\psi) = \sigma_c^2 \quad \forall x \in X. \qquad (11.2.16)$$

Thus the proof, given below, of Theorem 11.2.4 essentially reduces to verify (a) and the first equality in (11.2.14). To prove the latter we will repeatedly use the following elementary properties of conditional expectations.

11.2.5 Remark. Let ξ and ξ' be integrable random variables on a probability space $(\Omega, \mathcal{F}, \gamma)$, and let $\mathcal{G}$ and $\mathcal{G}'$ be sub-σ-algebras of $\mathcal{F}$.

(a) $E(\xi) = E[E(\xi|\mathcal{G})]$.

(b) If ξ is $\mathcal{G}$-measurable, then

   (b$_1$) $E(\xi\xi'|\mathcal{G}) = \xi E(\xi'|\mathcal{G})$, and

   (b$_2$) $E(\xi|\mathcal{G}) = \xi$.

(c) If $\mathcal{G} \subset \mathcal{G}'$ then $E[E(\xi|\mathcal{G})|\mathcal{G}'] = E[E(\xi|\mathcal{G}')|\mathcal{G}] = E(\xi|\mathcal{G})$.

(d) If ξ is square-integrable, then $[E(\xi|\mathcal{G})]^2 \le E(\xi^2|\mathcal{G})$. $\square$

Proof of Theorem 11.2.4. (a) As c^2 is in $\mathbb{B}_w(X)$ (by assumption), (7.3.8) and (7.5.30) give that h^2 is in $\mathbb{B}_w(X)$. This implies [by (7.3.8) again] that $P(h^2)$ is in $\mathbb{B}_w(X)$, and so does ψ since (11.2.12) gives that $0 \le \psi \le P(h^2)$.

   (b) By a previous remark [see (11.2.15), (11.2.16)], it only remains to prove the first equality in (11.2.14). To do this we will first show that the random variables

$$Y_t := h(x_t) - Ph(x_{t-1}) = h(x_t) - E_x[h(x_t)|x_{t-1}], \quad t = 1, 2, \ldots, \quad (11.2.17)$$

satisfy that

$$E_x(Y_t^2) = E_x[\psi(x_{t-1})] \quad \forall x \in X, \; t = 1, 2, \ldots, \quad (11.2.18)$$

and also that the variance in (11.2.10) can be written as

$$\text{var}[s_n(x)] = E_x[s_n(x) - n\mu(c)]^2 + o_1(x, n) \quad (11.2.19)$$

and as

$$\text{var}[s_n(x)] = E_x\left[\sum_{t=1}^{n-1} Y_t^2\right] + o_2(x, n), \quad (11.2.20)$$

where, for $i = 1, 2$, $o_i(x, n)$ is a sequence such that, as $n \to \infty$,

$$o_i(x, n)/n \to 0 \quad \forall x \in X. \quad (11.2.21)$$

Hence, by the definition (11.2.9) of the limiting average variance the first equality in (11.2.14) will follow from (11.2.20), (11.2.21) and (11.2.18).

Proof of (11.2.18). Let

$$\mathcal{F}_t := \sigma\{x_0, \ldots, x_t\} \quad \text{for} \quad t = 0, 1, \ldots. \quad (11.2.22)$$

be the σ-algebra generated by $\{x_0, \ldots, x_t\}$. Then, by Remark 11.2.5(a) and the Markov property, for $t \ge 1$,

$$\begin{aligned}
E_x(Y_t^2) &= E_x[E_x(Y_t^2|\mathcal{F}_{t-1})] \\
&= E_x[E_x(Y_t^2|x_{t-1})] \\
&= E_x[\psi(x_{t-1})],
\end{aligned}$$

where the latter equality, which gives (11.2.18), follows from (11.2.17) and the fact that $\psi(x)$ in (11.2.12) can also be written as

$$\psi(x) = \int_X \left[h(y) - \int_X h(z)P(dz|x) \right]^2 P(dy|x). \tag{11.2.23}$$

Proof of (11.2.19). From (11.2.7) and (11.2.8)

$$s_n(x) - E_x s_n(x) = [s_n(x) - n\mu(c)] - [h(x) - E_x h(x_n)]. \tag{11.2.24}$$

Thus, squaring and taking expectations $E_x(\cdot)$, we see that (11.2.10) can be written as in (11.2.19) with

$$o_1(x, n) := -[h(x) - E_x h(x_n)]^2. \tag{11.2.25}$$

Now, to prove (11.2.21) with $i = 1$, use the elementary inequality

$$(a + b)^2 \le 2(a^2 + b^2) \quad \forall a, b \in \mathbb{R} \tag{11.2.26}$$

and Remark 11.2.5(d) to get

$$|o_1(x, n)| \le 2[h^2(x) + E_x h^2(x_n)],$$

which yields (11.2.21) with $i = 1$ because, as h^2 is in $\mathbb{B}_w(X)$, (7.3.8) gives

$$E_x h^2(x_n) \to \mu(h^2) \quad \forall x \in X.$$

Proof of (11.2.20). We begin by noting that the random variables Y_t in (11.2.17) satisfy

$$E_x(Y_t Y_s) = 0 \quad \forall 1 \le t < s. \tag{11.2.27}$$

Indeed, with $\mathcal{F}_t$ as in (11.2.22), the Markov property and Remark 11.2.5(b$_2$) give

$$E_x(Y_s | \mathcal{F}_{s-1}) = E_x[h(x_s)|x_{s-1}] - E_x[h(x_s)|x_{s-1}] = 0 \quad \forall s,$$

and, therefore, (11.2.27) follows from Remark 11.2.5(a), (b$_1$) because

$$\begin{aligned} E_x(Y_t Y_s) &= E_x[E_x(Y_t Y_s | \mathcal{F}_{s-1})] \\ &= E_x[Y_t E_x(Y_s | \mathcal{F}_{s-1})] \\ &= 0. \end{aligned}$$

We also note that (11.2.5) yields

$$c(x_t) - \mu(c) = h(x_t) - Ph(x_t) \quad \forall t = 0, 1, \ldots, \tag{11.2.28}$$

and so we get

$$c(x_t) - \mu(c) = Y_t - [Ph(x_t) - Ph(x_{t-1})] \quad \forall t = 1, 2, \ldots,$$

and

$$s_n(x) - n\mu(c) = M_{n-1} + [h(x_0) - Ph(x_{n-1})] \quad \forall n \geq 2 \qquad (11.2.29)$$

where, for every $n \geq 1$,

$$M_n := \sum_{t=1}^{n} Y_t. \qquad (11.2.30)$$

Observe that (11.2.27) implies that

$$E_x(M_n^2) = E_x\left(\sum_{t=1}^{n} Y_t^2\right), \qquad (11.2.31)$$

and, on the other hand,

$$E_x(M_n) = 0 \qquad (11.2.32)$$

because [by Remark 11.2.5(a)]

$$E_x(Y_t) = E_x h(x_t) - E_x h(x_t) = 0. \qquad (11.2.33)$$

Moreover, using (11.2.29) we can rewrite (11.2.24) as

$$s_n(x) - E_x s_n(x) = M_{n-1} + [h(x_0) - Ph(x_{n-1})] - [h(x) - E_x h(x_n)].$$

From this relation, together with (11.2.31) and (11.2.32), we obtain (11.2.20) with

$$o_2(x,n) := o_1(x,n) + A(x,n) + 2B(x,n),$$

with $o_1(x,n)$ as in (11.2.25), and

$$
\begin{aligned}
A(x,n) &:= E_x[h(x_0) - Ph(x_{n-1})]^2, \\
B(x,n) &:= E_x\{[h(x_0) - Ph(x_{n-1})] \cdot M_{n-1}\}.
\end{aligned}
$$

Thus, to complete the proof of Theorem 11.2.4, it only remains to prove (11.2.21) with $i = 2$. In fact, as $o_1(x,n)/n \to 0$, we only need to show that, for every initial state x, as $n \to \infty$

$$A(x,n)/n \to 0 \qquad (11.2.34)$$

and

$$B(x,n)/n \to 0 \qquad (11.2.35)$$

To prove (11.2.34), use (11.2.26) and Remark 11.2.5(a), (d) to get

$$A(x,n) \leq 2[h^2(x) + E_x h^2(x_n)].$$

This yields (11.2.34), as in the proof of (11.2.19).

Finally, to obtain (11.2.35) we may use the Cauchy-Schwartz inequality to see that

$$B(x,n)^2 \leq E_x[h(x_0) - Ph(x_{n-1})]^2 \cdot E_x(M_n^2)$$
$$= A(x,n) \cdot E_x \left(\sum_{t=1}^{n-1} Y_t^2 \right) \quad \text{[by (11.2.31)]}.$$

Therefore, (11.2.35) follows from (11.2.34), (11.2.18), and the second equality in (11.2.14).

This completes the proof of Theorem 11.2.4. □

11.2.6 Remark. (a) Let M_n and $\mathcal{F}_n$ be as in (11.2.30) and (11.2.22), respectively. Then it is clear that $\{M_n, \mathcal{F}_n\}$ is a martingale. In §11.3 and §11.4 we will use a martingale of this form in combination with the following result.

(b) **(The Martingale Stability Theorem.)** Let $\{Y_t\}$ be a sequence of random variables on a probability space $(\Omega, \mathcal{F}, P)$ and let $\{\mathcal{F}_t\}$ be a nondecreasing sequence of sub-σ-algebras of $\mathcal{F}$ such $\{M_n, \mathcal{F}_n, n = 1, 2, \ldots\}$, with $M_n := \sum_{t=1}^{n} Y_t$, is a martingale. If $1 \leq q \leq 2$ and

$$\sum_{t=1}^{\infty} t^{-q} E(|Y_t|^q | \mathcal{F}_{t-1}) < \infty \quad P\text{-a.s.,} \tag{11.2.36}$$

then

$$\lim_{n \to \infty} \frac{1}{n} M_n = 0 \quad P\text{-a.s.}$$

For a proof of this fact see, for instance, Hall and Heyde [1], Theorem 2.18.

(c) **(Alternative expressions for σ_c^2.)** Letting $\widehat{c}$ be the centered function

$$\widehat{c}(\cdot) := c(\cdot) - \mu(c),$$

we can also write (11.2.11) as

$$\sigma_c^2 = \mu(2\widehat{c}h - \widehat{c}^2), \tag{11.2.37}$$
$$\sigma_c^2 = \mu(\widehat{c}^2) + 2\mu \left[\sum_{t=1}^{\infty} \widehat{c}P^t\widehat{c} \right], \tag{11.2.38}$$

and

$$\sigma_c^2 = E_\mu[\widehat{c}^2(x_0)] + 2 \sum_{t=1}^{\infty} E_\mu[\widehat{c}(x_0)\widehat{c}(x_t)]. \tag{11.2.39}$$

Indeed, to obtain (11.2.37), we write the Poisson equation (11.2.5) as $Ph = h - \widehat{c}$, so that

$$h^2 - (Ph)^2 = 2\widehat{c}h - \widehat{c}^2. \tag{11.2.40}$$

This equality and (11.2.11) yield (11.2.37). On the other hand, to get (11.2.38) first write (7.5.30) as

$$h = \sum_{t=0}^{\infty} P^t \widehat{c} = \widehat{c} + \sum_{t=1}^{\infty} P^t \widehat{c}.$$

Then the right-hand side of (11.2.40) becomes

$$2\widehat{c}h - \widehat{c}^2 = \widehat{c}^2 + 2\sum_{t=1}^{\infty} \widehat{c}P^t\widehat{c},$$

and so (11.2.38) follows from (11.2.37). Finally, note that for any initial state $x_0 = x$ and $t = 0, 1, \ldots,$

$$E_x[\widehat{c}(x_0)\widehat{c}(x_t)] = \widehat{c}(x)E_x[\widehat{c}(x_t)] = \widehat{c}(x)P^t\widehat{c}(x)$$

and integration with respect to μ gives

$$E_\mu[\widehat{c}(x_0)\widehat{c}(x_t)] = \mu(\widehat{c}P^t\widehat{c}).$$

Thus (11.2.39) follows from (11.2.38). $\square$

Notes on §11.2

1. The conclusion of Theorem 11.2.1(c) remains valid if the hypothesis "$\{x_t\}$ is positive Harris recurrent" is replaced by "the state space X is a *locally compact* separable metric space and $\{x_t\}$ has a *unique* i.p.m. μ". (See Hernández-Lerma and Lasserre [13], Lemma 4.2.) On the other hand, it follows from Theorem 11.2.1(b) that if $\{x_t\}$ is positive Harris recurrent with i.p.m. μ and g is a *bounded measurable* function on X, then

$$\lim_{n\to\infty} \frac{1}{n} E_x \left[\sum_{t=0}^{n-1} g(x_t)\right] = \mu(g) \quad \text{for all} \quad x \in X. \tag{11.2.41}$$

This can be used to obtain additional results. For instance, if g is a nonnegative function in $L_1(\mu)$, then

$$\liminf_{n\to\infty} \frac{1}{n} E_x \left[\sum_{t=0}^{n-1} g(x_t)\right] \geq \mu(g) \quad \text{for all} \quad x \in X.$$

This follows from (11.2.41) and the fact that $g \in L_1(X)^+$ is the pointwise limit of a nondecreasing sequence of bounded measurable functions.
On the other hand, writing $g(x_n)$ as

$$\sum_{t=0}^{n} g(x_t) - \sum_{t=0}^{n-1} g(x_t),$$

it is easily deduced from (11.2.1) that if $\{x_t\}$ is positive Harris recurrent with i.p.m. μ and g is in $L_1(\mu)$, then

$$\lim_{n\to\infty} |g(x_n)|/n = 0 \quad P_\nu\text{-a.s. for each } \nu \text{ in } \mathcal{P}(X). \qquad (11.2.42)$$

In fact, in (11.2.42) we may replace $|g(x_n)|$ by $\max_{1\le k\le n} |g(x_k)|$; see, for instance, Meyn and Tweedie [1, Theorem 17.3.3].

2. (The Central Limit Theorem from Markov chains.) Suppose that the hypotheses of Theorem 11.2.4 are satisfied, and let $Z(\cdot)$ denote the Gaussian distribution with mean 0 and variance 1, i.e.,

$$Z(r) := (2\pi)^{-1/2} \int_{-\infty}^{r} e^{-y^2/2} dy, \quad r \in \mathbb{R}.$$

If $\sigma_c^2 > 0$, then for each initial state $x \in X$

$$\lim_{n\to\infty} P_x\{(n\sigma_c^2)^{-1/2}[s_n(x) - n\mu(c)] \le r\} = Z(r) \quad \forall r \in \mathbb{R}. \qquad (11.2.43)$$

[For a proof see, for example, the references given after (11.2.11).]

3. The reader should be warned that there are *different definitions* of limiting average variance. For instance, Baykal-Gürsoy and Ross [1], Filar *et al.* [1], Puterman [1, p. 408], etc., define the "limiting average variance" as

$$\lim_{n\to\infty} \frac{1}{n} \sum_{t=0}^{n-1} E_x[c(x_t) - \mu(c)]^2, \qquad (11.2.44)$$

which in general [despite (11.2.19)] does not coincide with (11.2.9). In particular, observe that under the hypotheses of Theorem 11.2.4 the limiting value in (11.2.44) is

$$\mu[(c - \mu(c))^2] = \mu[(h - Ph)^2],$$

which is not the same as σ_c^2 in (11.2.11) [or (11.2.37)].

11.3 The w-geometrically ergodic case

Let $w \ge 1$ be a given weight function. In this section we study the optimality criteria in Definition 11.1.1 for a w-geometrically ergodic control model, by which we mean the following.

11.3.1 Definition. A Markov control model $\mathcal{M} = (X, A, \{A(x)|x \in X\}, Q, c)$ is said to be w-**geometrically ergodic** if Assumptions 10.2.1 and 10.2.2 are both satisfied.

We already have a lot of information on such a model. For instance, by Proposition 10.2.3(b), the "lim sup" and "lim inf" average costs [see

(11.1.5), (11.1.7)] coincide on the class Π_{DS} of deterministic stationary policies; that is, for each f^∞ in Π_{DS} (equivalently, for each decision function $f \in \mathbb{F}$)

$$J(f^\infty, x) = J^I(f^\infty, x) = J(f) \quad \forall x \in X, \tag{11.3.1}$$

where $J(f) = \mu_f(c_f)$ is the constant in (10.2.18). Moreover, in (11.3.1) we may replace the initial state $x_0 = x$ by any initial distribution ν in the family

$$\mathcal{P}_w(X) := \{\nu \in \mathcal{P}(X) | \nu(w) := \int_X w d\nu < \infty\}. \tag{11.3.2}$$

That is, instead of (11.3.1) we have

$$J(f^\infty, \nu) = \int_X J(f^\infty, x)\nu(dx) = J(f) \quad \forall \nu \in \mathcal{P}_w(X). \tag{11.3.3}$$

Indeed, by (10.4.3), the n-step cost $J_n(f^\infty, x)$ satisfies

$$|J_n(f^\infty, x)/n| \le kw(x) \quad \forall x \in X,$$

where $k := \bar{c}[1 + b/(1-\beta)]$. Hence, if ν is in $\mathcal{P}_w(X)$, from (11.3.1) and the Dominated Convergence Theorem we obtain

$$
\begin{aligned}
J(f) &= \int_X J(f^\infty, x)\nu(dx) \\
&= \lim_{n\to\infty} \int_X n^{-1} J_n(f^\infty, x)\nu(dx) \\
&= \lim_{n\to\infty} n^{-1} J_n(f^\infty, \nu) \quad [\text{by (11.1.2)}] \\
&= J(f^\infty, \nu),
\end{aligned}
$$

which gives (11.3.3).

A result similar to (11.3.1) and (11.3.3) is obtained if we replace the *expected* AC $J(f^\infty, \cdot)$ by the *sample path* AC $J^0(f^\infty, \cdot)$ in (11.1.6):

11.3.2 Proposition. *If the Markov control model is w-geometrically ergodic, then for each deterministic stationary policy $f^\infty \in \Pi_{DS}$*

(a) *the state (Markov) process is positive Harris recurrent, and*

(b) *for each initial distribution ν in $\mathcal{P}(X)$,*

$$J^0(f^\infty, \nu) = \lim_{n\to\infty} J_n^0(f^\infty, \nu)/n = J(f) \quad P_\nu^{f^\infty}\text{-a.s.} \tag{11.3.4}$$

Proof. In (10.2.9) replace u by an indicator function I_B, with B in $\mathcal{B}(X)$. This gives

$$\lim_{t\to\infty} Q_f^t(B|x) = \mu_f(B) \quad \forall x \in X, \ B \in \mathcal{B}(X),$$

which clearly implies (11.2.3) with $Q_f(\cdot|x)$ and $\mu_f(B)$ in lieu of $P(\cdot|x)$ and α_B, respectively. Thus part (a) follows from Theorem 11.2.1(b), and (b) follows from (11.2.1). $\square$

Despite these facts, however, w-geometric ergodicity is not enough to guarantee a "good behavior" of the Markov control model with respect to the optimality criteria in Definition 11.1.1. In particular, it does not ensure the existence of sample path AC-optimal policies [Definition 11.1.1(b)]. It is thus convenient to consider sample path AC-optimality in the restricted sense of Definition 11.1.3. We shall consider two cases:

(1) $\widehat{\Pi} := \Pi_{DS}$ and $\widehat{\mathcal{P}}(X) := \mathcal{P}(X)$;

(2) $\widehat{\Pi} := \Pi$ and $\widehat{\mathcal{P}}(X) := \mathcal{P}_\delta(X)$, the set defined in (11.1.19)

Case (1), which is dealt with in subsection A, is quite straightforward but is explicitly stated here (Theorem 11.3.3) mainly for the purpose of comparison with the more technically demanding case (2). The latter case is studied in subsection B, and, finally, in subsection C we consider a variance minimization problem.

A. Optimality in Π_{DS}

By Theorem 10.3.1 we know that for a w-geometrically ergodic control model there exists a deterministic stationary policy f_0^∞ which is AC-optimal in Π_{DS}, that is,

$$J(f_0) = \rho_0, \quad \text{with} \quad \rho_0 := \inf_{\mathbb{F}} J(f). \tag{11.3.5}$$

[Furthermore, f_0^∞ is AC-optimal (in all of Π) if the one-stage cost $c(x, a)$ is *nonnegative*—see Remark 10.3.2.] We now have the following.

11.3.3 Theorem. *Suppose that the Markov control model is w-geometrically ergodic and let f_0^∞ be as in (11.3.5). Then:*

(a) *f_0^∞ is sample path AC-optimal with respect to Π_{DS} and $\mathcal{P}(X)$; that is,*

$$J^0(f_0^\infty, \nu) = \rho_0 \quad P_\nu^{f_0^\infty}\text{-a.s.} \quad \forall \nu \in \mathcal{P}(X)$$

and

$$J^0(f_0^\infty, \nu) \geq \rho_0 \quad P_\nu^{f_0^\infty}\text{-a.s.} \quad \forall f \in \Pi_{DS}, \ \nu \in \mathcal{P}(X).$$

(b) *(f_0^∞, ν) is a "minimum pair in Π_{DS}" for each initial distribution ν in $\mathcal{P}_w(X)$, the set defined in (11.3.2); that is,*

$$J(f_0^\infty, \nu) = \rho_0 \quad \forall \nu \in \mathcal{P}_w(X).$$

Proof. Part (a) follows from (11.3.5) and Proposition 11.3.2(b), and part (b) from (11.3.3). □

Observe that we can write (11.3.1) as

$$J(f^\infty, \nu) = J(f) \quad \forall \nu \in \mathcal{P}_\delta(X) := \{\delta_x | x \in X\}. \tag{11.3.6}$$

Hence, as $\mathcal{P}_\delta(X)$ is contained in $\mathcal{P}_w(X)$, we can see that Theorem 11.3.3(b) gives a statement stronger than (11.3.5). More explicitly, the former is a result valid for any initial distribution in $\mathcal{P}_w(X)$, whereas in Theorem 10.3.1 we proved (11.3.5) for "initial states" only, that is, for "initial distributions" δ_x in $\mathcal{P}_\delta(X) \subset \mathcal{P}_w(X)$.

B. Optimality in Π

The following assumption is supposed to hold throughout the rest of this section.

11.3.4 Assumption. The hypotheses of Theorem 10.3.6(a) are satisfied and, in addition, there is a constant $r \geq 0$ such that

$$\sup_{A(x)} c^2(x, a) \leq rw(x) \quad \forall x \in X. \tag{11.3.7}$$

In other words, Assumption 11.3.4 combines Assumptions 10.2.1, 10.2.2 and 10.3.5, but in addition to Assumption 10.2.1(d), the cost-per-stage $c(x, a)$ is also required to satisfy the "second order" condition (11.3.7).

By Theorem 10.3.6(a) and Theorem 10.3.4, there exists a canonical triplet (ρ^*, h_*, f_*) with h_* in $\mathbb{B}_w(X)$; that is, $\rho^* \in \mathbb{R}$, $h_* \in \mathbb{B}_w(X)$, and $f_*^\infty \in \Pi_{DS}$ satisfy (10.3.13), (10.3.14) and (10.3.15). Moreover, from (10.3.15) it follows that ρ^* satisfies (11.1.23). The following theorem states, in particular, that under the additional condition (11.3.7) there is a policy π^* for which (11.1.24) and (11.1.25) also hold.

11.3.5 Theorem. *Suppose that Assumption 11.3.4 holds. Then*

(a) *For each policy $\pi \in \Pi$ and initial state $x \in X$*

$$J^0(\pi, x) \geq \liminf_{n \to \infty} \frac{1}{n} J_n^0(\pi, x) \geq \rho^* \quad P_x^\pi\text{-a.s.}, \tag{11.3.8}$$

and the following statements (b), (c) and (d) are equivalent for a deterministic stationary policy f^∞:

(b) *f^∞ is AC-optimal;*

(c) *$\pi^* := f^\infty$ satisfies (11.1.24) and (11.1.25)—that is, f^∞ is sample path AC-optimal with respect to Π and $\mathcal{P}_\delta(X)$ [see (11.1.19)]*

(d) $J(f^\infty, \nu) = \rho^*$ *for every initial distribution ν in $\mathcal{P}_w(X)$, the set defined in (11.3.2).*

Hence, by Theorem 10.3.6(a), there exists a deterministic stationary policy f^∞ that satisfies (b), (c), (d). Furthermore, if the cost-per-stage function $c(x, a)$ is such that

$$c(x, a) \quad \text{is bounded below}, \tag{11.3.9}$$

then (b), (c) and (d) are equivalent to:

(e) f^∞ *is strong AC-optimal* [Definition 11.1.1(c)].

The proof of Theorem 11.3.5 is given in subsection D.

11.3.6 Remark. (a) It is worth noting that Theorem 11.3.5(c) and the second inequality in (11.3.8) give that $\pi^* := f^\infty$ is in fact **strong** sample path AC-optimal in Π and $\mathcal{P}_\delta(X)$, where "strong" means that (11.1.24) and (11.1.25) are satisfied when the "lim sup" sample path cost $J^0(\pi, x)$ is replaced by the *lim inf sample path AC*

$$\underline{J}^0(\pi, x) := \liminf_{n \to \infty} \frac{1}{n} J_n^0(\pi, x). \tag{11.3.10}$$

(b) From (11.1.12) and (11.1.23) it is evident that

$$\rho_{\min} \leq \rho^*. \tag{11.3.11}$$

Let us now *suppose that (11.3.9) holds*, and let $\pi \in \Pi$ and $\nu \in \mathcal{P}(X)$ be arbitrary. Then, as

$$J_n^0(\pi, \nu) = \int_X J_n^0(\pi, x)\nu(dx),$$

Fatou's Lemma and (11.3.10) yield that P_ν^π-a.s.

$$
\begin{aligned}
\underline{J}^0(\pi, \nu) &:= \liminf_{n \to \infty} \frac{1}{n} J_n^0(\pi, \nu) \\
&\geq \int_X \underline{J}^0(\pi, x)\nu(dx) \\
&\geq \rho^* \quad \text{by (11.3.8).}
\end{aligned}
$$

From this inequality and (11.1.6) it follows that

$$J^0(\pi, \nu) \geq \underline{J}^0(\pi, \nu) \geq \rho^* \quad P_\nu^\pi\text{-a.s.} \tag{11.3.12}$$

for arbitrary $\pi \in \Pi$ and $\nu \in \mathcal{P}(X)$. On the other hand, from (11.1.5), (11.1.7), and again using Fatou's Lemma,

$$J(\pi, \nu) \geq J_I(\pi, \nu) \geq E_\nu^\pi[\underline{J}^0(\pi, \nu)] \geq \rho^*,$$

where the third inequality follows from (11.3.12). Thus

$$J(\pi,\nu) \geq \rho^* \quad \forall \pi \in \Pi, \; \nu \in \mathcal{P}(X), \tag{11.3.13}$$

which implies the reverse of inequality (11.3.11), that is, $\rho_{min} \geq \rho^*$. Hence, under (11.3.9),

$$\rho_{min} = \rho^*. \tag{11.3.14}$$

Combining these we can easily obtain the following corollary of Theorem 11.3.5. $\square$

11.3.7 Corollary. *If Assumption 11.3.4 and also (11.3.9) are satisfied, then there exists a deterministic stationary policy f^∞ such that*

(a) *(f^∞, ν) is a minimum pair for all ν in $\mathcal{P}_w(X)$, and*

(b) *f^∞ is sample path AC-optimal [in the sense of Definition 11.1.1(b)].*

Proof. By Theorem 11.3.5, there exists an AC-optimal deterministic stationary policy f^∞. Thus, by (11.3.14) and (11.3.3),

$$J(f^\infty, \nu) = \rho^* = \rho_{min} \quad \forall \nu \in \mathcal{P}_w(X),$$

and (a) follows. On the other hand, Proposition 11.3.2(b) yields

$$J^0(f^\infty, \nu) = \rho^* = \rho_{min} \quad P_\nu^{f^\infty}\text{-a.s.} \quad \forall \nu \in \mathcal{P}(X)$$

which together with (11.3.12) gives (b). $\square$

C. Variance minimization

To state the variance-minimization problem we are interested in, let (ρ^*, h_*) be a solution to the ACOE (10.3.13)–(10.3.14), with $h^* \equiv h_*$. Also, for each state $x \in X$, let $A^*(x) \subset A(x)$ be the set defined in (10.3.35), and recall (10.3.36): $f \in \mathbb{F}$ is a *canonical* decision function if and only if $f(x)$ is in $A^*(x)$ for all $x \in X$. Moreover, in analogy with (11.2.12) [or (11.2.23)], consider the function

$$\psi(x,a) := \int_X h_*^2(y)Q(dy|x,a) - \left[\int_X h_*(y)Q(dy|x,a)\right]^2. \tag{11.3.15}$$

As usual, for each $f \in \mathbb{F}$ we shall write

$$\psi_f(x) := \psi(x, f(x)). \tag{11.3.16}$$

Then under the same hypotheses of Theorem 11.3.5 we now get the following.

11.3.8 Theorem. (Existence of minimum-variance policies.) *If Assumption 11.3.4 is satisfied, then there exists a constant $\sigma_*^2 \geq 0$, a canonical* *decision function* $f_* \in \mathbb{F}_{ca}$, *and a function* $V_*(\cdot)$ *in* $\mathbb{B}_w(X)$ *such that for each* $x \in X$:

$$\sigma_*^2 + V_*(x) = \min_{A^*(x)} \left[\psi(x, a) + \int_X V_*(y)Q(dy|x, a) \right] \quad (11.3.17)$$

$$= \psi_{f_*}(x) + \int_X V_*(y)Q_{f_*}(dy|x).$$

Furthermore, f_*^∞ *minimizes the limiting average variance* $Var\,(f^\infty, x)$ *in (11.1.8) over the set of AC-optimal policies* f^∞, *and* $Var\,(f_*^\infty, \cdot) = \sigma_*^2$; *in fact,*

$$Var\,(f_*^\infty, x) = \mu_{f_*}(\psi_{f_*}) = \sigma_*^2 \quad \forall x \in X, \qquad (11.3.18)$$

and

$$\sigma_*^2 \le Var\,(f^\infty, x) \quad \forall f \in \mathbb{F}_{AC}, \; x \in X, \qquad (11.3.19)$$

where $\mathbb{F}_{AC}$ *is the set of AC-optimal decision functions [see (10.3.2)].*

Comparing (11.3.17) with the ACOE (10.3.13)–(10.3.14), we see that (11.3.17) looks very much as an "ACOE" for some Markov control problem. This is indeed the case, as shown in the proof of Theorem 11.3.8 (see subsection E).

D. Proof of Theorem 11.3.5

We shall begin with some preliminary results concerning the weight function

$$v(\cdot) := w(\cdot)^{1/2}. \qquad (11.3.20)$$

Consider the inequality (10.2.4)—which is equivalent to (10.2.1)—with $b(\cdot)$ a constant [for instance, replace $b(\cdot)$ by the constant $b := \|b(\cdot)\|$], i.e.,

$$\int_X w(y)Q_f(dy|x) \le \beta w(x) + b \quad \forall f \in \mathbb{F}, \; x \in X. \qquad (11.3.21)$$

Then, "taking the square root" of both sides of (11.3.21) and using Jensen's inequality we see that $v := w^{1/2}$ satisfies

$$\int_X v(y)Q_f(dy|x) \le \beta_0 v(x) + b_0 \quad \forall f \in \mathbb{F}, \; x \in X, \qquad (11.3.22)$$

with $\beta_0 := \beta^{1/2} < 1$ and $b_0 := b_0^{1/2}$.

Now, as in (7.2.1), consider the v-**norm**

$$\|u\|_v := \sup_X |u(x)|/v(x)$$

of a real-valued function u on X. Then, since $w(\cdot) \ge v(\cdot) \ge 1$, we have

$$\|u\|_v \ge \|u\|_w.$$

On the other hand, if $R(dy|x)$ denotes a signed kernel on X [Definition 7.2.3(b)], then its v-norm [see (7.2.8)]

$$\|R\|_v := \sup_X v(x)^{-1} \int_X v(y)|R(dy|x)|$$

satisfies

$$\|R\|_v^2 \le \|R\|_w,$$

by (11 3.20) and Jensen's inequality. Finally, as in the proof of Theorem 11.2.4(a), the inequality (11.3.7), which can also be written as

$$c_f^2(x) \le rw(x) \quad \forall x \in X, \ f \in \mathbb{F},$$

implies that h_f^2 is in $\mathbb{B}_w(X)$, where h_f is the function in (10.2.19); hence, h_f is in the space $\mathbb{B}_v(X)$ of measurable functions on X with a finite v-norm. From these remarks and (10.2.7) [or (10.2.9)] we obtain the following.

11.3.9 Lemma. *For each deterministic stationary policy f^∞:*

(a) *The state (Markov) process $\{x_t\}$ is v-geometrically ergodic, that is,*

$$\left| \int_X u(y)Q_f^t(dy|x) - \mu_f(u) \right| \le \|u\|_v R_0 \rho_0^t v(x) \quad [\text{cf. (10.2.9)}]$$

for all $x \in X$ and $t = 0, 1, \dots$, where $\rho_0 := \rho^{1/2} < 1$ and $R_0 := R^{1/2}$;

(b) *The unique solution $(J(f), h_f)$ of (10.2.21)–(10.2.22) is such that the bias function h_f is in $\mathbb{B}_v(X)$.*

The next two lemmas are crucial for the proof of Theorem 11.3.5.

11.3.10 Lemma. *For each policy $\pi \in \Pi$ and initial state $x \in X$:*

(a) $E_x^\pi \left[\sum_{t=1}^\infty t^{-2} w(x_t) \right] < \infty;$

hence the following statements hold P_x^π-a.s.:

(b) $\sum_{t=1}^\infty t^{-2} w(x_t) < \infty,$

(c) $t^{-2} w(x_t) \to 0$, *and*

(d) $t^{-1} v(x_t) \to 0.$

Proof. By (11.3.20), it is clear that (a) $\Rightarrow$ (b) $\Rightarrow$ (c) $\Rightarrow$ (d). Thus, it suffices to prove (a). This, however, follows directly from (10.4.2), which yields

$$E_x^\pi w(x_t) \le k \cdot w(x)$$

for some constant k. □

Choose an arbitrary policy $\pi \in \Pi$ and initial state $x \in X$, and let

$$\mathcal{F}_t \equiv \mathcal{F}_t(\pi, x) := \sigma(x_0, a_0, \ldots, x_t, a_t) \text{ for } t = 0, 1, \ldots \qquad (11.3.23)$$

be the σ-algebra generated by the state and control variables up to time t. Moreover, let h_* be as in (11.3.15) [that is, (ρ^*, h_*) is the solution to the ACOE (10.3.13)], and define the random variables, for $t = 1, 2, \ldots$,

$$Y_t(\pi, x) := h_*(x_t) - E_x^\pi[h_*(x_t)|\mathcal{F}_{t-1}] \qquad (11.3.24)$$

and

$$M_n(\pi, x) := \sum_{t=1}^n Y_t(\pi, x). \qquad (11.3.25)$$

11.3.11 Lemma. For each policy $\pi \in \Pi$ and initial state $x \in X$, the sequence $\{M_n(\pi, x), \mathcal{F}_n\}$ is a P_x^π-martingale, and

$$\lim_{n \to \infty} \frac{1}{n} M_n(\pi, x) = 0 \quad P_x^\pi\text{-a.s.} \qquad (11.3.26)$$

Proof. Choose an arbitrary policy $\pi \in \Pi$ and initial state $x \in X$. For notational ease, in (11.3.23)–(11.3.25) we shall write

$$\mathcal{F}_t := \mathcal{F}_t(\pi, x), \quad Y_t := Y_t(\pi, x), \quad M_t := M_t(\pi, x). \qquad (11.3.27)$$

Now note that (ρ^*, h_*) is a solution to the Poisson equation (10.3.14) and so, by Lemma 11.3.9(b),

$$h_* \text{ is in } \mathbb{B}_v(X), \text{ that is, } |h_*(\cdot)| \leq \|h_*\|_v v(\cdot). \qquad (11.3.28)$$

Hence, by (11.3.24), (11.3.27) and (11.3.28),

$$|Y_t| \leq |h_*(x_t)| + E_x^\pi[|h_*(x_t)| \,|\, \mathcal{F}_{t-1}],$$

so that, by (11.3.28),

$$|Y_t| \leq \|h_*\|_v\{v(x_t) + E_x^\pi[v(x_t)|\mathcal{F}_{t-1}]\}, \qquad (11.3.29)$$

and it follows that [by Remark 11.2.5(a)]

$$E_x^\pi|Y_t| \leq 2\|h_*\|_v E_x^\pi[v(x_t)].$$

This inequality and (11.3.25) show that M_n is P_x^π-integrable for each n. On the other hand, it is clear that M_n is $\mathcal{F}_n$-measurable and that [by Remark 11.2.5 (b$_2$)]

$$E_x^\pi(M_{n+1} - M_n|\mathcal{F}_n) = 0.$$

Therefore, $\{M_n, \mathcal{F}_n\}$ is a martingale, which proves the first part of the lemma.

To prove (11.3.26) we shall use the Martingale Stability Theorem in Remark 11.2.6(b). Hence, it suffices to show that [as in (11.2.36) with $q = 2$]

$$\sum_{t=1}^{\infty} t^{-2} E_x^{\pi}(Y_t^2|\mathcal{F}_{t-1}) < \infty \quad P_x^{\pi}\text{-a.s.} \tag{11.3.30}$$

To prove this, first we use (11.2.26) and the fact that $v^2 = w$ to see that (11.3.29) yields [using Remark 11.2.5(d)]

$$Y_t^2 \le 2\|h_*\|_v^2 \{w(x_t) + E_x^{\pi}[w(x_t)|\mathcal{F}_{t-1}]\}.$$

Hence [by Remark 11.2.5(b$_2$)]

$$\begin{aligned}
E(Y_t^2|\mathcal{F}_{t-1}) &\le 4\|h_*\|_v^2 E[w(x_t)|\mathcal{F}_{t-1}] \\
&\le 4\|h_*\|_v^2 (\beta + b)w(x_{t-1}),
\end{aligned} \tag{11.3.31}$$

where the second inequality is obtained from the first inequality in the proof of Lemma 10.4.1 together with the fact that $w(\cdot) \ge 1$. Finally, (11.3.30) follows from (11.3.31) and Lemma 11.3.10(a) because

$$\sum_{t=1}^{\infty} t^{-2} w(x_{t-1}) \le w(x_0) + \sum_{t=1}^{\infty} t^{-2} w(x_t) < \infty \quad P_x^{\pi}\text{-a.s.}$$

This completes the proof of Lemma 11.3.11. $\square$

We are now ready for the proof of Theorem 11.3.5.

Proof of Theorem 11.3.5. (a) Choose an arbitrary policy $\pi \in \Pi$ and initial state $x \in X$, and consider the so-called **AC-discrepancy function**

$$\widehat{D}(x, a) := c(x, a) + \int_X h_*(y)Q(dy|x, a) - h_*(x) - \rho^*. \tag{11.3.32}$$

Observe that the ACOE (10.3.13) can be written as

$$\min_{A(x)} \widehat{D}(x, a) = 0 \quad \forall x \in X,$$

and that $\widehat{D}$ is *nonnegative*. Moreover, using again the notation (11.3.27), we may write (11.3.24) as

$$\begin{aligned}
Y_t &= h_*(x_t) - \int_X h_*(y)Q(dy|x_{t-1}, a_{t-1}), \\
&= h_*(x_t) - h_*(x_{t-1}) - \widehat{D}(x_{t-1}, a_{t-1}) + c(x_{t-1}, a_{t-1}) - \rho^*,
\end{aligned}$$

and (11.3.25) becomes

$$M_n = h_*(x_n) - h_*(x_0) - \sum_{t=0}^{n-1} \widehat{D}(x_t, a_t) + J_n^0(\pi, x) - n\rho^*.$$

Thus, since $\widehat{D} \geq 0$,

$$J_n^0(\pi, x) \geq n\rho^* + M_n - h_*(x_n) + h_*(x_0). \tag{11.3.33}$$

Finally, note that (11.3.28) and Lemma 11.3.10(d) imply that $|h_*(x_n)|/n \to 0$ P_x^π-a.s. as $n \to \infty$, and, similarly, $M_n/n \to 0$ P_x^π-a.s. by (11.3.26). Therefore, multiplying by $1/n$ both sides of (11.3.33) and taking lim inf as $n \to \infty$ we obtain the second inequality in (11.3.8). As the first inequality is obvious, we thus have (11.3.8).

We will next prove the equivalence of (b), (c) and (d).

(b) $\Rightarrow$ (c). Let $f^\infty \in \Pi_{DS}$ be an AC-optimal policy, the existence of which is ensured by Theorem 10.3.6(a). Then, by Proposition 11.3.2(b)

$$J^0(f^\infty, x) = J(f) = \rho^* \quad P_x^{f^\infty}\text{-a.s.} \quad \forall x \in X.$$

This fact and (11.3.8) yield (c).

(c) $\Rightarrow$ (b). Suppose that f^∞ is sample path AC-optimal with respect to Π and $\mathcal{P}_\delta(X)$; that is, $\pi^* := f^\infty$ satisfies (11.1.24) and, moreover, (11.1.25) holds. Then, by (11.3.1),

$$J(f^\infty, x) = J(f) = \rho^* \quad \forall x \in X, \tag{11.3.34}$$

which together with (11.1.23) implies (b).

(b) $\Leftrightarrow$ (d). This follows from (11.3.34) and (11.3.3).

Finally, let us suppose that (11.3.9) is satisfied. Then, by Definition 11.1.1(c), we have (e) $\Rightarrow$ (b). Conversely, suppose that $f^\infty \in \Pi_{DS}$ is AC-optimal. Then f^∞ satisfies (11.3.34) and, on the other hand, (11.3.8) and Fatou's Lemma give

$$\rho^* \leq \liminf_{n\to\infty} E_x^\pi[J_n^0(\pi, x)]/n$$
$$= J^I(\pi, x) \quad \text{[by (11.1.7) and (11.1.3)]}$$

for any policy π and initial state x. Therefore, (b) $\Rightarrow$ (e).

This completes the proof of Theorem 11.3.5. $\square$

E. Proof of Theorem 11.3.8

We shall first state some preliminary facts.

Let $f \in \mathbb{F}$ be an arbitrary decision function. By Lemma 11.3.9(b), the bias function h_f is in $\mathbb{B}_v(X)$, and, therefore, the function

$$\Psi_f(x) := \int_X h_f^2(y) Q_f(dy|x) - \left[\int_X h_f(y) Q_f(dy|x)\right]^2 \tag{11.3.35}$$

for $x \in X$, belongs to $\mathbb{B}_w(X)$—recall (11.3.20). Furthermore, comparing (11.3.35) with (11.2.12), we obtain from Theorem 11.2.4(b)

$$\text{Var } (f^\infty, x) = \lim_{n \to \infty} \frac{1}{n} E_x^{f^\infty} \left[\sum_{t=0}^{n-1} \Psi_f(x_t) \right] =: \sigma^2(f), \tag{11.3.36}$$

where, by (11.2.11),

$$\sigma^2(f) := \mu_f(\Psi_f). \tag{11.3.37}$$

Let us now suppose that $f \in \mathbb{F}$ is a *canonical decision function*, that is, f is in $\mathbb{F}_{ca}$. Then $(J(f), h_f) = (\rho^*, h_f)$ satisfies the ACOE (10.3.13), (10.3.14), and, in addition, Theorem 10.3.7 yields that

$$h_f(\cdot) = h_*(\cdot) + k_f \quad (\text{if } f \in \mathbb{F}_{ca}) \tag{11.3.38}$$

for some constant k_f, where $h_* := h^*$ is the function in (10.3.13)–(10.3.14). Thus, from (11.3.38), (11.3.35) and (11.3.15)–(11.3.16) we obtain that

$$\psi_f(\cdot) = \Psi_f(\cdot) \quad \forall f \in \mathbb{F}_{ca}. \tag{11.3.39}$$

On the other hand, *if $f^\infty \in \Pi_{DS}$ is AC-optimal*, Theorem 10.3.6(b) gives that (11.3.38) holds μ_f-*almost everywhere*; that is, there exists a Borel set $N = N_f$ in X such that $\mu_f(N) = 0$ and

$$h_f(x) = h_*(x) + k_f \quad \forall x \in N^c, \tag{11.3.40}$$

where N^c denotes the complement of N. We also get the following.

11.3.12 Lemma. *Suppose that $f^\infty \in \Pi_{DS}$ is AC-optimal, but it is not canonical, that is, f is in $\mathbb{F}_{AC} \backslash \mathbb{F}_{ca}$ [see (10.3.2)]. Then there exists a canonical decision function $\widehat{f} \in \mathbb{F}_{ca}$ such that $\mu_{\widehat{f}} = \mu_f$, and*

$$\text{Var } (\widehat{f}^\infty, x) = \text{Var } (f^\infty, x) = \sigma^2(f) \quad \forall x \in X, \tag{11.3.41}$$

where $\sigma^2(f)$ is the constant in (11.3.36), (11.3.37).

Proof. Let $g \in \mathbb{F}_{ca}$ be a canonical decision function—whose existence is assured by Theorem 10.3.6(a)—and define a new function $\widehat{f}$ as

$$\widehat{f} := g \text{ on } N, \text{ and } \widehat{f} := f \text{ on } N^c,$$

where $N = N_f$ is the μ_f-null set in (11.3.40). Then $\widehat{f}$ is canonical, and

$$Q_{\widehat{f}}(\cdot|x) = Q_f(\cdot|x) \text{ on } N^c. \tag{11.3.42}$$

Hence, on the one hand, (11.3.42) yields (by definition of i.p.m.)

$$\mu_f(\cdot) = \mu_{\widehat{f}}(\cdot), \tag{11.3.43}$$

and, on the other hand, (11.3.42) and (11.3.40) give [by (11.3.35) and (11.3.16)]

$$\psi_f(\cdot) = \Psi_f(\cdot) \quad \mu_f\text{-a.e.} \tag{11.3.44}$$

Therefore, (11.3.41) follows from (11.3.43)–(11.3.44) and (11.3.36)–(11.3.37). □

With these preliminaries we can now easily prove Theorem 11.3.8.

Proof of Theorem 11.3.8. Let $A^*(x) \subset A(x)$ and $\psi(x, a)$ be as in (10.3.35) and (11.3.15), respectively, and recall (10.3.36) and (11.3.16). Consider the new Markov control model

$$\mathcal{M}_{\text{var}} := (X, A, \{A^*(x) | x \in X\}, Q, \widehat{c}),$$

where $\widehat{c}(x, a) := \psi(x, a)$. It is easily verified that $\mathcal{M}_{\text{var}}$ satisfies the hypotheses of Theorem 10.3.6(a), replacing $c(x, a)$ and $A(x)$ with $\widehat{c}(x, a)$ and $A^*(x)$. Therefore, by Theorem 10.3.6(a), there exists a canonical triplet (σ_*^2, V_*, f_*) for $\mathcal{M}_{\text{var}}$, with V_* in $\mathbb{B}_w(X)$; that is, there exists a constant $\sigma_*^2 \geq 0$, a function V_* in $\mathbb{B}_w(X)$, and a canonical decision function $f_* \in \mathbb{F}_{ca}$ that satisfy (11.3.17). Moreover, as in (10.3.15)

$$\sigma_*^2 = \mu_{f_*}(\psi_{f_*}) = \text{Var}\,(f_*^\infty, x) \quad \forall x \in X, \tag{11.3.45}$$

and

$$\sigma_*^2 \leq \mu_f(\psi_f) = \text{Var}\,(f^\infty, x) \quad \forall f \in \mathbb{F}_{ca},\ x \in X, \tag{11.3.46}$$

where the second equality in (11.3.45) and (11.3.46) follows from (11.3.39) and (11.3.36)–(11.3.37). Finally, to verify (11.3.19), observe that (11.3.46) and Lemma 11.3.12 yield

$$\sigma_*^2 \leq \sigma^2(f) = \text{Var}\,(f^\infty, x) \quad \forall f \in \mathbb{F}_{AC},\ x \in X.$$

This completes the proof of Theorem 11.3.8. □

Notes on §11.3

1. This section is based on Hernández-Lerma, Vega-Amaya and Carrasco [1]. For related results concerning Markov control models with a *finite* state space X, see Mandl [1], [2], and Mandl and Lausmanová [1]. The equality preceding (11.3.33) was perhaps first noted by Mandl [1]. Kurano [1] considers the variance-minimization problem for Markov control models with a transition law $Q(\cdot | x, a)$ that satisfies a Doeblin-like condition, and that are absolutely continuous with respect to some given reference measure. In the *finite state* case, the minimization of the "limiting average variance" defined as in (11.2.43) has been studied in the references given in Note 3 of §11.2. For additional references on sample path AC optimality see §5.7, and Note 1 in §11.4.

2. By (10.3.31), the optimal bias function $\widehat{h}$ [defined in (10.3.27)] and the function h_* $(:= h^*)$ in (11.3.15) satisfy that $\widehat{h}(\cdot) = h_*(\cdot) - k$ for some constant k. Hence, by (11.3.39) the variance $\sigma^2(f)$ in (11.3.36) or (11.3.37) can also be written as

$$\sigma^2(f) = \mu_f[\widehat{h}^2 - (Q_f\widehat{h})^2],$$

or

$$\sigma^2(f) = \int_X \left\{ \widehat{h}^2(x) - \left[\int_X \widehat{h}(y)Q_f(dy|x) \right]^2 \right\} \mu_f(dx),$$

for $f \in \mathbb{F}_{ca}$. Similarly, the "cost-per-stage" function $\psi(x, a)$ in (11.3.15) can be written as

$$\psi(x, a) = \int_X \left[\widehat{h}(y) - \int_X \widehat{h}(z)Q(dz|x, a) \right]^2 Q(dy|x, a).$$

These expressions suggest that the bias-optimal polices and the minimum-variance policies f_*^∞ in Theorem 11.3.8 should be related in some sense, but it is open question what this relation (if any) should be.

11.4   Strictly unbounded costs

In this section we consider another Markov control model $\mathcal{M}$ for which there exists a sample path AC-optimal policy [Definition 1.1.1(b)]. We shall suppose that $\mathcal{M}$ satisfies the following assumption, which was already used in §5.7 (Condition 5.7.4).

11.4.1 Assumption. (a) $J(\widehat{\pi}, \widehat{x}) < \infty$ for some policy $\widehat{\pi}$ and some initial state $\widehat{x}$.

(b) The one-stage cost function $c(x, a)$ is *nonnegative* and lower semicon-tinuous (l.s.c.), and the set $\{a \in A(x)|c(x, a) \le r\}$ is compact for each $x \in X$ and $r \in \mathbb{R}$.

(c) $c(x, a)$ is *strictly unbounded*, that is [with $\mathbb{K}$ as in (8.2.1)], there is a nondecreasing sequence of compact sets $K_n \uparrow \mathbb{K}$ such that

$$\lim_{n \to \infty} \inf\{c(x, a)|(x, a) \notin K_n\} = \infty.$$

(d) The transition law Q is *weakly continuous*, that is,

$$(x, a) \mapsto \int u(y)Q(dy|x, a)$$

is a continuous bounded function on $\mathbb{K}$ for every continuous bounded function u on X.

11.4.2 Remark. (a) A function that satisfies Assumption 11.4.1(c) is also known as a *norm-like* function or as a *moment* function. Each of the following conditions implies that $c(x, a)$ is strictly unbounded:

(a$_1$) The state space X is compact.

(a$_2$) c is *inf-compact*, that is, the level set $\{(x, a) \in \mathbb{K} \mid c(x, a) \leq r\}$ is compact for every number $r \geq 0$.

(a$_3$) X and A are σ-compact (Borel) spaces; $A(x)$ is compact for every $x \in X$, and the set-valued mapping $x \mapsto A(x)$ is u.s.c. Moreover, the function $c^i(x) := \inf_{A(x)} c(x, a)$ satisfies that for every $r \geq 0$ there is a compact set K_r for which

$$c^i(x) \geq r \quad \forall x \notin K_r.$$

(These conditions were already discussed in Remark 5.7.5.)

(b) We will use Assumption 11.4.1(c) here and in Chapter 12 in the same way we did in Chapters 5 and 6: If M is a family of probability measures on $\mathbb{K}$ such that

$$\sup_{\mu \in M} \int_{\mathbb{K}} c(x, a) \mu(d(x, a)) < \infty, \tag{11.4.1}$$

then by Theorems 12.2.15 and 12.2.16, for each sequence $\{\mu_n\}$ in M there is a subsequence $\{\mu_{n_i}\}$ and a p.m. μ on $\mathbb{K}$ (but not necessarily in M) such that $\{\mu_{n_i}\}$ *converges weakly* to μ, which means that

$$\lim_{i \to \infty} \int v \, d\mu_{n_i} = \int v \, d\mu \quad \forall v \in C_b(\mathbb{K}), \tag{11.4.2}$$

where $C_b(\mathbb{K})$ denotes the Banach space of continuous bounded function on $\mathbb{K}$ with the sup norm. In this section "weak" convergence refers to (11.4.2) but in Chapter 12 we consider other forms of "weak" convergence of measures. $\square$

Let Φ and Π_{RS} be as in Definitions 8.2.1 and 8.2.2. Instead of (8.2.5) sometimes we shall use the notation

$$c_\varphi(x) := \int_A c(x, a) \varphi(da|x), \quad Q_\varphi(\cdot|x) := \int_A Q(\cdot|x, a) \varphi(da|x). \tag{11.4.3}$$

For a decision function $f \in \mathbb{F}$ ($\subset \Phi$), the notation (11.4.3) reduces to (10.2.2), (10.2.15).

11.4.3 Definition. A randomized stationary policy φ^∞ (in particular, a deterministic stationary policy $f^\infty \in \Pi_{DS} \subset \Pi_{RS}$) is said to be

(a) **stable** if [using the notation (11.4.3)] the transition law Q_φ admits an i.p.m. p_φ, and the AC $J(\varphi^\infty, p_\varphi)$ is finite and such that

$$J(\varphi^\infty, p_\varphi) = \int_X c_\varphi(x) p_\varphi(dx); \tag{11.4.4}$$

(b) **positive Harris recurrent** if φ^∞ is stable and Q_φ is Harris recurrent.

The following proposition states some useful properties of stable policies. In part (b) of the proposition, $\rho_{\min}$ and ρ^* are the numbers defined in (11.1.12) and (11.1.23), respectively.

11.4.4 Proposition. (a) *If $\varphi^\infty \in \Pi_{RS}$ is stable, then*

$$J(\varphi^\infty, p_\varphi) = \int_X J(\varphi^\infty, x) p_\varphi(dx). \tag{11.4.5}$$

(b) *If there exists a stable policy $\varphi^\infty \in \Pi_{RS}$ such that $(\varphi^\infty, p_\varphi)$ is a minimum pair, that is [by Definition 11.1.1(a)]*

$$J(\varphi^\infty, p_\varphi) = \rho_{\min}, \tag{11.4.6}$$

then

$$\rho_{\min} = \rho^*. \tag{11.4.7}$$

(c) *Let $\varphi^\infty \in \Pi_{RS}$ be a stable policy. Then $(\varphi^\infty, p_\varphi)$ is a minimum pair if and only if*

$$J(\varphi^\infty, x) = \rho^* \quad p_\varphi\text{-a.a. } x \in X. \tag{11.4.8}$$

Proof. (a) By (11.1.5) and the Individual Ergodic Theorem (7.5.24), the limit

$$J(\varphi^\infty, x) = \lim_{n \to \infty} J_n(\varphi^\infty, x)/n$$

exists P_φ-a.e. and satisfies

$$\int_X J(\varphi^\infty, x) p_\varphi(dx) = \int_X c_\varphi(x) p_\varphi(dx).$$

The latter equality and (11.4.4) give (11.4.5).

(b) By the definitions (11.1.12) and (11.1.23) of $\rho_{\min}$ and ρ^*, it is evident that $\rho_{\min} \le \rho^*$. On the other hand, $J(\varphi^\infty, x) \ge \rho^*$ for all $x \in X$ and, therefore, (11.4.6) and (11.4.5) give the reverse inequality,

$$\rho_{\min} = \int_X J(\varphi^\infty, x) p_\varphi(dx) \ge \rho^*.$$

(c) This part is a direct consequence of (a) and (b). $\square$

Before stating the main result in this section, Theorem 11.4.6, let us recall (from Chapter 5) the facts in the following lemma. Observe that part (a) in the lemma ensures the existence of a policy φ^∞ that satisfies the hypotheses of Proposition 11.4.4(b)—hence (11.4.7) holds. Moreover, part (c) states that if, among other things, (11.4.8) *holds for all state x*, then there exists a *deterministic* stationary policy which is *AC-optimal* [see (11.1.10)].

11.4.5 Lemma. *Suppose that Assumption 11.4.1 is satisfied. Then:*

(a) *There exists a stable policy $\varphi_*^\infty \in \Pi_{RS}$ such that $(\varphi_*^\infty, p_{\varphi_*})$ is a minimum pair.*

(b) *If in addition the policy φ_*^∞ in (a) is positive Harris recurrent, then its sample path AC $J^0(\varphi_*^\infty, \cdot)$ satisfies that*

$$J^0(\varphi_*^\infty, \nu) = \rho^* \quad P_\nu^{\varphi_*^\infty}\text{-a.s.} \quad \forall \nu \in \mathcal{P}(X), \tag{11.4.9}$$

where $\rho^* = \rho_{\min}$ [by (a) and Proposition 11.1.4(b)].

(c) *Suppose that there exists a randomized stationary policy φ^∞ such that*

$$J(\varphi^\infty, x) = \rho^* \quad \forall x \in X, \tag{11.4.10}$$

and let

$$h(x) := \liminf_{N \to \infty} \frac{1}{N} \sum_{n=0}^{N-1} h_n(x) \quad (\geq 0) \tag{11.4.11}$$

where, for $n \geq 1$,

$$h_n(x) := J_n(\varphi^\infty, x) - M_n \quad and \quad M_n := \inf_X J_n(\varphi^\infty, x),$$

*and $h_0(\cdot) = M_0 := 0$. If h is a real-valued function, then there exists a **deterministic** stationary policy f^∞ which is AC-optimal with value function ρ^*, that is,*

$$J(f^\infty, x) = J^*(x) = \rho^* \quad \forall x \in X. \tag{11.4.12}$$

Proof. Part (a) is the same as Theorem 5.7.9(a), whereas (11.4.9) follows from Theorem 11.2.1(a). Furthermore, part (c) is a consequence of Theorem 5.4.3(ii), which states that (11.4.10) and (11.4.11) yield the Average Cost Optimality Inequality (ACOI)

$$\begin{aligned}
\rho^* + h(x) &\geq c_\varphi(x) + \int_X h(y) Q_\varphi(dy|x) \\
&\geq \min_{A(x)} \left[c(x,a) + \int_X h(y) Q(dy|x,a) \right].
\end{aligned} \tag{11.4.13}$$

Hence, by the usual argument, there exists a decision function $f \in \mathbb{F}$ such that $f(x) \in A(x)$ attains the minimum in (11.4.13) for every $x \in X$, that is,

$$\rho^* + h(x) \geq c_f(x) + \int_X h(y) Q_f(dy|x) \quad \forall x \in X,$$

which yields (11.4.12). $\square$

We shall now state our main result, which, in particular, in part (c) gives conditions for the existence of sample path AC-optimal policies. Also note that (11.4.16) is a statement stronger than (11.4.7) because it uses the *lim inf* expected AC in (11.1.7).

11.4.6 Theorem. *Suppose that Assumption 11.4.1 is satisfied. Then:*

(a) *For each policy π and initial distribution ν*

$$\liminf_{n\to\infty} J_n^0(\pi,\nu)/n \geq \rho^* \quad P_\nu^\pi\text{-a.s.;} \qquad (11.4.14)$$

hence

$$J^I(\pi,\nu) \geq \rho^*, \qquad (11.4.15)$$

and

$$\rho^* = \inf_{\mathcal{P}(X)} \inf_\Pi J^I(\pi,\nu). \qquad (11.4.16)$$

(b) *If $\pi^* \in \Pi$ is an AC-optimal policy and the AC-value function $J^*(\cdot)$ equals ρ^*, that is*

$$J(\pi^*,x) = J^*(x) = \rho^* \quad \forall x \in X, \qquad (11.4.17)$$

then π^* is strong AC-optimal and

$$\liminf_{n\to\infty} J_n^0(\pi^*,\nu)/n = \rho^* \quad P_\nu^{\pi^*}\text{-a.s. } \forall \nu \in \mathcal{P}(X). \qquad (11.4.18)$$

(c) *If the policy φ_*^∞ in Lemma 11.4.5 is positive Harris recurrent, then it is sample path AC-optimal; in fact, every positive Harris recurrent policy φ^∞ in Π_{RS} (or in Π_{DS}) for which $(\varphi^\infty, p_\varphi)$ is a minimum pair, is also sample path AC-optimal.*

Proof. As the proof of (11.4.14) is quite "technical", to simplify the exposition we will first *suppose* that it holds and prove the remaining parts of the theorem; then we will prove (11.4.14).

Suppose that (11.4.14) is satisfied, and choose an arbitrary policy π and initial distribution ν. Then (11.1.3), (11.1.7), and Fatou's Lemma yield

$$J^I(\pi,\nu) \geq E_\nu^\pi \left[\liminf_{n\to\infty} J_n^0(\pi,\nu)/n\right] \geq \rho^*; \qquad (11.4.19)$$

that is (11.4.15) holds. Moreover, as [by (11.1.5)]

$$J(\pi,\nu) \geq J^I(\pi,\nu), \qquad (11.4.20)$$

(11.4.16) follows from (11.4.19) and the definition (11.1.12) of ρ^*.

Proof of (b). If π^* satisfies (11.4.17), then (11.4.16) yields that π^* is strong AC-optimal, and, on the other hand, using (11.4.20) and (11.4.19) with $\nu = \delta_x$,

$$\rho^* = J(\pi^*,x) \geq E_x^{\pi^*} \left[\liminf_{n\to\infty} J_n^0(\pi^*,x)/n\right] \geq \rho^*,$$

i.e.,

$$E_x^{\pi^*} \left[\liminf_{n\to\infty} J_n^0(\pi^*,x)/n\right] = \rho^* \quad \forall x \in X.$$

This equality and (11.4.14) give (11.4.18).

Proof of (c). Part (c) follows from (11.4.14) and (11.4.8).

We have now completed the proof of Theorem 11.4.6 except for the key fact (11.4.14). The proof of the latter is based on Remark 11.4.2(b), Lemma 9.4.4, and the next two lemmas. (Lemma 11.4.7 combines Propositions A.2 and E.2 in Appendices A and E, respectively.)

11.4.7 Lemma. *Let S be a metric space with Borel σ-algebra $\mathcal{B}(S)$, and let $C_b(S)$ be the Banach space of real-valued continuous bounded functions on S.*

(a) *A function $u : S \to \mathbb{R}$ is l.s.c. and bounded below if and only if there exists a nondecreasing sequence of functions u_n in $C_b(S)$ such that $u_n \uparrow u$ pointwise.*

(b) *Let $u : S \to \mathbb{R}$ be l.s.c. and bounded below, and let $\mu, \mu_n (n = 1, 2, \ldots)$ be probability measures on $\mathcal{B}(S)$ such that μ_n converges weakly to μ, that is*

$$\int_S g d\mu_n \to \int_S g d\mu \quad \forall g \in C_b(S).$$

Then

$$\liminf_{n \to \infty} \int_S u d\mu_n \geq \int_S u d\mu.$$

In Lemma 11.4.8 below we use the following terminology and notation. Let (S, τ) be a separable metrizable space, that is, a separable topological space for which there exists a metric d on S consistent with the topology τ. For each metric d on S we denote by $\mathcal{U}(S, d)$ the subfamily of functions in $C_b(S)$ which are *uniformly continuous* with respect to d. We take $\mathcal{U}(S, d)$ to have the relative topology of $C_b(S)$.

11.4.8 Lemma. *Let (S, τ) be a separable metrizable space. Then there exists a metric d^* on S consistent with τ such that:*

(a) *the family $\mathcal{U}(S, d^*)$ is separable;*

(b) *for each function u in $C_b(S)$ there exist sequences $\{u_n^0\}$ and $\{u_n^1\}$ in $\mathcal{U}(S, d^*)$ such that $u_n^0 \uparrow u$ and $u_n^1 \downarrow u$ pointwise as $n \to \infty$.*

Proof. See Bertsekas and Shreve [1], Corollary 7.6.1 (p. 113), Proposition 7.9 (p. 116), and Lemma 7.7 (p. 125). □

Proof of (11.4.14). Choose an arbitrary policy π and initial distribution ν, and let $(\Omega, \mathcal{F}, P_\nu^\pi)$ be the "canonical" probability space in Remark 8.2.3(c). Furthermore, define on Ω a random variable $\widehat{J}$ as in the left-hand side of (11.4.14), that is

$$\widehat{J} := \liminf_{n \to \infty} J_n^0(\pi, \nu)/n \tag{11.4.21}$$

with $J_n^c(\pi, \nu)$ as in (11.1.4). If for some sample path $\omega = (x_0, a_0, x_1, a_1, \dots)$ of the state-action process it occurs that $\widehat{J}(\omega) = +\infty$, then (11.4.14) trivially holds. Thus without loss of generality we may restrict to sample paths in the set $\Omega' := \{\omega | J(\omega) < \infty\}$. Now consider the *empirical measures*

$$\gamma_n(\Gamma) := \frac{1}{n} \sum_{t=0}^{n-1} I_\Gamma(x_t, a_t) \quad \text{for} \quad \Gamma \in \mathcal{B}(X \times A), \ n = 1, 2, \dots.$$

By (8.2.3), each γ_n is a (random) probability measure on $X \times A$ concentrated on the set $\mathbb{K}$ defined in (8.2.1), and, moreover, we can write $\widehat{J}$ as

$$\widehat{J} = \liminf_{n \to \infty} \int_{\mathbb{K}} c(x, a) \gamma_n(d(x, a)).$$

To prove (11.4.14) we will proceed in two steps. First we will show that:

(i) *For each $\omega \in \Omega'$ there exist a probability measure γ^ω on $X \times A$, contrated on $\mathbb{K}$, such that*

$$\widehat{J}(\omega) \geq \int_{\mathbb{K}} c(x, a) \gamma^\omega(d(x, a)). \tag{11.4.22}$$

Thus, by Lemma 9.4.4, there exists a stochastic kernel $\varphi_\omega \in \Phi$ such that

$$\gamma^\omega(B \times C) = \int_B \varphi_\omega(C|x) \widehat{\gamma}^\omega(dx) \quad \forall B \in \mathcal{B}(X), \ C \in \mathcal{B}(A), \tag{11.4.23}$$

where $\widehat{\gamma}^\omega(\cdot) := \gamma^\omega(\cdot \times A)$ is the marginal of γ^ω on X.

In the second step we will show that.

(ii) *For P_ν^π-almost all ω, the randomized stationary policy φ_ω^∞ is stable, with i.p.m. $p_{\varphi_\omega} = \widehat{\gamma}^\omega$.*

Therefore, by Definition 11.4.3(a) and (11.4.22),

$$\widehat{J}(\omega) \geq J(\varphi_\omega^\infty, \widehat{\gamma}^\omega) \geq \rho^* \quad P_\nu^\pi\text{-a.s.,}$$

and so the proof of (11.4.14) will be complete.

Proof of (i). Choose an arbitrary $\omega \in \Omega'$ and a sequence $\{n_i\}$ such that

$$\widehat{J}(\omega) = \lim_{i \to \infty} \int_{\mathbb{K}} c(x, a) \gamma_{n_i}^\omega(d(x, a)).$$

Then

$$\sup_i \int_{\mathbb{K}} c(x, a) \gamma_{n_i}^\omega(d(x, a)) < \infty,$$

which [as in Remark 11.4.2(b)] implies the existence of a p.m. γ^ω on $\mathbb{K}$ and a subsequence $\{m_i\}$ of $\{n_i\}$ such that $\{\gamma^\omega_{m_i}\}$ converges weakly to γ^ω, that is, as $i \to \infty$

$$\int_{\mathbb{K}} v \, d\gamma^\omega_{m_i} \to \int_{\mathbb{K}} v \, d\gamma^\omega \quad \forall v \in C_b(\mathbb{K}). \tag{11.4.24}$$

Thus, since $c(x,a)$ is l.s.c. and nonnegative [Assumption 11.4.1(b)], we obtain (11.4.22) from Lemma 11.4.7(b).

Proof of (ii). In Lemma 11.4.8 take $S = X$, the state space, and let $\mathcal{U}$ be a countable dense subset of $\mathcal{U}(X, d^*)$. For each function $u \in \mathcal{U}$ define on $\mathbb{K}$ the function

$$Lu(x,a) := u(x) - \int_X u(y) Q(dy|x,a),$$

and let us consider two random sequences $\{M_n(u)\}$ and $\{Y_t(u)\}$ as in (11.3.24) and (11.3.25), that is,

$$M_n(u) := \sum_{t=1}^{n} Y_t(u),$$

with

$$Y_t(u) := u(x_t) - E^\pi_\nu[u(x_t)|x_{t-1}, a_{t-1}].$$

Observe that we can also write $Y_t(u)$ as

$$Y_t(u) = Lu(x_t, a_t) + E^\pi_\nu[u(x_{t+1})|x_t, a_t] - E^\pi_\nu[u(x_t)|x_{t-1}, a_{t-1}],$$

so that

$$M_n(u) = \sum_{t=1}^{n} Lu(x_t, a_t) + E^\pi_\nu[u(x_{n+1})|x_n, a_n] - E^\pi_\nu[u(x_1)|x_0, a_0]. \tag{11.4.25}$$

As in Lemma 11.3.11, one can show that $\{M_n(u), \mathcal{F}_n\}$ is a martingale and, on the other hand, as u is bounded, so is the sequence

$$E^\pi_\nu[Y_t(u)^2|\mathcal{F}_{t-1}], \quad t = 1, 2, \ldots.$$

Hence, by the Martingale Stability Theorem in Remark 11.2.6(b),

$$\lim_{n \to \infty} \frac{1}{n} M_n(u) = 0 \quad P^\pi_\nu\text{-a.s.}$$

Equivalently, from (11.4.25) and noting that

$$\frac{1}{n} \sum_{t=0}^{n-1} Lu(x_t, a_t) = \int_{\mathbb{K}} Lu(x,a) \gamma_n(d(x,a)),$$

we have

$$\lim_{n\to\infty} \int_{\mathbb{K}} Lu(x,a)\gamma_n^\omega(d(x,a)) = 0 \quad \forall \omega \in \Omega_u,$$

where Ω_u is a measurable subset of Ω such that $P_\nu^\pi(\Omega_u) = 1$. Thus

$$\lim_{n\to\infty} \int_{\mathbb{K}} Lu(x,a)\gamma_n^\omega(d(x,a)) = 0 \quad \forall u \in \mathcal{U}, \ \omega \in \Omega^*, \qquad (11.4.26)$$

where

$$\Omega^* := \cap_{u\in\mathcal{U}}\Omega_u.$$

Moreover, by Assumption 11.4.1(c), the function Lu is in $C_b(\mathbb{K})$ for every u in $\mathcal{U}$. Hence, for each ω in Ω^* there is a sequence $\{m_i(\omega)\}$ as in (11.4.24), so that, by (11.4.26),

$$\int_{\mathbb{K}} Lu(x,a)\gamma^\omega(d(x,a)) = 0 \quad \forall u \in \mathcal{U}.$$

In fact, by Lemma 11.4.8(b), the latter equality holds for all u in $C_b(X)$, i.e.,

$$\int_{\mathbb{K}} Lu(x,a)\gamma^\omega(d(x,a)) = 0 \quad \forall u \in C_b(X),$$

which [by (11.4.23)] can also be written as

$$\int_X \int_X u(y)Q_{\varphi_\omega}(dy|x)\widehat{\gamma}^\omega(dx) = \int_X u(x)\widehat{\gamma}^\omega(dx) \quad \forall u \in C_b(X),$$

and (ii) follows. The proof of Theorem 11.4.6 is now complete. $\square$

Notes on §11.4

1. Theorem 11.4.6 comes from Vega-Amaya [2, 3]. Related results are obtained by Lasserre [3] using a different approach. In addition to these works and the paper by Hernández-Lerma, Vega-Amaya and Carrasco [1] mentioned in Note 1 of §11.3, we know of no previous works on sample path AC-optimality for MCPs on general (uncountable) Borel spaces.

2. Vega-Amaya [2, Theorem 6.3.1] gives a proof of Lemma 11.4.5(a) different from our proof in §5.7.

11.5   Examples

11.5.1 Example. (Examples 10.9.3 and 8.6.2, continued.) Let us consider the inventory-production system (8.6.5) with cost-per-stage (8.6.7), namely,

$$x_{t+1} = (x_t + a_t - z_t)^+ \quad \text{for } t = 0, 1, \ldots, \qquad (11.5.1)$$

and

$$c(x, a) := p \cdot a + m \cdot (x + a) - s \cdot E[\min(x + a, z_0)]. \tag{11.5.2}$$

The state and control spaces are $X := [0, \infty)$ and $A = A(x) = [0, \theta]$ for all x in X. In Example 10.9.3 we saw that Assumption 8.6.1, together with the condition 8.6.3 and (10.9.8), implies that the system is w-geometrically ergodic with respect to the weight function

$$w(x) = \exp[r_*(x + 2\bar{z})], \tag{11.5.3}$$

where $\bar{z} := E(z_0)$, and r_* is a positive number such that

$$\psi(r_*) < 1,$$

with $\psi(r) := \exp[r(\theta - z_0)]$, the moment generating function of $\theta - z_0$. Moreover, (10.9.21) implies that Assumption 10.3.5 is satisfied. Therefore, to complete the verification of, say, Assumption 11.3.4, it only remains to check that (11.3.7) holds.

To verify that (11.3.7) is satisfied first note that, as $A := [0, \theta]$,

$$0 \le E[\min(x + a), z_0)] \le x + \theta \quad \forall x \in X. \tag{11.5.4}$$

Using this inequality and (11.5.2), a straightforward calculation shows that there are constants k_i $(i = 0, 1, 2)$ such that

$$c^2(x, a) \le k_0 + k_1 x + k_2 x^2 \quad \forall x \in X.$$

This yields (11.3.7) for some r sufficiently large, with $w(x)$ as in (11.5.3).

In other words, Assumption 11.3.4 holds, and so are all of the results in §11.3 except perhaps for the last statement in Theorem 11.3.5. The problem in the latter case is that (11.3.9) might not be true. In fact, from (11.5.4) and (11.5.2) it can be seen that $c(x, a)$ is minorized by $(m - s)x + s\theta$, which is not bounded below unless $m = s$; see (8.6.8).

Concerning the inventory system (11.5.1), see also Remark 11.5.3, below.
□

11.5.2 Example. (Examples 10.9.5 and 8.6.4, continued.) We shall consider again the controlled queueing system (8.6.16), namely

$$x_{t+1} = (x_t + a_t \eta_t - \xi_t)^+ \quad \text{for } t = 0, 1, \ldots, \tag{11.5.5}$$

under the Assumptions 8.6.5. In Examples 8.6.4 and 10.9.5 we already verified Assumption 11.3.4 except for (11.3.7), which in the present case refers to the weight function w in (8.6.18) or (10.9.23), that is,

$$w(x) := e^{\widehat{r}x} \quad \text{for } x \in X := [0, \infty), \tag{11.5.6}$$

for some positive number $\widehat{r}$.

The hypotheses in Example 8.6.4 and 10.9.5 imply the hypotheses of Theorem 10.3.6(a). However, since we did not give a specific form to the cost-per-stage $c(x, a)$, in addition to (8.6.19) we shall suppose that

$$c(x, a) \text{ is nonnegative and satisfies (11.3.7).} \qquad (11.5.7)$$

With this additional requirement, the queueing system (11.5.5) satisfies Assumption 11.3.4 and (11.3.9); hence all of the results in §11.3 hold in this case.

On the other hand, the hypotheses in the previous paragraphs also imply parts (a) and (d) *in Assumption 11.4.1*. To obtain (b) and (c) in that assumption, we may further suppose that, for instance, $c(x, a)$ is continuous and satisfies either (a$_2$) or (a$_3$) in Remark 11.4.2. [Note that in the present example the control set $A \equiv A(x)$ is compact and independent of x, and so the requirement on $A(\cdot)$ in Remark 11.4.2(a$_3$) trivially holds.] In other words, in the latter case, Assumption 11.4.1 is valid, and, therefore, so are the corresponding results in Lemma 11.4.5 and Theorem 11.4.6. In particular, in lieu of the minimum pair $(\varphi_*^\infty, p_{\varphi_*})$ in Lemma 11.4.5(a) we may take a pair (f^∞, μ_f) consisting of an AC-optimal deterministic stationary policy f^∞—as in Theorem 11.3.6(a)—and the associated i.p.m. μ_f. Then f^∞ is a positive Harris recurrent policy [see Definition 11.4.3(b) and Proposition 11.3.2(a)] for which the conclusion of Theorem 11.4.6(c) is true. This would be, in other words, an alternative proof (instead of using Theorem 11.3.5) that f^∞ is sample path AC-optimal. $\square$

11.5.3 Remark. In the last part of the previous example we easily verified Assumption 11.4.1 by imposing suitable conditions on the cost-per-stage $c(x, a)$. This can also be done in Example 11.5.1. For instance, consider the system (11.5.1), with the same control sets, $A = A(x) = [0, \theta]$ for all $x \in X$, but the cost function $c(x, a)$ in (11.5.2) is replaced by, say,

$$c'(x, a) := c_1 \cdot (x - x_*)^2 + c_2 \cdot (a - a_*)^2 \quad \forall x \in X, \ a \in A, \qquad (11.5.8)$$

where c_1 and c_2 are given positive constants. In (11.5.8), $x_* \in X$ and $a_* \in A$ are fixed nominal values of the inventory level and production rate, respectively, and the interpretation of an AC-optimal control policy $\pi = \{a_t\}$ is that, in the long run, it minimizes the mean-square distance of the inventory level $\{x_t\}$ and production rates $\{a_t\}$ to the given nominal values x_* and a_*. We can now see that part (b) in Assumption 11.4.1 is trivially satisfied, whereas part (c) follows from either (a$_2$) or (a$_3$) in Remark 11.4.2. Furthermore, since we already verified (d) in Example 8.6.2, to complete the proof that Assumption 11.4.1 holds we only need to check part (a). To do this, take for instance the initial state $\hat{x} := 0$, and let $\hat{\pi}$ be the stationary policy f^∞ such that $f(x) := 0$ for all $x \in X$. Then a straightforward calculation using (10.9.2) shows that

$$J_n(f^\infty, 0) = n(c_1 x_*^2 + c_2 a_*^2) \quad \forall n = 0, 1, \ldots,$$

and, therefore, the corresponding average cost is

$$J(f^\infty, 0) = c_1 x_*^2 + c_2 a_*^2 < \infty.$$

This completes the verification of Assumption 11.4.1 for the system (11.5.1) with the cost-per-stage in (11.5.8). □

In Example 11.5.6 below we wish to illustrate the results in §11.4 for an $\mathbb{R}^d$-valued additive-noise control system

$$x_{t+1} = F(x_t, a_t) + z_t \tag{11.5.9}$$

with a quadratic cost-per-stage function

$$c(x, a) = x' \Delta x + a' \Theta a, \tag{11.5.10}$$

where a is in $\mathbb{R}^q$ and Δ and Θ are suitable matrices; x' and a' denote the transpose of x and a, respectively. Since the analysis of this control system relies on the properties of the noncontrolled Markov chain

$$x_{t+1} = F(x_t) + z_t, \tag{11.5.11}$$

we shall first present the following result, which is a variant of Example 7.4.6.

11.5.4 Proposition. *Consider the Markov chain (11.5.11) and the following conditions:*

(a) *$F : \mathbb{R}^d \to \mathbb{R}^d$ is locally bounded; that is, for each compact set $K \subset \mathbb{R}^d$ there is a constant $m = m(K)$ such that $|F(x)| \le m$ for all $x \in K$.*

(b) *The disturbance sequence $\{z_t\}$ consists of i.i.d. random vectors in $\mathbb{R}^d$, independent of the initial state x_0, and whose common distribution has a density g which is positive λ-a.e., where λ denotes Lebesgue measure. In addition, $E(z_0) = 0$.*

(c) *There exist positive constants $s \ge 1$, M_1, and b such that*

$$E|F(x) + z_0|^s \le |x|^s - b \quad \forall |x| > M_1. \tag{11.5.12}$$

(d) *There exist positive constants $s \ge 1$, $\beta < 1$, and M_2 such that $E|z_0|^s < \infty$ and*

$$E|F(x) + z_0|^s \le \beta |x|^s \quad \forall |x| > M_2. \tag{11.5.13}$$

Then the following holds:

(i) *Under (a) and (b), the Markov chain is aperiodic, λ-irreducible and Harris recurrent with respect to λ.*

(ii) *Under (a), (b) and (c), the chain has a unique i.p.m.—hence, by (i), it is positive Harris recurrent.*

(iii) *Under (a), (b) and (d), there exist a p.m.* μ, *and positive numbers* $\rho < 1$ *and* R *such that*

$$\|P^n(\cdot|x) - \mu(\cdot)\|_w \le R\rho^n \quad for \quad \mu\text{-a.a. } x \qquad (11.5.14)$$

[cf. (7.3.7)], where w *is the weight function* $w(x) = 1 + |x|^s$, *and*

$$\int w d\mu < \infty. \qquad (11.5.15)$$

Proof. Part (i) follows from (7.4.18). For the proof of (ii) see Tweedie [1] or Mokkadem [1, Proposition 1], and for the proof of (iii) see Tweedie [2] or Mokkadem [1, Proposition 3]. $\square$

11.5.5 Remark. (a) The main difference between Proposition 11.5.4(iii) and Example 7.4.6 is that the latter requires F to be *continuous*. Moreover, from (i), and comparing (11.5.14) with (7.3.7), it can be seen that the measure in (11.5.14) is the *unique i.p.m.* for the Markov chain. This can also be deduced from the conclusion (ii) by noting that the inequality (11.5.13) implies that (11.5.12) holds with $M_1 := \max\{M_2, [b/(1-\beta)]^{1/s}\}$, because

$$\beta|x|^s \le |x|^s - b \Leftrightarrow |x| \ge [b/(1-\beta)]^{1/s}.$$

(b) Using Minkowski's inequality it can be seen that (11.5.12) and (11.5.13) are both satisfied if there are positive constants $\beta^* < 1$ and M such that

$$|F(x)| + \|z_0\|_s \le \beta^*|x| \quad \forall |x| > M, \qquad (11.5.16)$$

where $\|z_0\|_s := (E|z_0|^s)^{1/s}$.

(c) Proposition 11.5.4 is of course applicable to the *linear case* in which $F(x) = \Gamma x$ for some matrix Γ. $\square$

In the following example we use Proposition 11.5.4 (with $s = 2$) as a guide to derive conditions ensuring that (11.5.9) and (11.5.10) satisfy the hypotheses of §11.4.

11.5.3 Example. (An additive-noise quadratic cost system.) Consider the stochastic control system (11.5.9) with cost-per-stage (11.5.10). The state space is $X = \mathbb{R}^d$, and we shall assume that the control constraint sets $A(x)$, for each $x \in X$, are closed subsets of (say) $\mathbb{R}^q$ and such that

$$\mathbb{K} := \{(x,a)|x \in X, \ a \in A(x)\}$$

is a convex set. Firstly, we will also assume:

(a) *The matrices* Δ *and* Θ *in (11.5.10) are symmetric and positive definite.*

This condition clearly implies that the quadratic cost (11.5.10) satisfies Assumptions 11.4.1(b) and 11.4.1(c).

To verify Assumption 11.4.1(d) we suppose:

(b) $\{z_t\}$ *satisfies the condition (b) in Proposition 11.5.4, and*

(c) $F : \mathbb{K} \to X$ *is continuous.*

Then, since

$$\int_X u(y)Q(dy|x, a) = E(u[F(x, a) + z_0]), \qquad (11.5.17)$$

the Bounded Convergence Theorem and (c) yield Assumption 11.4.1(d). [For this to be true we only need the first part of (b)—we do not require the density g to be positive λ-a.e.]

To derive a sufficient condition for Assumption 11.4.1(a) we will use Proposition 11.5.4(iii) and the following notation: If $\varphi^\infty \in \Pi_{RS}$ is a randomized stationary policy we write

$$F(x, \varphi) := \int_A F(x, a)\varphi(da|x) \quad \text{for} \quad x \in X, \qquad (11.5.18)$$

where $A := \mathbb{R}^q$. [By Definitions 8.2.1 and 8.2.2(b), in (11.5.18) we may replace A by $A(x)$, if necessary.] Let us now consider the conditions:

(d) $F(\cdot, \varphi)$ *is locally bounded for each randomized stationary policy* φ^∞.

(e) $E|z_0|^2 < \infty$ *and, moreover, there exists a randomized stationary policy* $\widehat{\varphi}^\infty$ *and positive constants* $\beta < 1$, M, *and* k *such that*

(e$_1$) $E|F(x, \widehat{\varphi}) + z_0|^2 \le \beta|x|^2 \quad \forall |x| > M$;

(e$_2$) $\int_A (a'\Theta a)\widehat{\varphi}(da|x) \le k|x|^2 \quad \forall x \in X$.

From (b), (d) and (e), we can see that $\widehat{\varphi}^\infty$ is *stable* [Definition 11.4.3(a)]. Indeed, by Proposition 11.5.4(iii), the transition law $Q_{\widehat{\varphi}}$ has an i.p.m. $\widehat{p} := p_{\widehat{\varphi}}$ that, in particular, satisfies (11.5.15) with $w(x) = 1 + |x|^2$ and $\mu = \widehat{p}$. Therefore, from (11.5.10) and (e$_2$), there is a constant $\widehat{k}$ such that

$$J(\widehat{\varphi}^\infty, \widehat{p}) \le \widehat{k} \int_X |x|^2 \widehat{p}(dx) < \infty.$$

Thus (11.4.4) holds, which in turn [by (11.4.5)] gives Assumption 11.4.1(a) with $\widehat{\pi} = \widehat{\varphi}^\infty$ and some initial state $\widehat{x}$.

Summarizing, the current conditions (a) to (e) imply that (11.5.9) and (11.5.10) satisfy Assumption 11.4.1, and so the corresponding results in §11.4 are applicable. Furthermore, if we wish to use, for instance, Lemma 11.4.5(b) or Theorem 11.4.6(c), we then need conditions for the policy φ^∞_* in those results to be positive Harris recurrent. One way of getting this is to assume (or to verify, when a specific control system is given) that (e) holds for every randomized stationary policy φ^∞, with constants β, M, and k

that may depend on φ^∞—note that we may replace (e_1) by the analogues of (11.5.12) or (11.5.16), namely,

$$E|F(x,\varphi) + z_0|^2 \leq |x|^2 - b \quad \forall |x| > M_1$$

or

$$|F(x,\varphi)| + \|z_0\|_2 \leq \beta^*|x| \quad \forall |x| > M,$$

with constants b, M_1 β^*, and M that may depend on φ^∞. $\square$

Notes on §11.5

The control model in Example 11.5.1 and Remark 11.5.3 has been studied by Vega-Amaya [2, 3]. These references contain other examples related to §11.3 and §11.4.

Example 11.5.6 comes from Hernández-Lerma and Lasserre [14], where the reader can find additional references on results related to Proposition 11.5.4 (on which Example 11.5.6 is based).

12

The Linear Programming Approach

12.1 Introduction

In this chapter we study the *linear programming* (LP) approach to Markov control problems. Our ultimate goal is to show how a Markov control problem can be approximated by *finite* linear programs.

To reach this goal, we shall first proceed to find a suitable linear program associated to the Markov control problem. Here, by a "suitable" linear program we mean a linear program (P) that together with its dual (P*) satisfies that

$$\sup(P^*) \le (MCP)^* \le \inf(P), \qquad (12.1.1)$$

where (using terminology specified in the following section)

$$\begin{aligned}
\inf(P) \quad &:= \quad \text{value of the primal program (P),} \\
\sup(P^*) \quad &:= \quad \text{value of the dual program (P*),} \\
(MCP)^* \quad &:= \quad \text{value function of the Markov control problem.}
\end{aligned}$$

In particular, if there is *no duality gap* for (P), so that

$$\sup(P^*) = \inf(P), \qquad (12.1.2)$$

then of course the values of (P) and of (P*) yield the desired value function $(MCP)^*$.

However, to find an *optimal policy* for the Markov control problem, (12.1.1) and (12.1.2) are not good enough because they do not guarantee that (P) or (P*) are *solvable*. If it can be ensured that, say, the primal

(P) is solvable—in which case we write its value as min (P)—and that

$$\min(P) = (MCP)^*, \tag{12.1.3}$$

then an optimal solution for (P) can be used to determine an optimal policy for the Markov control problem. Likewise, if the dual (P*) is solvable and its value—which in this case is written as max (P*)—satisfies

$$\max(P^*) = (MCP)^*, \tag{12.1.4}$$

then we can use an optimal solution for (P*) to find an optimal policy for the Markov control problem. In fact, one of the main results in this chapter (Theorem 12.4.2) gives conditions under which (12.1.3) and (12.1.4) are both satisfied, so that in particular *strong duality* for (P) holds, that is,

$$\max(P^*) = \min(P). \tag{12.1.5}$$

The LP approach to Markov control problems was already studied in Chapter 6, but from a very different viewpoint. Namely, in Chapter 6 we first introduced linear programs (P_α) associated to α-discounted MCPs, with $0 < \alpha < 1$, and derived properties such as (12.1.1)–(12.1.5). Then to study the average cost (AC) problem we suitably modified (P_α) to obtain "modified" programs (MP_α), and finally we obtained and analized the AC-related linear program (MP_1) as the "limit" of (MP_α) as $\alpha \uparrow 1$. In the present chapter, however, we do not use α-discounted programs to study (MP_1), which we now call (P) [as in (12.1.1)]. Instead, we go directly to $(P) \equiv (MP_1)$ and analize it without using α-discounted programs.

Another key difference with respect to Chapter 6 is that here we obtain *necessary and sufficient* conditions for (P) to be consistent (Theorem 12.3.7), as opposed to Chapter 6 that only gives sufficient conditions. Moreover, we study minimizing sequences for (P) and maximizing sequences for (P*), and, more importantly, we prove the convergence of an approximation scheme for (P) based on *finite* linear programs (Theorem 12.5.7).

A. Outline of the chapter

Section 12.2 presents background material that can be omitted on a first reading. It contains, in particular, a brief introduction to infinite LP and some important facts on the concept of "tightness". In §12.3 we introduce the program (P) associated to the AC problem, and we show that (P) is solvable and that there is no duality gap, so that (12.1.2) becomes

$$\sup(P^*) = \min(P).$$

Several equivalent formulations of the consistency of (P) are also proved.

Section 12.4 deals with approximating sequences for (P) and its dual (P*) In particular, it is shown that if a suitable maximizing sequence for

(P*) exists, then the strong duality condition (12.1.5) is satisfied. Section 12.5 presents an approximation scheme for (P) using finite-dimensional programs. The scheme consists of three main steps. In step 1 we introduce an "increasing" sequence of *aggregations* of (P), each one with finitely many constraints. In step 2 each aggregation is *relaxed* (from an equality to an inequality), and, finally, in step 3, each aggregation-relaxation is combined with an *inner approximation* that has a finite number of decision variables. Thus the resulting aggregation-relaxation-inner approximation turns out to be a finite linear program, that is, a program with finitely many constraints and decision variables. The corresponding convergence theorems are stated in §12.5, and they are all proved in the final section 12.6.

To fix ideas, we shall consider only the so-called "unichain" AC problem. However, from the proof of our main results it should be clear that similar results are valid for other Markov control problems, in particular for discounted and for "multichain" AC MCPs.

12.2   Preliminaries

This section contains background material that can be omitted on a first reading; the reader may refer to it as needed.

The material is divided into four subsections. Subsection A reviews some basic definitions and facts related to dual pairs of vector spaces and linear operators. Subsections B and C summarize the main results on infinite LP needed in later sections. Finally, Subsection D reviews the notion of "tightness" and its connection to the existence of i.p.m.'s for Markov chains.

A. Dual pairs of vector spaces

Let $\mathcal{X}$ and $\mathcal{Y}$ be two arbitrary (real) vector spaces, and let $\langle \cdot, \cdot \rangle$ be a **bilinear form** on $\mathcal{X} \times \mathcal{Y}$, that is, a real-valued function on $\mathcal{X} \times \mathcal{Y}$ such that

- the map $x \mapsto \langle x, y \rangle$ is linear on $\mathcal{X}$ for every $y \in \mathcal{Y}$, and

- the map $y \mapsto \langle x, y \rangle$ is linear on $\mathcal{Y}$ for every $x \in \mathcal{X}$.

Then the pair $(\mathcal{X}, \mathcal{Y})$ is called a **dual pair** if the bilinear form "separates points" in x and y, that is,

- for each $x \neq 0$ in $\mathcal{X}$ there is some $y \in \mathcal{Y}$ with $\langle x, y \rangle \neq 0$, and

- for each $y \neq 0$ in $\mathcal{Y}$ there is some $x \in \mathcal{X}$ with $\langle x, y \rangle \neq 0$.

If $(\mathcal{X}, \mathcal{Y})$ is a dual pair, then so is $(\mathcal{Y}, \mathcal{X})$.

If $(\mathcal{X}_1, \mathcal{Y}_1)$ and $(\mathcal{X}_2, \mathcal{Y}_2)$ are two dual pairs of vector spaces with bilinear forms $\langle \cdot, \cdot \rangle_1$ and $\langle \cdot, \cdot \rangle_2$, respectively, then the product $(\mathcal{X}_1 \times \mathcal{X}_2, \mathcal{Y}_1 \times \mathcal{Y}_2)$ is endowed with the bilinear form

$$\langle (x_1, x_2), (y_1, y_2) \rangle := \langle x_1, y_1 \rangle_1 + \langle x_2, y_2 \rangle_2. \qquad (12.2.1)$$

this definition can be extended to the product of three or more dual pairs.

12.1.1 Examples. (a) If $\mathcal{X} = \mathcal{Y} = \mathbb{R}^n$ for some $n = 1, 2, \ldots$, then $\langle x, y \rangle$ will denote the usual "inner product" $x \cdot y$ of the vectors x, y. that is,

$$\langle x, y \rangle := x \cdot y = x_1 y_1 + \cdots + x_n y_n.$$

(b) Let S be a Borel space with Borel σ-algebra $\mathcal{B}(S)$, and let $\mathcal{X} := M(S)$ be a vector space of finite signed measures on $\mathcal{B}(S)$. In the following sections $M(S)$ will be the Banach space $\mathbb{M}(S)$ of finite signed measures on $\mathcal{B}(S)$, endowed the *total variation norm* $\| \cdot \|_{TV}$, or the subspace $\mathbb{M}_w(S)$ of finite signed measures μ with finite *w-norm*

$$\|\mu\|_w := \int_S w \, d|\mu|, \qquad (12.2.2)$$

for some weight function $w \geq 1$. (See §7.2.)

Now let $\mathcal{Y} := F(S)$ be a vector space of real-valued measurable functions on S. In the following sections $F(S)$ will be one of the Banach spaces

$$\mathbb{B}_w(S) \supset \mathbb{B}(S) \supset C_b(S) \supset C_0(S), \qquad (12.2.3)$$

where

- $\mathbb{B}_w(S)$ is the Banach space of measurable functions u with finite *w-norm*

$$\|u\|_w := \sup_S |u(s)| / w(s), \qquad (12.2.4)$$

for some weight function $w \geq 1$ (see §7.2.);

- $\mathbb{B}(S)$ is the subspace of measurable *bounded* functions u with finite *supremum* (or *sup*) *norm*

$$\|u\| := \sup_S |u(s)| \qquad (12.2.5)$$

[obtained from (12.2.4) with $w(\cdot) \equiv 1$];

- $C_b(S) \subset \mathbb{B}(S)$ is the subspace of *continuous bounded* functions, and

- $C_0(S) \subset C_b(S)$ is the subspace of continuous functions u *vanishing at infinity*, that is, for each $\varepsilon > 0$ there is a compact set K_ε such that

$$|u(s)| < \varepsilon \quad \forall s \notin K_\varepsilon. \qquad (12.2.6)$$

In the latter case sometimes we shall simply write

$$\lim_{s \to \infty} u(s) = 0. \qquad (12.2.7)$$

The spaces $C_b(S)$ and $C_0(S)$ coincide if S is *compact*.

In any of the above cases, the dual pair $(\mathcal{X}, \mathcal{Y}) = (M(S), F(S))$ is endowed with the bilinear form

$$\langle \mu, u \rangle := \int_S u d\mu. \qquad (12.2.8)$$

Thus, by (12.2.1) and part (a), the bilinear form corresponding to the dual pair $(\mathbb{R}^n \times M(S), \mathbb{R}^n \times F(S))$ is

$$\langle (x, \mu), (y, u) \rangle = x \cdot y + \langle \mu, u \rangle. \square \qquad (12.2.9)$$

Given a dual pair $(\mathcal{X}, \mathcal{Y})$, we denote by $\sigma(\mathcal{X}, \mathcal{Y})$ the **weak topology** on $\mathcal{X}$ (also referred to as the σ-**topology** on $\mathcal{X}$), namely, the coarsest—or weakest—topology on $\mathcal{X}$ under which all the elements of $\mathcal{Y}$ are continuous when regarded as linear forms $\langle \cdot, y \rangle$ on $\mathcal{X}$. Equivalently, the base of neighborhoods of the origin of the σ-topology is the family of all sets of the form

$$N(I, \varepsilon) := \{ x \in \mathcal{X} | |\langle x, y \rangle| \le \varepsilon \quad \forall y \in I \}, \qquad (12.2.10)$$

where $\varepsilon > 0$ and I is a *finite* subset of $\mathcal{Y}$. (See, for instance, Robertson and Robertson [1], p. 32.)

Let $\{x_n\}$ be a sequence or a net in $\mathcal{X}$. (See Note 1 at the end of this section for the definition of "net".) Then x_n *converges to x in the weak topology* $\sigma(\mathcal{X}, \mathcal{Y})$ if

$$\langle x_n, y \rangle \to \langle x, y \rangle \quad \forall y \in \mathcal{Y}. \qquad (12.2.11)$$

For instance, for the dual pair $(M(S), F(S))$ in Example 12.2.1(b), a sequence or a net of measures μ_n converges to μ in the weak topology $\sigma(M(S), F(S))$ if

$$\langle \mu_n, u \rangle \to \langle \mu, u \rangle \quad \forall u \in F(S), \qquad (12.2.12)$$

where $\langle \cdot, \cdot \rangle$ stands for the bilinear form in (12.2.8).

12.2.2 Remark. (a) Let $(\mathcal{X}, \mathcal{Y})$ be a dual pair such that $\mathcal{Y}$ is a *Banach space* and $\mathcal{X} = \mathcal{Y}^*$ is the *topological dual* of $\mathcal{Y}$. In this case, the weak topology $\sigma(\mathcal{X}, \mathcal{Y})$ is called the **weak*** (weak-star) **topology** on $\mathcal{X}$, and so (12.2.11) is referred to as the **weak* convergence** of x_n to x.

(b) For instance, let $\mathcal{X} = \mathbb{M}(S)$ and $\mathcal{Y} := C_0(S)$ be as in Example 12.2.1. If S is a *locally compact separable metric* (LCSM) *space*, then—by the Riesz Representation Theorem (see, for example, Rudin [1])—$\mathbb{M}(S)$ is the topological dual of the (*separable*) Banach space $C_0(S)$, and so the weak topology $\sigma(\mathbb{M}(S), C_0(S))$ on $\mathbb{M}(S)$ is in fact the weak* topology.

The fact that $C_0(S)$ is a *separable* Banach space will be used later on in conjunction with the following result.

(c) (The *Alaoglu* or *Banach-Alaoglu-Bourbaki Theorem*; see Ash [1] or Brezis [1]). Let $\mathcal{Y}$ be a Banach space with topological dual $\mathcal{Y}^*$. Then:

(c_1) the closed unit sphere $U := \{y \in \mathcal{Y}^* | \|y\| \leq 1\}$ in $\mathcal{Y}^*$ is compact in the weak* topology $\sigma(\mathcal{Y}^*, \mathcal{Y})$.

(c_2) If in addition $\mathcal{Y}$ is *separable*, then the weak* topology of U is *metrizable*; hence U is weakly* *sequentially* compact.

A consequence of the metrizability in (c_2) is, in particular, that to verify that a subset, say V, of U is closed in the weak* topology, it suffices to use the usual criterion in metric spaces (if y_n is in V and $y_n \to y$ in the weak* topology, then y is in V) for *sequences* y_n, rather than nets.

An advantage of using sequences instead of nets is shown in the following proposition. □

12.2.3 Proposition. *Let* $(\mathcal{X}, \mathcal{Y})$ *be a dual pair of normed vector spaces. If* $\{x_n\}$ *is a net converging to x in the weak topology* $\sigma(\mathcal{X}, \mathcal{Y})$, *then*

$$\|x\| \leq \liminf \|x_n\|. \tag{12.2.13}$$

If $\{x_n\}$ *is a* **sequence**, *instead of a net, then (12.2.13) holds, and in addition the sequence* $\{\|x_n\|\}$ *is* **bounded.**

Proof. (12.2.13) follows from the fact that the map $x \mapsto \|x\|$ is lower semicontinuous (l.s.c.) in the weak topology. The result for sequences is well known—see, for instance, Ash [1, p. 145] or Brezis [1, p. 41]. □

12.2.4 Definition. Let $(\mathcal{X}, \mathcal{Y})$ and $(\mathcal{Z}, \mathcal{W})$ be two dual pairs of vector spaces and $G : \mathcal{X} \to \mathcal{Z}$ a linear map.

(a) G is said to be **weakly continuous** if it is continuous with respect to the weak topologies $\sigma(\mathcal{X}, \mathcal{Y})$ and $\sigma(\mathcal{Z}, \mathcal{W})$; that is, if $\{x_n\}$ is a net in $\mathcal{X}$ such that $x_n \to x$ in the weak topology $\sigma(\mathcal{X}, \mathcal{Y})$ [see (12.2.11)], then $Gx_n \to Gx$ in the weak topology $\sigma(\mathcal{Z}, \mathcal{W})$, i.e.,

$$\langle Gx_n, w \rangle \to \langle Gx, w \rangle \quad \forall w \in \mathcal{W}. \tag{12.2.14}$$

(b) The **adjoint** G^* of G is defined by the relation

$$\langle Gx, w \rangle = \langle x, G^*w \rangle \quad \forall x \in \mathcal{X}, \ w \in \mathcal{W}. \tag{12.2.15}$$

The following proposition gives a well-known (easy-to-use) criterion for the map G in Definition 12.2.4 to be weakly continuous—for a proof see, for instance, Robertson and Robertson [1], p. 38.

12.2.5 Proposition. *the linear map G is weakly continuous if and only if its adjoint G^* maps $\mathcal{W}$ into $\mathcal{Y}$, that is, $G^*(\mathcal{W}) \subset \mathcal{Y}$.*

12.2.6 Example. Let X and Y be two Borel spaces, and let $w_0(x)$ and $w(x,y)$ be weight functions on X and $X \times Y$, respectively, such that

$$1 \le w_0(x) \le w(x,y) \qquad \forall x \in X, \ y \in Y. \tag{12.2.16}$$

We shall consider spaces $\mathbb{B}_w(X \times Y)$, $\mathbb{M}_w(X \times Y)$, $\mathbb{B}_{w_0}(X)$, and $\mathbb{M}_{w_0}(X)$ as in Examples 12.2.1(b).

(a) Consider the dual pairs $(\mathbb{M}_w(X \times Y), \mathbb{B}_w(X \times Y))$ and $(\mathbb{R}, \mathbb{R})$, and the linear map

$$L_0 : \mathbb{M}_w(X \times Y) \to \mathbb{R}, \quad \mu \mapsto L_0\mu := \langle \mu, 1 \rangle. \tag{12.2.17}$$

By (12.2.8) with $u(x,y) \equiv 1$,

$$L_0\mu = \langle \mu, 1 \rangle = \mu(X \times Y) \quad \forall \mu \in \mathbb{M}_w(X \times Y).$$

In particular,

$$L_0\mu = \|\mu\|_{TV} \quad \text{if} \quad \mu \in \mathbb{M}_w(X \times Y)_+, \tag{12.2.18}$$

where $\mathbb{M}_w(X \times Y)_+$ stands for the convex cone of *nonnegative* measures in $\mathbb{M}_w(X \times Y)$ and $\| \cdot \|_{TV}$ denotes the total variation norm.

Since $\mathbb{B}_w(X \times Y)$ contains the constant functions [see (12.2.3)], the adjoint

$$r \mapsto (L_0^* r)(x,y) \equiv r \quad \forall r \in \mathbb{R},$$

obviously maps $\mathbb{R}$ into $\mathbb{B}_w(X \times Y)$, and so L_0 is weakly continuous, by Proposition 12.2.5.

(b) Consider the dual pairs $(\mathbb{M}_w(X \times Y), \mathbb{B}_w(X \times Y))$ and $(\mathbb{M}_{w_0}(X), \mathbb{B}_{w_0}(X))$, and the linear map

$$G_1 : \mathbb{M}_w(X \times Y) \to \mathbb{M}_{w_0}(X), \quad \mu \mapsto G_1\mu := \hat{\mu},$$

where $\hat{\mu}$ denotes the **marginal** (also known as the **projection**) of μ on X, that is,

$$\hat{\mu}(B) := \mu(B \times Y) \quad \forall B \in \mathcal{B}(X). \tag{12.2.19}$$

The adjoint $u \mapsto G_1^* u$, with

$$(G_1^* u)(x,y) := u(x) \quad \forall u \in \mathbb{B}_{w_0}(X), \ (x,y) \in X \times Y, \tag{12.2.20}$$

maps $\mathbb{B}_{w_0}(X)$ into $\mathbb{B}_w(X \times Y)$, because (12.2.16) gives

$$\frac{|G_1^* u|}{w} = \frac{|u|}{w_0} \cdot \frac{w_0}{w} \le \frac{|u|}{w_0},$$

and so $\|G_1^* u\|_w \le \|u\|_{w_0} < \infty$. Thus G_1 is weakly continuous, by Proposition 12.2.5.

(c) Consider the dual spaces in part (b), and let $P(B|x,y)$ be a *stochastic kernel* on X given $X \times Y$ (see Definition 7.2.3). Moreover, assume that

$$\int_X w_0(x') P(dx'|\cdot) \quad \text{is in} \quad \mathbb{B}_w(X \times Y),$$

that is, there is a constant k such that

$$\int_X w_0(x')P(dx'|x,y) \le kw(x,y) \quad \forall (x,y) \in X \times Y. \tag{12.2.21}$$

Then the linear map $G_2 : \mathbb{M}_w(X \times Y) \to \mathbb{M}_{w_0}(X)$, $\mu \mapsto G_2\mu$, defined by

$$(G_2\mu)(B) := \int_{X \times Y} P(B|x,y)\mu(d(x,y)) \quad \text{for} \quad B \in \mathcal{B}(X) \tag{12.2.22}$$

is weakly continuous. Indeed, for each function u in $\mathbb{B}_{w_0}(X)$, (12.2.21) yields

$$\left| \int_X u(x')P(dx'|x,y) \right| \le \|u\|_{w_0} \int_X w_0(x')P(dx'|x,y)$$
$$\le \|u\|_{w_0} kw(x,y),$$

which means that the adjoint G_2^*, $u \mapsto G_2^*u$, given by

$$(G_2^*u)(x,y) := \int_X u(x')P(dx'|x,y), \tag{12.2.23}$$

maps $\mathbb{B}_{w_0}(X)$ into $\mathbb{B}_w(X \times Y)$. Thus, the weak continuity of G_2 follows from Proposition 12.2.5.

(d) As a consequence of (b) and (c), if the inequality (12.2.21) holds, then the linear map

$$L_1 := G_1 - G_2 : \mathbb{M}_w(X \times Y) \to \mathbb{M}_{w_0}(X),$$

i.e.,

$$(L_1\mu)(B) := \hat{\mu}(B) - \int_{X \times Y} P(B|x,y)\mu(d(x,y)) \quad \text{for} \quad B \in \mathcal{B}(X), \tag{12.2.24}$$

is weakly continuous. Furthermore, with L_0 as in (12.2.17), the linear map

$$L := (L_0, L_1) : \mathbb{M}_w(X \times Y) \to \mathbb{R} \times \mathbb{M}_{w_0}(X),$$

i.e.,

$$L\mu := (L_0\mu, L_1\mu) \quad \text{for} \quad \mu \text{ in } \mathbb{M}_w(X \times Y), \tag{12.2.25}$$

is weakly continuous.

Note that the adjoints

$$L_1^* : \mathbb{B}_{w_0}(X) \to \mathbb{B}_w(X \times Y), \quad L^* : \mathbb{R} \times \mathbb{B}_{w_0}(X) \to \mathbb{B}_w(X \times Y)$$

of L_1 and L are given by $L_1^* = G_1^* - G_2^*$ and $L^* = L_0^* + L_1^*$; that is,

$$(L_1^*u)(x,y) = u(x) - \int_X u(x')P(dx'|x,y) \tag{12.2.26}$$

[see (12.2.20) and (12.2.23)] and

$$L^*(r,u)(x,y) = (L_0^*r)(x,y) + (L_1^*u)(x,y)$$
$$= r + u(x) - \int_X u(x')P(dx'|x,y), \tag{12.2.27}$$

respectively. □

12.2.7 Remark. (a) Let $P(B|x,y)$ be the stochastic kernel in Example 12.2.6(c), and consider the Banach spaces $C_b(\cdot)$ and $C_0(\cdot)$ in Example 12.2.1(b). By a standard abuse of terminology, the kernel P is said to be **weakly continuous** if the adjoint G_2^* in (12.2.23) maps $C_b(X)$ into $C_b(X \times Y)$, that is,

$$G_2^*u \text{ is in } C_b(X \times Y) \quad \text{if} \quad u \text{ is in } C_b(X). \tag{12.2.28}$$

[Observe that this is a Feller-like condition; see (12.2.47).]
   (b) Suppose, on the other hand, that P is weakly continuous and, moreover,

$$P(K|\cdot) \text{ vanishes at infinity for each compact } K \subset X; \tag{12.2.29}$$

that is [as in (12.2.6)], for each $\varepsilon > 0$ there is a compact set $K' = K'(\varepsilon, K)$ in $X \times Y$ such that

$$P(K|x,y) \le \varepsilon \quad \forall (x,y) \notin K'.$$

Then a straightforward calculation shows that, in addition to (12.2.28), G_2^* maps $C_0(X)$ into $C_0(X \times Y)$, that is

$$G_2^*u \text{ is in } C_0(X \times Y) \text{ if } u \text{ is in } C_0(X). \tag{12.2.30}$$

In other words, suppose that (12.2.28) and (12.2.29) are satisfied, and that X and Y—hence the product $X \times Y$—are *locally compact separable metric spaces*. Then, in view of Remark 12.2.2(b) and Proposition 12.2.5, the condition (12.2.30) states that the map $G_2 : \mathbb{M}(X \times Y) \to \mathbb{M}(X)$ defined by (12.2.22) is *weakly* continuous*, that is, continuous with respect to the weak* topologies $\sigma(\mathbb{M}(X \times Y), C_0(X \times Y))$ and $\sigma(\mathbb{M}(X), C_0(X))$. □

12.2.8 Remark. (Positive and dual cones.) (a) Let $(\mathcal{X}, \mathcal{Y})$ be a dual pair of vector spaces, and K a *convex cone* in $\mathcal{X}$, that is, $x + x'$ and λx belong to K whenever x and x' are in K and $\lambda > 0$. Unless explicitly stated otherwise, we shall assume that $K \ne X$ and the origin (that is, the zero vector, 0) is in K. In this case, K defines a partial order $\ge$ on X such that

$$x \ge x' \quad \Leftrightarrow \quad x - x' \in K,$$

and K will be referred to as a *positive cone*. The *dual cone* of K is the convex cone K^* in $\mathcal{Y}$ defined by

$$K^* := \{y \in \mathcal{Y} | \langle x, y \rangle \ge 0 \quad \forall x \in K\}. \tag{12.2.31}$$

(b) If $\mathcal{X} = M(S)$ is any of the measure spaces in Example 12.2.1(b), we will denote by $M(S)_+$ the "natural" *positive cone in* $M(S)$, which consists of all the *nonnegative* measures in $M(S)$, that is,

$$M(S)_+ := \{\mu \in M(S) | \mu \geq 0\}.$$

The corresponding dual cone $M(S)_+^*$ in (any of the spaces) $F(S)$ coincides with the "natural" positive cone

$$F(S)_+ := \{u \in F(S) | u \geq 0\}. \quad \square$$

B. Infinite linear programming

An infinite linear program requires the following components:

- two dual pairs $(\mathcal{X}, \mathcal{Y})$ and $(\mathcal{Z}, \mathcal{W})$ of real vector spaces;
- a weakly continuous linear map $L : \mathcal{X} \to \mathcal{Z}$, with adjoint $L^* : \mathcal{W} \to \mathcal{Y}$;
- a positive cone K in $\mathcal{X}$, with dual cone K^* in $\mathcal{Y}$ [see (12.2.31)]; and
- vectors $b \in \mathcal{Z}$ and $c \in \mathcal{Y}$.

Then the **primal** linear program is

$$\mathbb{P} : \quad \text{minimize } \langle x, c \rangle$$
$$\text{subject to: } Lx = b, \ x \in K. \tag{12.2.32}$$

The corresponding **dual** problem is

$$\mathbb{P}^* : \quad \text{maximize } \langle b, w \rangle$$
$$\text{subject to: } c - L^*w \in K^*, \ w \in \mathcal{W}. \tag{12.2.33}$$

An element x of $\mathcal{X}$ is called **feasible** for $\mathbb{P}$ if it satisfies (12.2.32), and $\mathbb{P}$ is said to be **consistent** if it has a feasible solution. If $\mathbb{P}$ is consistent, then its **value** is defined as

$$\inf \mathbb{P} := \inf\{\langle x, c \rangle | x \text{ is feasible for } \mathbb{P}\}; \tag{12.2.34}$$

otherwise, $\inf \mathbb{P} := +\infty$. The program $\mathbb{P}$ is **solvable** if there is a feasible solution x^* that achieves the infimum in (12.2.34). In this case, x^* is called an **optimal solution** for $\mathbb{P}$ and, instead of $\inf \mathbb{P}$, the value of $\mathbb{P}$ is written as

$$\min \mathbb{P} = \langle x^*, c \rangle.$$

Similarly, $w \in \mathcal{W}$ is **feasible** for the dual program $\mathbb{P}^*$ if it satisfies (12.2.33), and $\mathbb{P}^*$ is said to be **consistent** if it has a feasible solution. If $\mathbb{P}^*$ is consistent, then its **value** is defined as

$$\sup \mathbb{P}^* := \sup\{\langle b, w \rangle | w \text{ is feasible for } \mathbb{P}^*\}; \tag{12.2.35}$$

otherwise, $\sup \mathbb{P}^* := -\infty$. The dual $\mathbb{P}^*$ is **solvable** if there is a feasible solution w^* that attains the supremum in (12.2.35), in which case we write the value of $\mathbb{P}^*$ as

$$\max \mathbb{P}^* = \langle b, w^* \rangle.$$

The next theorem can be proved as in elementary (finite-dimensional) LP.

12.2.9 Theorem.

(a) **(Weak duality.)** *If $\mathbb{P}$ and $\mathbb{P}^*$ are both consistent, then their values are finite and satisfy*

$$\sup \mathbb{P}^* \leq \inf \mathbb{P}. \qquad (12.2.36)$$

(b) **(Complementary slackness.)** *If x is feasible for $\mathbb{P}$, w is feasible for $\mathbb{P}^*$, and*

$$\langle x, c - L^*w \rangle = 0, \qquad (12.2.37)$$

then x is optimal for $\mathbb{P}$ and w is optimal for $\mathbb{P}^*$.

The converse of Theorem 12.2.9(b) does not hold in general [as shown in Example 6.2.5(b)]. It does hold, however, if there is **no duality gap** for $\mathbb{P}$, which means that equality holds in (12.2.36), i.e.,

$$\sup \mathbb{P}^* = \inf \mathbb{P}. \qquad (12.2.38)$$

On the other hand, it is said that the **strong duality** condition for $\mathbb{P}$ holds if $\mathbb{P}$ and its dual are both solvable and

$$\max \mathbb{P}^* = \min \mathbb{P}. \qquad (12.2.39)$$

The following theorem gives conditions under which $\mathbb{P}$ is solvable and there is no duality gap—for a proof see Anderson and Nash [1, Theorem 3.9].

12.2.10 Theorem. *Let H be the set in $\mathcal{Z} \times \mathbb{R}$ defined as*

$$H := \{(Lx, \langle x, c \rangle + r | x \in K, \ r \geq 0\}.$$

If $\mathbb{P}$ is consistent and H is weakly closed [that is, closed in the weak topology $\sigma(\mathcal{Z} \times \mathbb{R}, \mathcal{W} \times \mathbb{R})$], then $\mathbb{P}$ is solvable and there is no duality gap, so that (12.2.38) becomes

$$\sup \mathbb{P}^* = \min \mathbb{P}.$$

The following *Generalized Farkas Theorem* of Craven and Koliha [1, Theorem 2] gives a necessary and sufficient condition for $\mathbb{P}$ to be consistent. The result is similar to Theorem 12.2.10 in that it also requires a certain set to be weakly closed.

12.2.11 Theorem. (Craven and Koliha [1].) *If L is weakly continuous and $L(K) \subset \mathcal{Z}$ is weakly closed then the following conditions are equivalent:*

(a) *(12.2.32) is satisfied; that is, the equation $Lx = b$ has a solution x in* K

(b) $L^* w \in K^* \Rightarrow \langle b, w \rangle \geq 0$.

We can view Theorem 12.2.11 as an "alternative theorem". Namely, if L is weakly continuous and $L(K)$ is weakly closed, then either

(i) (12.2.32) is satisfied, or

(ii) there exists $w \in W$ such that: $L^* w$ is in K^* and $\langle b, w \rangle < 0$.

Due to this fact, the Generalized Farkas Theorem 12.2.11 is sometimes referred to as the *Farkas Alternative Theorem*.

C. Approximation of linear programs

An important practical question is how to obtain—or at least estimate—the value of a linear program. In later sections we shall consider two approaches related to the following definitions.

12.2.12 Definition. (Minimizing and maximizing sequences.)

(a) A sequence $\{x_n\}$ in $\mathcal{X}$ is called a **minimizing sequence** for $\mathbb{P}$ if each x_n is feasible for $\mathbb{P}$ and $\langle x_n, c \rangle \downarrow \inf \mathbb{P}$.

(b) A sequence $\{w_n\}$ in $\mathcal{W}$ is called a **maximizing sequence** for the dual problem $\mathbb{P}^*$ if each w_n is feasible for $\mathbb{P}^*$ and $\langle b, w_n \rangle \uparrow \sup \mathbb{P}^*$.

Note that if $\mathbb{P}$ is consistent with a finite value $\inf \mathbb{P}$, then [by definition (12.2.34) of $\inf \mathbb{P}$] there exists a minimizing sequence. A similar remark holds for $\mathbb{P}^*$.

12.2.13 Definition. (Aggregations and inner approximations.)

(a) Let W be a subset of $\mathcal{W}$. Then the linear program

$$\mathbb{P}(W) \ : \ \text{minimize } \langle x, c \rangle$$
$$\text{subject to: } \langle Lx - b, w \rangle = 0 \quad \forall w \in W, \ x \in K, \quad (12.2.40)$$

is called an **aggregation** (of constraints) of $\mathbb{P}$.

(b) If $K' \subset K$ is a subset of the positive cone $K \subset \mathcal{X}$, then the program

$$\mathbb{P}(K') \ : \ \text{minimize } \langle x, c \rangle$$
$$\text{subject to: } Lx = b, \ x \in K', \quad (12.2.41)$$

is called an **inner approximation** of $\mathbb{P}$.

As K' is contained in K, we have $\inf \mathbb{P} \leq \inf \mathbb{P}(K')$. On the other hand, if x satisfies (12.2.32), then it satisfies (12.2.40), and so $\inf \mathbb{P}(W) \leq \inf \mathbb{P}$. Hence

$$\inf \mathbb{P}(W) \leq \inf \mathbb{P} \leq \inf \mathbb{P}(K').$$

Thus, we can use an aggregation (of constraints) to approximate $\inf \mathbb{P}$ *from below*, whereas an inner approximation can be used to approximate $\inf \mathbb{P}$ *from above*. One can also easily get the following (for a proof see Hernández-Lerma and Lasserre [15]):

12.2.14 Proposition. *Suppose that $\mathbb{P}$ is solvable.*

(a) *If W is weakly dense in $\mathcal{W}$, then $\mathbb{P}(W)$ is equivalent to $\mathbb{P}$ in the sense that $\mathbb{P}(W)$ is also solvable and*

$$\min \mathbb{P}(W) = \min \mathbb{P}.$$

(b) *If K' is weakly dense in K, then there is a sequence $\{x_n\}$ in K' such that*

$$\langle x_n, c \rangle \to \min \mathbb{P}.$$

D. Tightness and invariant measures

The Alaoglu Theorem in Remark 12.2.2(c) gives conditions for a certain set to be compact in the *weak* topology*. This is particularly useful when dealing, for example, with the dual pair $(\mathbb{M}(S), C_0(S))$ in Remark 12.2.2(b). However, we will also need to consider compact sets in the *weak topology* $\sigma(\mathbb{M}(S), C_b(S))$, and so in this subsection we will briefly review some related notions.

Let S be a Borel space, and let $\mathbb{M}(S)_+$ be the positive cone of (finite) nonnegative measures in $\mathbb{M}(S)$ [see Example 12.2.1(b) and Remark 12.2.8(b)]. A measure γ in $\mathbb{M}(S)_+$ is said to be **tight** if for each $\varepsilon > 0$ there is a compact set $K \subset S$ such that $\gamma(K^c) < \varepsilon$, where K^c denotes the complement of K.

For example, suppose that either (i) S is σ-compact, or (ii) S is a Polish space. Then any finite measure γ on the Borel σ-algebra $\mathcal{B}(S)$ is tight.

Similarly, a family Γ in $\mathbb{M}(S)_+$ is said to be **tight** if for each $\varepsilon > 0$ there is a compact set K such that $\gamma(K^c) < \varepsilon$ for all γ in Γ.

Tightness turns out to be closely related to the existence of a **strictly unbounded** (also known as a *moment* or *norm-like*) function $g \geq 0$ on S, which means that there exists an increasing sequence of compact sets $K_n \uparrow S$ such that

$$\lim_{n \to \infty} \inf_{x \notin K_n} g(x) = +\infty. \tag{12.2.42}$$

For instance, if $g \geq 0$ is **inf-compact**—that is, the level set $\{x | g(x) \leq r\}$ is compact for every number r—then g is strictly unbounded.

The connection between tightness and strictly unbounded functions is provided by the following theorem (see, for instance, Balder [1, §2] or Bourbaki [1, p. 109]).

12.2.15 Theorem. *Let Γ be a bounded family of measures in $\mathbb{M}(S)_+$. Then Γ is tight if and only if there is a strictly unbounded function $g \geq 1$ such that [using the notation (12.2.8)]*

$$\sup_{\Gamma} \langle \gamma, g \rangle < \infty. \tag{12.2.43}$$

Moreover, if Γ consists of probability measures only, then the condition $g \geq 1$ can be replaced by $g \geq 0$.

As an elementary application of Theorem 12.2.15, if S is a *compact* metric space, then each bounded family Γ of measures in $\mathbb{M}(S)_+$ is tight. Indeed, let $g(\cdot) \equiv \bar{g} > 0$ be a constant function on S, and M a constant such that $\|\gamma\|_{TV} \leq M$ for all γ in Γ. Then (12.2.42) trivially holds (because the infimum over the empty set S^c is $+\infty$), and (12.2.43) becomes

$$\sup_{\Gamma} \langle \gamma, g \rangle \leq \bar{g}M < \infty.$$

On the other hand, the connection between tightness and compactness in the weak topology $\sigma(\mathbb{M}(S), C_b(S))$ is provided by Prohorov's Theorem (see Billingsley [1] or Parthasarathy [1]):

12.2.16 Theorem. (Prohorov's Theorem.) *Let $\mathcal{P}(S)$ be the family of probability measures in $\mathbb{M}(S)_+$, and Γ a subset of $\mathcal{P}(S)$. If Γ is tight, then it is sequentially relatively compact in the weak topology $\sigma(\mathbb{M}(S), C_b(S))$; that is for each sequence $\{\mu_n\}$ in Γ there is a subsequence $\{\mu_m\}$ and a probability measure μ (not necessarily in Γ) such that*

$$\langle \mu_m, u \rangle \to \langle \mu, u \rangle \quad \forall u \in C_b(S). \tag{12.2.44}$$

Prohorov's Theorem 12.2.16 is true for a general *metric*—not necessarily Borel—space S. Furthermore, *the converse holds* (weak sequential relative compactness implies tightness) *if S is a Polish space.*

Finally, we will present a theorem of Benes [1, 2] that relates the concepts of tightness, strictly unbounded functions, and invariant probability measures (i.p.m.'s) for a Markov chain $\{x_n\}$ on the Borel space S, with transition kernel $P(B|x)$. Recall that if ν is a measure on S, then νP^n denotes the measure

$$(\nu P^n)(B) := \int_S P^n(B|x)\nu(dx) \quad \forall n = 0, 1, \ldots, B \in \mathcal{B}(S), \tag{12.2.45}$$

whereas if u is a function on X, then Pu stands for the function

$$(Pu)(x) := \int_S u(y)P(dy|x) \quad \text{for} \quad x \in S. \tag{12.2.46}$$

Moreover, the chain $\{x_n\}$, or the transition kernel P, is said to satisfy the (weak) **Feller property** if

$$u \in C_b(S) \Rightarrow Pu \in C_b(S). \qquad (12.2.47)$$

12.2.17 Theorem. *Suppose that the Borel space S is σ-compact, and that $\{x_n\}$ is a Markov chain on S that satisfies the Feller property. Then the following conditions are equivalent:*

(a) *$\{x_n\}$ has an i.p.m.*

(b) *there is a p.m. ν such that the sequence $\{\nu P^n, n = 0, 1, \ldots\}$ is tight.*

(c) *There is a p.m. ν and a strictly unbounded function $g \geq 0$ such that*

$$\sup_n \langle \nu P^n, g \rangle < \infty.$$

(d) *There is a p.m. ν and a compact set K in S such that*

$$\limsup_{N \to \infty} \frac{1}{N} \sum_{n=0}^{N-1} \nu P^n(K) > 0.$$

12.2.18 Remark. Benes [1] proves Theorem 12.2.17 under the following, stronger, assumptions:

(i) S is a LCSM space;

(ii) P satisfies the Feller property; and

(iii) For each compact set K, the function $x \mapsto P(K|x)$ vanishes at infinity.

As in Remark 12.2.7(b), it is easy to see that (ii) and (iii) imply that

$$Pu \text{ is in } C_0(S) \text{ if } u \text{ is in } C_0(S) \text{ [cf. (12.2.30)]}, \qquad (12.2.48)$$

and so (12.2.45), with $n = 1$, defines a map $P : \mathbb{M}(S) \to \mathbb{M}(S)$ which is continuous in the weak* topology $\sigma(\mathbb{M}(S), C_0(S))$. Benes uses this fact and the Alaoglu Theorem [Remark 12.2.2(c)] to relate (a) and (d) in Theorem 12.2.17, as well as a fifth condition (e) not included here. Without the latter condition (e), it can be verified that the proof by Benes also yields Theorem 12.2.17 in its present form, assuming σ-compactness and the Feller property, rather than (i), (ii), (iii). In fact, the relations

$$(a) \Rightarrow (b) \Leftrightarrow (c) \Rightarrow (d)$$

are immediate. Indeed, if (a) holds and γ denotes an i.p.m., then taking $\nu = \gamma$ in (b), the sequence $\nu P^n = \gamma$ $(n = 0, 1, \ldots)$ is tight because any single finite measure γ on a σ-compact metric space is tight. Hence (a) implies (b). On the other hand, the equivalence of (b) and (c) follows from Theorem 12.2.15, wheres (b) $\Rightarrow$ (d) follows from the definition of tightness. $\square$

Notes on §12.2

1. A partially ordered set $(D, \leq)$ is said to be **directed** if every finite subset of D has an upper bound; that is, if a, b are in D, then there exists $c \in D$ such that $a \leq c$ and $b \leq c$. A **net** (also known as a **generalized sequence**) in a topological space $\mathcal{X}$ is a function from a directed set D into $\mathcal{X}$ and it is denoted as $\{x_n, n \in D\}$ or simply $\{x_n\}$. The net $\{x_n\}$ is said to converge to the point x if for every neighborhood N of x there is an n_0 in D such that $x_n \in D$ for all $n \in D$ such that $n \geq n_0$. (For further properties of nets see, for instance, Ash [1].)

2. The material on infinite LP (subsection B) is borrowed from Anderson and Nash [1], except for the Generalized Farkas Theorem 12.2.11, which is due to Craven and Koliha [1]. For applications of Theorem 12.2.11 see, for instance, Hernández-Lerma and Lasserre [3, 5, 7], and Theorem 12.3.7 below. A Farkas-like result different from Theorem 12.2.11 can be found in Hernández-Lerma and Lasserre [4].

3. For additional comments and references related to Theorem 12.2.17 see Hernández-Lerma and Lasserre [2], which also gives necessary and sufficient conditions for existence of i.p.m.'s.

12.3   Linear programs for the AC problem

We will now consider the average cost (AC) criterion $J(\pi, \nu)$ from the viewpoint of LP. *Throughout the rest of this chapter we suppose that Assumption 11.4.1 is satisfied.* Thus, by Lemma 11.4.5(a) and Proposition 11.4.4(b), we already have in particular the existence of a stable randomized stationary policy $\varphi^\infty \in \Pi_{RS}$ such that (φ, p_φ) is a minimum pair, namely [by Definition 11.1.1(a)],

$$J(\varphi^\infty, p_\varphi) = \rho_{\min}, \tag{12.3.1}$$

where

$$\rho_{\min} := \inf_{\mathcal{P}(X)} J^*(\nu) = \inf_{\mathcal{P}(X)} \inf_{\Pi} J(\pi, \nu) \tag{12.3.2}$$

In this section we begin by introducing a linear program (P) that satisfies (12.1.1) with $(MCP)^* = \rho_{\min}$, that is,

$$\sup(P^*) \leq \rho_{\min} \leq \inf(P). \tag{12.3.3}$$

Then we will show that (P) is *solvable* and that there is *no duality gap* [see (12.2.38)], so that instead of (12.3.3) we will actually have the stronger relation

$$\sup(P^*) = \rho_{\min} = \min(P). \tag{12.3.4}$$

Finally, we will use the Generalized Farkas Theorem 12.2.11 to obtain *necessary and sufficient conditions for* (P) *to be consistent*, which will require a set of hypotheses different from Assumption 11.4.1.

A. The linear programs

We will first proceed as at the beginning of subsection 12.2.B to introduce the components of the linear program we are interested in.

The dual pairs. Let $\mathbb{K} \subset X \times A$ be the set defined in (8.2.1), and let $w(x, a)$ and $w_0(x)$ be the weight functions on $\mathbb{K}$ and X, respectively, defined as

$$w(x, a) := 1 + c(x, a), \quad w_0(x) := \min_{A(x)} w(x, a). \tag{12.3.5}$$

(By a well-known result of Rieder [1], also stated in Proposition D.6(a) of Volume I, Assumption 11.4.1(b) implies that $w_0(x)$ is measurable.) Then the dual pairs we are concerned with are

$$(\mathcal{X}, \mathcal{Y}) := (\mathbb{M}_w(\mathbb{K}), \mathbb{B}_w(\mathbb{K})) \tag{12.3.6}$$

and

$$(\mathcal{Z}, \mathcal{W}) := (\mathbb{R} \times \mathbb{M}_{w_0}(X), \mathbb{R} \times \mathbb{B}_{w_0}(X)), \tag{12.3.7}$$

where $\mathbb{M}_w(\mathbb{K})$ and $\mathbb{B}_w(\mathbb{K})$ are the weighted-norm spaces in Example 12.2.1(b), and similarly for $\mathbb{M}_{w_0}(X)$ and $\mathbb{B}_{w_0}(X)$. In particular, the bilinear form on $(\mathbb{M}_w(\mathbb{K}), \mathbb{B}_w(\mathbb{K}))$ is [as in (12.2.8)]

$$\langle \mu, u \rangle := \int_{\mathbb{K}} u \, d\mu, \tag{12.3.8}$$

and on $(\mathbb{R} \times \mathbb{M}_{w_0}(X), \mathbb{R} \times \mathbb{B}_{w_0}(X))$ is [by (12.2.1)]

$$\langle (r, \nu), (\rho, v) \rangle := r \cdot \rho + \int_X v \, d\nu. \tag{12.3.9}$$

Note that, since $c(x, a)$ is nonnegative [Assumption 11.4.1(b)], (12.3.5) yields

$$0 \leq c(x, a) \leq w(x, a) \quad \forall (x, a) \in \mathbb{K},$$

which implies that *the cost-per-stage function c is in* $\mathbb{B}_w(\mathbb{K})$, and, on the other hand,

$$1 \leq w_0(x) \leq w(x, a) \quad \forall (x, a) \in \mathbb{K}, \tag{12.3.10}$$

which is the same as (12.2.16) with $\mathbb{K}$ in lieu of $X \times Y$. Moreover, the policy $\widehat{\pi}$ and the initial state $\widehat{x}$ in Assumption 11.4.1(a) satisfy

$$\limsup_{n \to \infty} \frac{1}{n} \sum_{t=0}^{n-1} E_{\widehat{x}}^{\widehat{\pi}} [w(x_t, a_t)] = 1 + J(\widehat{\pi}, \widehat{x}) < \infty. \tag{12.3.11}$$

We will suppose that w and w_0 satisfy a condition of the form (12.2.21), with the kernel $P(\cdot | x, y)$ being replaced by the transition law $Q(\cdot | x, a)$;

namely:

12.3.1 Assumption. There is a constant k such that

$$\int_X w_0(y)Q(dy|x,a) \le kw(x,a) \quad \forall (x,a) \in \mathbb{K}.$$

This assumption is equivalent to saying that [as in Example 12.2.6(c)] the function

$$(x,a) \mapsto \int_X w_0(y)Q(dy|x,a) \quad \text{is in} \quad \mathbb{B}_w(\mathbb{K}).$$

The linear maps. In (12.2.17) and (12.2.24) replace $X \times Y$ by $\mathbb{K}$. This yields the linear map

$$L_0 : \mathbb{M}_w(\mathbb{K}) \to \mathbb{R} \quad \text{and} \quad L_1 : \mathbb{M}_w(\mathbb{K}) \to \mathbb{M}_{w_0}(X),$$

with

$$L_0\mu := \langle \mu, 1 \rangle = \mu(\mathbb{K}) \tag{12.3.12}$$

and

$$(L_1\mu)(B) := \widehat{\mu}(B) - \int_{\mathbb{K}} Q(B|x,a)\mu(d(x,a)) \quad \text{for} \quad B \in \mathcal{B}(X), \tag{12.3.13}$$

where $\widehat{\mu}$ denotes the marginal of μ on X—see (12.2.19). Finally, let

$$L : \mathbb{M}_w(\mathbb{K}) \to \mathbb{R} \times \mathbb{M}_{w_0}(X)$$

be the linear map in (12.2.25), i.e.,

$$L\mu := (L_0\mu, L_1\mu) \quad \text{for} \quad \mu \in \mathbb{M}_w(\mathbb{K}). \tag{12.3.14}$$

As in (12.2.27), the adjoint

$$L^* : \mathbb{R} \times \mathbb{B}_{w_0}(X) \to \mathbb{B}_w(\mathbb{K})$$

of L is given by

$$L^*(\rho, u)(x,a) := \rho + u(x) - \int_X u(y)Q(dy|x,a) \tag{12.3.15}$$

for every pair (ρ, u) in $\mathbb{R} \times \mathbb{B}_{w_0}(X)$ and (x,a) in $\mathbb{K}$. Hence, Assumption 12.3.1 and Proposition 12.2.5 yield [as in Example 12.2.6(d)] that

the linear map L in (12.3.14) is weakly continuous, $\qquad$ (12.3.16)

that is, continuous with respect to the weak topologies

$$\sigma(\mathbb{M}_w(\mathbb{K}), \mathbb{B}_w(\mathbb{K})) \quad \text{and} \quad \sigma(\mathbb{R} \times \mathbb{M}_{w_0}(X), \mathbb{R} \times \mathbb{B}_{w_0}(X)).$$

The linear programs. To complete the description of our linear program as at the beginning of Subsection 12.2.B, we introduce the "vectors"

$$b := (1,0) \text{ in } \mathbb{R} \times \mathbb{M}_{w_0}(X), \text{ and } c \text{ in } \mathbb{B}_w(\mathbb{K}),$$

where c is the cost-per-stage function, as well as the positive cone

$$K := \mathbb{M}_w(\mathbb{K})_+, \qquad (12.3.17)$$

whose dual cone is

$$K^* := \mathbb{B}_w(\mathbb{K})_+. \qquad (12.3.18)$$

[See Remark 12.2.8(b).] Then the **primal** linear program is

(P)   minimize $\langle \mu, c \rangle$

subject to: $L\mu = (1,0), \quad \mu \in \mathbb{M}_w(\mathbb{K})_+.$ (12.3.19)

More explicitly, by (12.3.12)–(12.3.14), the constraint (12.3.19) is satisfied if

$$\mu(\mathbb{K}) = 1 \quad \text{with} \quad \mu \in \mathbb{M}_w(\mathbb{K})_+, \qquad (12.3.20)$$

and $L_1\mu = 0$, i.e.,

$$\widehat{\mu}(B) - \int_{\mathbb{K}} Q(B|x,a)\mu(d(x,a)) = 0 \quad \forall B \in \mathcal{B}(X), \text{ with } \mu \in \mathbb{M}_w(\mathbb{K})_+. \qquad (12.3.21)$$

Observe that, in particular, (12.3.20) requires μ to be a *probability measure* (p.m.). Moreover, recalling Lemma 9.4.4 [see also Remark 12.3.2(a) below], (12.3.21) can be written as

$$\widehat{\mu}(B) = \int_X Q(B|x,\varphi)\widehat{\mu}(dx) \quad \forall B \in \mathcal{B}(X),$$

for some stochastic kernel $\varphi \in \Phi$, which means that μ *is feasible for* (P) *if* μ *is a p.m. on* $\mathbb{K}$ *such that its marginal* $\widehat{\mu}$ *on* X *is an i.p.m. for the transition kernel* $Q(\cdot|x,\varphi)$.

On the other hand, observe that

$$\langle b, w \rangle = \langle (1,0), (\rho, u) \rangle = \rho \quad \forall w = (\rho, u) \in \mathbb{R} \times \mathbb{B}_{w_0}(X).$$

Hence, by (12.3.18) and (12.3.15), the **dual** of (P) is [as in (12.2.33)]

(P*)   maximize ρ

subject to : $\rho + u(x) - \int_X u(y)Q(dy|x,a) \leq c(x,a)$ (12.3.22)

$\forall (x,a) \in \mathbb{K}, \text{ with } (\rho, u) \in \mathbb{R} \times \mathbb{B}_{w_0}(X).$

This completes the specification of the linear programs associated to the AC problem.

B. Solvability of (P)

Before proceeding to verify (12.3.3) and (12.3.4), let us note the following.

12.3.2 Remark. We will use the following *conventions*:

(a) A measure μ on $\mathbb{K} \subset X \times A$ may (and will) be viewed as a measure on all of $X \times A$ by defining $\mu(\mathbb{K}^c) := 0$, where $\mathbb{K}^c$ stands for the complement of $\mathbb{K}$ in $X \times A$.

(b) We will regard $c : \mathbb{K} \to \mathbb{R}_+$ as a function on all of $X \times A$ with $c(x, a) := +\infty$ if (x, a) is in $\mathbb{K}^c$. Observe that this convention is consistent with Assumption 11.4.1(c), and, moreover, by (12.3.5), the weight function $w = +\infty$ on $\mathbb{K}^c$. Any other function u in $\mathbb{B}_w(\mathbb{K})$ can be arbitrarily extended to $X \times A$, for example, as $u := 0$ on $\mathbb{K}^c$.

(c) $0 \cdot (+\infty) := 0$

(d) As in (12.2.20), a function u in $\mathbb{B}_{w_0}(X)$ will also be seen a function in $\mathbb{B}_w(\mathbb{K})$ given by $u(x, a) := u(x)$ for all (x, a) in $\mathbb{K}$.

Then, in particular, we may write the bilinear form in (12.3.8) as

$$\langle \mu, u \rangle = \int_{X \times A} u \, d\mu$$

for any measure μ in $\mathbb{M}_w(\mathbb{K})$ and any function u in $\mathbb{B}_w(\mathbb{K})$ or in $\mathbb{B}_{w_0}(X)$. $\square$

We will next show that (P) is **solvable** and that instead of (12.3.3) we have

$$\sup(\mathrm{P}^*) \le \rho_{\min} = \min(\mathrm{P}). \tag{12.3.23}$$

[Concerning part (b) in the next theorem, see Note 2 at the end of this section.]

12.3.3 Theorem. *Suppose that Assumptions 11.4.1 and 12.3.1 are satisfied. Then:*

(a) **[Solvability of (P)].** *There exists an optimal solution μ^* for (P), and*

$$\min(\mathrm{P}) = \rho_{\min} = \langle \mu^*, c \rangle. \tag{12.3.24}$$

(b) **[Consistency of (P*)].** *The dual problem (P*) is consistent and it satisfies the inequality in (12.3.23).*

Proof. (a) By Lemma 11.4.5(a), there exists a stable randomized policy φ_*^∞ such that $(\varphi_*^\infty, p_{\varphi_*})$ is a minimum pair. That is, by Definitions 11.4.3(a) and 11.1.1(a), p_{φ_*} is an i.p.m. for the transition kernel

$$Q_{\varphi_*}(B|x) := \int_A Q(B|x, a)\varphi_*(da|x),$$

and
$$J(\varphi^*, p_{\varphi_*}) = \int_X c_{\varphi_*}(x) p_{\varphi_*}(dx) = \rho_{\min} < \infty, \tag{12.3.25}$$
where
$$c_{\varphi_*}(x) := \int_A c(x, a) \varphi_*(da|x).$$

Furthermore, as p_{φ_*} is an i.p.m. for Q_{φ_*}, for every B in $\mathcal{B}(X)$ we have

$$p_{\varphi_*}(B) = \int_X Q_{\varphi_*}(B|x) p_{\varphi_*}(dx),$$

i.e.,

$$p_{\varphi_*}(B) = \int_X \int_A Q(B|x, a) \varphi_*(da|x) p_{\varphi_*}(dx). \tag{12.3.26}$$

Now let μ^* be the measure on $X \times A$ defined as

$$\mu^*(B \times C) := \int_B \varphi_*(C|x) p_{\varphi_*}(dx) \quad \forall B \in \mathcal{B}(X), \ C \in \mathcal{B}(A).$$

Then, by Definition 8.2.1 (of Φ), μ^* is a p.m. on $X \times A$, concentrated on $\mathbb{K}$ [that is, $\mu^*(\mathbb{K}) = 1$], and its marginal on X coincides with p_{φ_*}:

$$\widehat{\mu}^*(B) := \mu^*(B \times A) = p_{\varphi_*}(B) \quad \forall B \in \mathcal{B}(X).$$

It follows that we may rewrite (12.3.26) and (12.3.25) as

$$\widehat{\mu}^*(B) - \int_{\mathbb{K}} Q(B|x, a) \mu^*(d(x, a)) = 0 \quad \forall B \in \mathcal{B}(X),$$

and
$$J(\varphi^*, p_{\varphi_*}) = \langle \mu^*, c \rangle = \rho_{\min} < \infty, \tag{12.3.27}$$
which means that we already have the second equality in (12.3.24), as well as the equalities $\mu^*(\mathbb{K}) = 1$ and $L_1 \mu^* = 0$ in (12.3.20) and (12.3.21).

Therefore, to complete the proof of part (a) it suffices to show that

(i) μ^* is $\mathbb{M}_w(\mathbb{K})$ [see (12.2.2)], so that μ^* is indeed feasible for (P); and

(ii) $\langle \mu, c \rangle \geq \rho_{\min}$ for any feasible solution μ for (P), which would yield $\inf(P) \geq \rho_{\min}$.

In other words, (i), (ii) and (12.3.27) will give that μ^* is feasible for (P) and

$$\rho_{\min} = \langle \mu^*, c \rangle \geq \inf(P) \geq \rho_{\min}, \quad \text{i.e.,} \quad \langle \mu^*, c \rangle = \rho_{\min}.$$

Let us, then, prove (i), (ii).

Proof of (i). This is easy because, by (12.3.5) and (12.3.27),

$$\langle \mu^*, w \rangle = 1 + \langle \mu^*, c \rangle < \infty.$$

Proof of (ii). If μ satisfies (12.3.20) and (12.3.21), then, in particular, μ is a probability measure on $X \times A$ concentrated on $\mathbb{K}$; see Remark 12.3.2(a). Thus, by Lemma 9.4.4, there is a stochastic kernel $\varphi \in \Phi$ such that

$$\mu(B \times C) = \int_B \varphi(C|x)\widehat{\mu}(dx) \quad \forall B \in \mathcal{B}(X), \ C \in \mathcal{B}(A).$$

Furthermore, taking $(\varphi^\infty, p_\varphi) := (\varphi^\infty, \widehat{\mu})$, (12.3.21) gives that φ^∞ is a stable randomized policy, and, therefore, by (11.4.4) and the definition of $\rho_{\min}$,

$$\langle \mu, c \rangle = J(\varphi^\infty, \widehat{\mu}) \geq \rho_{\min}.$$

This proves (ii), which completes the proof of part (a).

(b) By (a) and the weak duality property (12.2.36), to prove (b) it suffices to show that (P*) is consistent. This, however, is obvious: for example, the pair (ρ, u) with $\rho = u(\cdot) \equiv 0$ satisfies (12.3.22). $\square$

C. Absence of duality gap

To continue with the program for this section, we now turn our attention to proving (12.3.4).

12.3.4 Theorem. (Absence of duality gap.) *If Assumptions 11.4.1 and 12.3.1 are satisfied, then (12.3.4) holds.*

Proof. We wish to use Theorem 12.2.10 with $\mathcal{Z}$ and L as in (12.3.7) and (12.3.14), respectively. Hence, we wish to show that the set

$$H := \{(L\mu, \langle \mu, c \rangle + r)|\mu \in \mathbb{M}_w(\mathbb{K})_+, \ r \geq 0\}$$

is closed in the weak topology

$$\sigma(\mathbb{R} \times \mathbb{M}_{w_0}(X) \times \mathbb{R}, \ \mathbb{R} \times \mathbb{B}_{w_0}(X) \times \mathbb{R}).$$

[See (12.2.1) for the bilinear form on a product of dual pairs, and Note 1 in §12.2 for the definition of "nets", which are used next.] Thus, let $(D, \leq)$ be a directed set, and consider a net $\{(\mu_\alpha, r_\alpha), \alpha \in D\}$ in $\mathbb{M}_w(\mathbb{K})_+ \times \mathbb{R}_+$ such that

$$L_0\mu_\alpha := \mu_\alpha(\mathbb{K}) \to r_* \tag{12.3.28}$$
$$\langle L_1\mu_\alpha, u \rangle \to \langle \nu_*, u \rangle \quad \forall u \in \mathbb{B}_{w_0}(X), \ \text{and} \tag{12.3.29}$$
$$\langle \mu_\alpha, c \rangle + r_\alpha \to \rho_*. \tag{12.3.30}$$

We will show that $((r_*, \nu_*), \rho_*)$ is in H; that is, there exists a measure μ in $\mathbb{M}_w(\mathbb{K})_+$ and a number $r \geq 0$ such that

$$r_* = L_0\mu := \mu(\mathbb{K}), \tag{12.3.31}$$
$$\nu_* = L_1\mu, \ \text{and} \tag{12.3.32}$$
$$\rho^* = \langle \mu, c \rangle + r. \tag{12.3.33}$$

We shall consider two cases, $r_* = 0$ and $r_* > 0$.

Case 1: $r_* = 0$. As μ_α is *nonnegative*,

$$L_0 \mu_\alpha = \mu_\alpha(\mathbb{K}) = \|\mu_\alpha\|_{TV}. \tag{12.3.34}$$

[See (12.2.18).] Therefore, if $r_* = 0$ in (12.3.28), it follows easily that (12.3.31)–(12.3.33) hold with $\mu(\cdot) = 0$ and $r = \rho_*$.

Case 2: $r_* > 0$. By (12.3.28) [together with (12.3.34)] and (12.3.30), there exists α_0 in D such that

$$\|\mu_\alpha\|_{TV} \le 2r_* \quad \text{and} \quad \langle \mu_\alpha, c \rangle \le 2\rho_* \quad \forall \alpha \ge \alpha_0. \tag{12.3.35}$$

Hence, as $\langle \mu_\alpha, c+1 \rangle = \langle \mu_\alpha, c \rangle + \|\mu_\alpha\|_{TV}$, we get that $\Gamma := \{\mu_\alpha, \alpha \ge \alpha_0\}$ is a *bounded* set of measures, which combined with Assumption 11.4.1(c) and Theorem 12.2.15 yields that Γ is *tight*. Moreover, if $\mu_\alpha(\mathbb{K}) > 0$, we may "normalize" μ_α rewriting it as $\mu_\alpha(\cdot)/\mu_\alpha(\mathbb{K})$, and so we may assume that Γ is a (tight) family of probability measures. Then, by Prohorov's Theorem 12.2.16, for each sequence $\{\mu_n\}$ in Γ there is a subsequence $\{\mu_m\}$ and a p.m. μ on $\mathbb{K}$ such that

$$\langle \mu_m, v \rangle \to \langle \mu, v \rangle \quad \forall v \in C_b(\mathbb{K}). \tag{12.3.36}$$

In particular, taking $v(\cdot) \equiv 1$, (12.3.28) yields that μ *satisfies (12.3.31)*. We will next show that

(i) μ is in $\mathbb{M}_w(\mathbb{K})_+$, that is, $\|\mu\|_w := \langle \mu, w \rangle < \infty$ [see (12.2.2)], and

(ii) μ satisfies (12.3.32).

Proof of (i). As $w := 1 + c$, to prove (i) we need to show that $\langle \mu, c \rangle$ is finite. We will prove the latter by showing that

$$[(12.3.36),\ c \ge 0\ \text{and l.s.c.}] \Rightarrow \liminf \langle \mu_m, c \rangle \ge \langle \mu, c \rangle. \tag{12.3.37}$$

Indeed, if $c \ge 0$ and l.s.c. [as in Assumption 11.4.1(b)], then there exists an increasing sequence of functions v_k in $C_b(\mathbb{K})$ such that $v_k \uparrow c$. It follows from (12.3.36) that for each k

$$\liminf_{m \to \infty} \langle \mu_m, c \rangle \ge \liminf_{m \to \infty} \langle \mu_m, v_k \rangle = \langle \mu, v_k \rangle.$$

Thus, letting $k \to \infty$, the Monotone Convergence Theorem gives (12.3.37).

Proof of (ii). As in Example 12.2.6(d), the weak continuity condition on Q [Assumption 11.4.1(d)] implies that the adjoint of L_1, namely,

$$(L_1^* u)(x, a) := u(x) - \int_X u(y) Q(dy|x, a),$$

maps $C_b(X)$ into $C_b(\mathbb{K})$. Therefore, (12.3.36) and (12.3.29) yield that for any function u in $C_b(X)$

$$
\begin{aligned}
\langle L_1\mu, u\rangle = \langle \mu, L_1^*u\rangle &= \lim_{m\to\infty} \langle \mu_m, L_1^*u\rangle \quad [\text{by } (12.3.36)]\\
&= \lim_{m\to\infty} \langle L_1\mu_m, u\rangle\\
&= \langle \nu_*, u\rangle \quad [\text{by } (12.3.29)].
\end{aligned}
$$

That is, $\langle L_1\mu, u\rangle = \langle \nu_*, u\rangle$ for any function u in $C_b(X)$, which implies (12.3.32). This proves (ii).

Summarizing, we have shown that μ is a measure in $\mathbb{M}_w(\mathbb{K})_+$ that satisfies (12.3.31) and (12.3.32). Finally, from (12.3.37) and (12.3.30) we see that

$$
\rho_* \geq \langle \mu, c\rangle + \liminf_{m\to\infty} r_m \geq \langle \mu, c\rangle \quad \text{as} \quad r_m \geq 0 \ \forall m.
$$

Thus, defining $r := \rho_* - \langle \mu, c\rangle$ (≥ 0), we conclude that μ and r satisfy (12.3.31), (12.3.32) and (12.3.33). This shows that H is indeed weakly closed, and so (12.3.4) follows. $\square$

Having (12.3.4), in the following sections we consider conditions for the solvability of the dual problem (P*) and for the convergence of approximations to the optimal values max(P*) and min(P). First, however, we shall conclude this section by showing a different approach to obtain the consistency of (P).

D. The Farkas alternative

Theorem 12.2.11 is important because, in particular, it can be used to obtain *necessary and sufficient conditions* for (P) to be consistent. It requires, however, that a certain set should be weakly closed, which turns out to be technically demanding. Hence, to apply Theorem 12.2.11 to our linear program (P) we shall proceed first to introduce a set of hypotheses, Assumption 12.3.5, different from Assumption 11.4.1, and then we shall proceed to "perturbate" (P) in a suitable form.

12.3.5 Assumption. Assumption 12.3.1 is satisfied and, in addition:

(a) X and $\mathbb{K}$ are locally compact separable metric (LCSM) spaces.

(b) The one-stage cost $c(x, a)$ is nonnegative and l.s.c.

(c) Q is weakly continuous [see Assumption 11.4.1(d)].

(d) For every compact subset C of X, the function $(x, a) \mapsto Q(C|x, a)$ vanishes at infinity; that is, for each $\varepsilon > 0$ there is a compact set $C' = C'(\varepsilon, C)$ such that $Q(C|x, a) \leq \varepsilon$ for all $(x, a) \notin C'$.

12.3.6 Remark. (a) Observe that Assumption 12.3.5 and Assumption 11.4.1 are not directly "comparable". For instance, the latter requires X

and $\mathbb{K}$ to be *Borel spaces*, which is a condition weaker than Assumption 12.3.5(a). We now need X and $\mathbb{K}$ to be *locally compact separable metric* (LCSM) *spaces* because we wish to consider dual pairs $(\mathcal{X}, \mathcal{Y})$ as in Remark 12.2.2(b), (c), and Remark 12.2.18. A sufficient condition for $\mathbb{K} \subset X \times A$ to be as in Assumption 12.3.5(a) is that X and A are both LCSM (which implies that $X \times A$ is LCSM) and that $\mathbb{K}$ is either open or closed in $X \times A$. For a proof of the latter fact see, for instance, Dieudonné [1, pp. 66, 75].

Similarly, Assumption 12.3.5(b) is weaker that 11.4.1(b), and, furthermore, Assumption 12.3.5 requires neither 11.4.1(a) nor 11.4.1(c), but it does require the "vanishing-at-infinity" condition 12.3.5(d).

(b) As in Remark 12.2.7(b) [see (12.2.29) and (12.2.30)], Assumptions 12.3.5(c) and 12.3.5(d) imply that

$$(x, a) \mapsto \int_X u(y)Q(dy|x, a) \text{ is in } C_0(\mathbb{K}) \text{ for each } u \in C_0(X). \quad (12.3.38)$$

For example, consider a general discrete-time system

$$x_{t+1} = F(x_t, a_t, z_t), \quad t = 0, 1, \ldots,$$

with values in, say, $X = \mathbb{R}^d$, and i.i.d. disturbances in $Z = \mathbb{R}^m$. Then, since

$$\int_X u(y)Q(dy|x, a) = Eu[F(x, a, z_0)],$$

Assumptions 12.3.5(c) and 12.3.5(d) are both satisfied if, for every z in Z, the function $F(x, a, z)$ is continuous in (x, a) and $F(x, a, z) \to \infty$ as $(x, a) \to \infty$. $\square$

A "perturbation" of (P). Let v_0 be a strictly positive function in $C_0(X)$, and consider the linear operator

$$\mathcal{L} : \mathbb{M}_w(\mathbb{K}) \times \mathbb{R}^2 \to \mathbb{M}_{w_0}(X) \times \mathbb{R}^2$$

defined as

$$\mathcal{L}(\mu, r_1, r_2) := (L_1\mu, \langle \mu, 1 \rangle + r_1, \langle \mu, v_0 \rangle - r_2), \quad (12.3.39)$$

with L_1 as in (12.3.13). [To write $\langle \mu, v_0 \rangle$ in (12.3.39) we have used Remark 12.3.2(d).] The adjoint

$$\mathcal{L}^* : \mathbb{B}_{w_0}(X) \times \mathbb{R}^2 \to \mathbb{B}_w(\mathbb{K}) \times \mathbb{R}^2$$

is given by

$$\mathcal{L}^*(u, \rho_1, \rho_2) = (L_1^*u + \rho_1 + \rho_2 v_0, \rho_1, -\rho_2). \quad (12.3.40)$$

We will next use $\mathcal{L}$ and the Generalized Farkas Theorem 12.2.11 to obtain the following.

12.3.7 Theorem [Equivalent formulations of the consistency of (P).] *If Assumption 12.3.5 holds, then the following statements are equivalent:*

(a) (P) *is consistent, that is, there is a measure μ that satisfies (12.3.19).*

(b) *The linear equation*

$$\mathcal{L}(\mu, r_1, r_2) = (0, 1, \varepsilon) \text{ has a solution } (\mu, r_1, r_2) \text{ in } \mathbb{M}_w(\mathbb{K})_+ \times \mathbb{R}_+^2$$

$$(12.3.41)$$

for some $\varepsilon > 0$.

(c) *The condition*

$$L_1^* u + \rho_1 + \rho_2 v_0 \geq 0 \text{ with } u \in \mathbb{B}_{wo}(X), \ \rho \geq 0, \text{ and } \rho_2 \leq 0 \quad (12.3.42)$$

implies

$$\rho_1 + \varepsilon \rho_2 \geq 0 \text{ for some } \varepsilon > 0. \tag{12.3.43}$$

In the proof of Theorem 12.3.7 we will use the following lemma, where we use the notation in Remark 12.2.2(a), (b), and Remark 12.3.2(d).

12.3.8 Lemma. *Suppose that Assumption 12.3.5(a) holds, and let $\{\mu_n\}$ be a bounded sequence of measures on $\mathbb{K}$. If μ_j converges to μ in the weak* topology $\sigma(\mathbb{M}(\mathbb{K}), C_0(\mathbb{K}))$, then the marginals $\widehat{\mu}_j$ on X converge to $\widehat{\mu}$ in the weak* topology $\sigma(\mathbb{M}(X), C_0(X))$; that is, if*

$$\langle \mu_j, v \rangle \to \langle \mu, v \rangle \quad \forall v \in C_0(\mathbb{K}), \tag{12.3.44}$$

then

$$\langle \widehat{\mu}_j, u \rangle \to \langle \widehat{\mu}, u \rangle \quad \forall u \in C_0(X). \tag{12.3.45}$$

Proof. Under Assumption 12.3.5(a), $\mathbb{K}$ is σ-compact; that is, there is an increasing sequence of compact sets $K_n \uparrow \mathbb{K}$. Moreover, by Urysohn's Lemma (see, for instance, Rudin [1], p. 39), for any given $\varepsilon > 0$ and each $n = 1, 2, \ldots$, there is a function α_n in $C_0(\mathbb{K})$ such that $0 \leq \alpha_n \leq 1$, with $\alpha_n = 1$ on K_n and $\alpha_n(x, a) = 0$ if the distance from (x, a) to K_n is $\geq \varepsilon$.

Now choose an arbitrary function u in $C_0(X)$ and define the functions

$$v_n(x, a) := \alpha_n(x, a) u(x) \quad \text{for} \quad (x, a) \in \mathbb{K}.$$

Then v_n is in $C_0(\mathbb{K})$ and for (x, a) in $\mathbb{K}$,

$$|v_n(x, a)| \leq |u(x)| \leq \|u\| (:= \text{sup norm of } u), \text{ and}$$
$$v_n(x, a) \to u(x) \text{ as } n \to \infty.$$

Hence, by the Bounded Convergence Theorem, for every fixed j,

$$\lim_{n \to \infty} \langle \mu_j, v_n \rangle = \langle \mu_j, u \rangle = \langle \widehat{\mu}_j, u \rangle \tag{12.3.46}$$

and, on the other hand,

$$\lim_{n \to \infty} \langle \mu, v_n \rangle = \langle \mu, u \rangle = \langle \widehat{\mu}, u \rangle, \tag{12.3.47}$$

where in the latter equality we have used Remark 12.3.2(d) to write

$$\langle \mu, u \rangle = \int_{\mathbb{K}} u d\mu = \int_X u d\widehat{\mu} = \langle \widehat{\mu}, u \rangle \text{ for } u \in C_0(X), \qquad (12.3.48)$$

and similarly for the second equality in (12.3.46). Moreover, for every fixed n, (12.3.44) yields

$$\lim_{j \to \infty} \langle \mu_j, u_n \rangle = \langle \mu, u_n \rangle. \qquad (12.3.49)$$

Finally, the desired conclusion (12.3.45) follows from (12.3.46)–(12.3.49) and the inequality [which uses (12.3.48) again]

$$|\langle \widehat{\mu}_j, u \rangle - \langle \widehat{\mu}, u \rangle| = |\langle \mu_j, u \rangle - \langle \mu, u \rangle|$$
$$\leq \ |\langle \mu_j, u \rangle - \langle \mu_j, v_n \rangle| + |\langle \mu_j, v_n \rangle - \langle \mu, v_n \rangle| + |\langle \mu, v_n \rangle - \langle \mu, u \rangle|. \ \square$$

We are now ready for the proof of Theorem 12.3.7.

Proof of Theorem 12.3.7. (a) $\Leftrightarrow$ (b). If μ satisfies (12.3.19), then

$$L\mu := (\langle \mu, 1 \rangle, L_1\mu) = (1, 0), \text{ i.e., } \langle \mu, 1 \rangle = 1 \text{ and } L_1\mu = 0,$$

and so $(\mu, 0, 0)$ satisfies (12.3.41) with $\varepsilon := \langle \mu, v_0 \rangle$. Conversely, suppose that (μ, r_1, r_2) satisfies (12.3.41), that is,

$$L_1\mu = 0, \ \langle \mu, 1 \rangle + r_1 = 1, \text{ and } \langle \mu, v_0 \rangle - r_2 = \varepsilon.$$

Then, in particular, $\langle \mu, v_0 \rangle \geq \varepsilon$, which implies that $\langle \mu, 1 \rangle = \mu(\mathbb{K}) > 0$. Therefore, the measure $\mu^* := \mu/\langle \mu, 1 \rangle$ satisfies (12.3.19).

(b) $\Leftrightarrow$ (c). In this proof we use the Generalized Farkas Theorem 12.2.11 with the following identifications:

$$\begin{aligned}
(\mathcal{X}, \mathcal{Y}) &:= (\mathbb{M}_w(\mathbb{K}) \times \mathbb{R}^2, \mathbb{B}_w(\mathbb{K}) \times \mathbb{R}^2), \\
(\mathcal{Z}, \mathcal{W}) &:= (\mathbb{M}_{w_0}(X) \times \mathbb{R}^2, \mathbb{B}_{w_0}(X) \times \mathbb{R}^2), \\
K &:= \mathbb{M}_w(\mathbb{K})_+ \times \mathbb{R}^2_+, \qquad\qquad\qquad (12.3.50) \\
L &:= \mathcal{L} \text{ [in (12.3.39)], and } b := (0, 1, \varepsilon).
\end{aligned}$$

Then condition (a) in Theorem 12.2.11 turns out to be precisely (12.3.41), whereas condition (b) in Theorem 12.2.11 becomes

$$\mathcal{L}^*(u, \rho_1, \rho_2) \geq 0 \Rightarrow \langle (0, 1, \varepsilon), (u, \rho_1, \rho_2) \rangle \geq 0,$$

which is the same as "the condition (12.3.42) implies (12.3.43)". Therefore, to prove the equivalence of (b) and (c) in Theorem 12.3.7 we only need to verify the hypotheses of Theorem 12.2.11, namely:

(a) $\mathcal{L}$ *is weakly continuous*,

(b) $\mathcal{L}(K)$ *is weakly closed*.

In fact, part (i) is obvious because the adjoint $\mathcal{L}^*$ [in (12.3.40)] maps $\mathcal{W}$ into $\mathcal{Y}$—see Proposition 12.2.5. Thus, it only remains to prove (ii).

Proof of (ii). To prove that $\mathcal{L}(K)$ is closed, with K as in (12.3.50), consider a directed set $(D, \leq)$, and a net $\{(\mu^\alpha, r_1^\alpha, r_2^\alpha), \alpha \in D\}$ in K such that $\mathcal{L}(\mu_1^\alpha, r_1^\alpha, r_2^\alpha)$ converges weakly to, say, (ν, ρ_1, ρ_2) in $\mathbb{M}_{w_0}(X) \times \mathbb{R}^2$; that is,

$$\langle L_1\mu^\alpha, u \rangle \to \langle \nu, u \rangle \quad \forall u \in \mathbb{B}_{w_0}(X), \qquad (12.3.51)$$

$$\langle \mu^\alpha, 1 \rangle + r_1^\alpha \to \rho_1, \text{ and} \qquad (12.3.52)$$

$$\langle \mu^\alpha, v_0 \rangle - r_2^\alpha \to \rho_2. \qquad (12.3.53)$$

We wish to show that the limiting triplet (ν, ρ_1, ρ_2) is in $\mathcal{L}(K)$; that is, there exists (μ^0, r_1^0, r_2^0) in K such that

$$L_1\mu^0 = \nu, \ \langle \mu^0, 1 \rangle + r_1^0 = \rho_1, \ \langle \mu^0, v_0 \rangle - r_2^0 = \rho_2. \qquad (12.3.54)$$

We shall consider two cases, $\rho_1 = 0$ and $\rho_1 > 0$.

If $\rho_1 = 0$, then (12.3.52) yields, in particular,

$$\langle \mu^\alpha, 1 \rangle = \mu^\alpha(\mathbb{K}) \to 0.$$

which in turn yields $\nu = 0$, by (12.3.51) and the weak continuity of L_1. Hence (12.3.54) holds for $(\mu^0, r_1^0, r_2^0) = (0, 0, -\rho_2)$. Let us now consider the case $\rho_1 > 0$.

Suppose that $\rho_1 > 0$. Then, by (12.3.52), there exists $\alpha_0 \in D$ such that

$$0 \leq \langle \mu^\alpha, 1 \rangle = \mu^\alpha(\mathbb{K}) = \|\mu^\alpha\|_{TV} \leq 2\rho_1 \quad \forall \alpha \geq \alpha_0. \qquad (12.3.55)$$

Hence, by Remark 12.2.2(b), (c) there is a (nonnegative) measure μ^0 on $\mathbb{K}$ and a sequence $\{j\}$ in D such that $\mu^j \to \mu^0$ in the weak* topology $\sigma(\mathbb{M}(\mathbb{K}), C_0(\mathbb{K}))$, i.e.,

$$\langle \mu^j, v \rangle \to \langle \mu^0, v \rangle \quad \forall v \in C_0(\mathbb{K}). \qquad (12.3.56)$$

Moreover, μ^0 is in $\mathbb{M}(\mathbb{K})_+$ since, by (12.3.55) and Proposition 12.2.3,

$$\mu^0(\mathbb{K}) = \|\mu^0\|_{TV} \leq \liminf_{j\to\infty} \|\mu^j\|_{TV} \leq 2\rho_1.$$

On the other hand, from (12.3.56), Lemma 12.3.8 and (12.3.8) we get that $L_1\mu^j$ converges to $L_1\mu^0$ in the weak* topology $\sigma(\mathbb{M}(X), C_0(X))$, i.e.,

$$\langle L_1\mu^j, u \rangle = \langle \mu^j, L_1^* u \rangle \to \langle \mu^0, L_1^* u \rangle = \langle L_1\mu^0, u \rangle \quad \forall u \in C_0(X).$$

This fact and (12.3.51) yield the first equality in (12.3.54), $L_1\mu^0 = \nu$. Finally, the second and third equalities in (12.3.54) hold with

$$r_1^0 := \rho_1 - \langle \mu^0, 1 \rangle \text{ and } r_2^0 := \langle \mu^0, v_0 \rangle - \rho_2,$$

which concludes the proof of (ii), and Theorem 12.3.7 follows. □

From Theorem 12.3.7 we can obtain a sufficient condition for (P) to be consistent under Assumption 12.3.5, as well as a connection with Theorem 12.2.17—see Remark 12.3.10.

12.3.9 Corollary. [A sufficient condition for the consistency of (P).] *Suppose that Assumption 12.3.5 is satisfied, and let v_0 be as in (12.3.39), a strictly positive function in $C_0(X)$. Let w_0 be the weight function on X and suppose, in addition, that there exists $\varepsilon > 0$, a randomized stationary policy φ^∞, and an initial state $\bar{x} \in X$ such that*

$$\liminf_{n \to \infty} E_{\bar{x}}^{\varphi^\infty}[w_0(x_n)]/n = 0 \qquad (12.3.57)$$

and

$$\liminf_{n \to \infty} \frac{1}{n} \sum_{t=0}^{n-1} E_{\bar{x}}^{\varphi^\infty}[v_0(x_t)] \geq \varepsilon. \qquad (12.3.58)$$

Then (P) *is consistent.*

Proof. The idea is to prove that part (c) in Theorem 12.3.7 is satisfied. More precisely, we wish to show that, under (12.3.57) and (12.3.58), the condition (12.3.42) implies (12.3.43). To do this, let us first note that (12.3.57) yields

$$\liminf_{n \to \infty} E_{\bar{x}}^{\varphi^\infty}|u(x_n)|/n = 0 \quad \forall u \in \mathbb{B}_{w_0}(X), \qquad (12.3.59)$$

which is obvious because $|u(\cdot)| \leq \|u\|_{w_0} w_0(\cdot)$ for u in $\mathbb{B}_{w_0}(X)$.

Let us now use the definition of L_1^* [see (12.2.26)] to rewrite (12.3.42) as

$$u(x) \geq \int_X u(y)Q(dy|x,a) - \rho_1 - \rho_2 v_0(x) \quad \forall x \in X,\ a \in A(x),$$

which integrated with respect to $\varphi(\cdot|x)$ yields

$$u(x) \geq \int_X u(y)Q(dy|x,\varphi) - \rho_1 - \rho_2 v_0(x) \quad \forall x \in X.$$

Iteration of the latter inequality gives, for all $x \in X$ and $n = 1, 2, \ldots,$

$$u(x) \geq E_x^{\varphi^\infty} u(x_n) - n\rho_1 - \rho_2 \sum_{t=0}^{n-1} E_x^{\varphi^\infty} v_0(x_t),$$

i.e.,

$$u(x) + n\rho_1 + \rho_2 \sum_{t=0}^{n-1} E_x^{\varphi^\infty} v_0(x_t) \geq E_x^{\varphi^\infty} u(x_n).$$

Thus, multiplying by $1/n$ and taking $\liminf$ as $n \to \infty$, (12.3.59) and (12.3.58) give (12.3.43) as $\rho_2 \leq 0$. □

12.3.10 Remark. The connection between Corollary 12.3.9 and Theorem 12.2.17 (see also Remark 12.2.18) is as follows. Let $\varepsilon > 0$ in (12.3.58) be such that $\varepsilon \leq \|v_0\|$, and choose $0 < \varepsilon_0 < \varepsilon$. As $v_0(\cdot) > 0$ is in $C_0(X)$, there is a compact set C in X such that $0 < v_0(x) \leq \varepsilon_0$ for all x not in C. Then, writing X as the union of C and its complement, for each $x \in X$ and $t = 0, 1, \ldots$ we obtain

$$
\begin{aligned}
E_x^{\varphi^\infty}[v_0(x_t)] &= \int_X v_0(y) Q^t(dy|x, \varphi) \\
&\leq (\|v_0\| - \varepsilon_0) Q^t(C|x, \varphi) + \varepsilon_0.
\end{aligned}
$$

Therefore, with $\bar{x}$ as in (12.3.58),

$$
\liminf_{n \to \infty} \frac{1}{n} \sum_{t=0}^{n-1} Q^t(C|\bar{x}, \varphi) \geq (\varepsilon - \varepsilon_0)/(\|v_0\| - \varepsilon_0) > 0, \qquad (12.3.60)
$$

which gives part (d) in Theorem 12.2.17 for the transition kernel $P(\cdot|x) := Q(\cdot|x, \varphi)$ on $S := X$, with $\nu := \delta_{\bar{x}}$ and the compact set $K := C$. Thus, part (a) in Theorem 12.2.17 implies the existence of an i.p.m. $\hat{\mu}$ for $Q(\cdot|x, \varphi)$, and if we could show that the p.m. $\mu(d(x, a)) := \varphi(da|x)\hat{\mu}(dx)$ is in $\mathbb{M}_w(\mathbb{K})$, then we would have the same conclusion of Corollary 12.3.9, the consistency of (P), by a quite different approach. Finally, it is worth noting (and easy to prove) that (12.3.57) and (12.3.58) are also *necessary* for (P) to be consistent.

For further comments on—and references related to—Theorem 12.2.17 see Hernández-Lerma and Lasserre [2]. □

12.3.11 Remark. (Absence of duality gap.) Theorem 12.3.4 remains valid if Assumption 12.3.1 holds, but Assumption 11.4.1 is replaced with:

(a) Assumption 12.3.5 is satisfied;

(b) $c(x, a)$ is inf-compact [see Remark 11.4.2(a$_2$)];

(c) (P) is consistent.

The proof of Theorem 12.3.4 is the same under this new set of hypotheses. □

Notes on §12.3

1. In subsections A, B, C we essentially followed Hernández-Lerma and Lasserre [14]. Subsection D comes from Hernández-Lerma and González-Hernández [1]. The latter reference and also Hordijk and Lasserre [1] deal with AC problems in the *multichain* case.

The approach in this section to prove (12.3.4) is quite different from the approach in Chapter 6, where we used a variant of the "vanishing discount" approach. Namely, for each discount factor $0 < \alpha < 1$ we introduced a linear program, say (P_α), related to the α-discount Markov control problem, and then we studied (P) as the "limit" of (P_α) as $\alpha \uparrow 1$.

2. Historically speaking, it is interesting to note that the LP formulation of the AC problem—as well as to other Markov control problems—was born trying to solve the corresponding dynamic programming equation, which in our present case is the *Average Cost Optimality Equation* (ACOE)

$$\rho^* + h^*(x) = \min_{A(x)} \left[c(x,a) + \int_X h^*(y)Q(dy|x,a) \right], \tag{12.3.61}$$

where [by (11.4.7) and (11.1.23)]

$$\rho^* := \inf_X J^*(x) = \rho_{\min}. \tag{12.3.62}$$

Observe that if h^* is a function in $\mathbb{B}_{w_0}(X)$, then the ACOE (12.3.61) implies that the pair (ρ^*, h^*) satisfies (12.3.22), that is, (ρ^*, h^*) is feasible for the dual program (P*). On the other hand, if (ρ, u) satisfies (12.3.22), then using straightforward arguments [or using (12.3.23)] one can see that

$$\rho \le \rho^* (= \rho_{\min}).$$

It is due to the latter inequality that the pairs (ρ, u) that are feasible for (P*) are also called **subsolutions to the ACOE**. Thus the equality

$$\sup(\mathrm{P}^*) = \rho^* (= \rho_{\min})$$

in (12.3.4) is sometimes stated in the stochastic control literature saying that "ρ^* is the supremum of the subsolutions to the ACOE".

See Chapter 6 for early references (going back to about 1960) on the LP formulation of Markov control problems.

12.4 Approximating sequences and strong duality

In the rest of this chapter we are mainly interested in the approximation of the AC-related linear program (P) and its dual (P*). In this section we first study minimizng sequences for (P), and then maximizing sequences for (P*).

A. Minimizing sequences for (P)

By Definition 12.2.12(a), a sequence of measures μ_n in $\mathbb{M}_w(\mathbb{K})_+$ is said to be a **minimizing sequence** for (P) if each μ_n is feasible for (P), that is, it satisfies (12.3.19), and in addition

$$\langle \mu_n, c \rangle \downarrow \min(\mathrm{P}), \tag{12.4.1}$$

where we have used that (P) is *solvable* [Theorem 12.3.3(a)] to write its value as min (P) rather than inf (P).

12.4.1 Theorem. *Suppose that Assumptions 11.4.1 and 12.3.1 are satisfied. I' $\{\mu_n\}$ is a minimizing sequence for* (P), *then there exists a subsequence $\{j\}$ of $\{n\}$ such that $\{\mu_j\}$ converges in the weak topology $\sigma(\mathbb{M}(\mathbb{K}),$ $C_b(\mathbb{K}))$ to an optimal solution for* (P).

Proof. let $\{\mu_n\}$ be a minimizing sequence for (P); that is [by (12.3.19)],

$$\langle \mu_n, 1 \rangle = 1 \quad \text{and} \quad L_1 \mu_n = 0 \quad \forall n, \tag{12.4.2}$$

and (12.4.1) holds. In particular, (12.4.1) implies that for any given $\varepsilon > 0$ there exists $n(\varepsilon)$ such that

$$\min(P) \le \langle \mu_n, c \rangle \le \min(P) + \varepsilon \quad \forall n \ge n(\varepsilon). \tag{12.4.3}$$

By the second inequality [together with Assumption 11.4.1(c) and Theorems 12.2.15 and 12.2.16], there exists a p.m. μ^* on $\mathbb{K}$ and a subsequence $\{j\}$ of $\{n\}$ such that

$$\langle \mu_j, v \rangle \to \langle \mu^*, v \rangle \quad \forall v \in C_b(\mathbb{K}). \tag{12.4.4}$$

Moreover, by (12.3.37),

$$\langle \mu^*, c \rangle \le \liminf_{j \to \infty} \langle \mu_j, c \rangle \le \min(P) + \varepsilon. \tag{12.4.5}$$

Thus, as ε was arbitrary, the latter inequality and (12.4.3) yield

$$\min(P) = \langle \mu^*, c \rangle. \tag{12.4.6}$$

This will prove that μ^* is optimal for (P) provided that μ^* is *feasible* for (P); in other words, provided that μ^* is a measure in $\mathbb{M}_w(\mathbb{K})_+$ and that

$$L\mu^* = (\langle \mu^*, 1 \rangle, L_1 \mu^*) = (1, 0). \tag{12.4.7}$$

This, however, is obvious because (12.4.5) yields $\langle \mu^*, w \rangle = 1 + \langle \mu^*, c \rangle < \infty$, whereas (12.4.7) follows from (12.4.2) and (12.4.4). $\square$

B. Maximizing sequences for (P*)

By Definition 12.2.12(b) and the definition of the dual program (P*) [see (12.3.22)], a sequence (ρ_n, u_n) in $\mathbb{R} \times \mathbb{B}_{w_0}(X)$ is a maximizing sequence for (P*) if

$$\rho_n + u_n(x) \le c(x, a) + \int_X u_n(y) Q(dy | x, a) \tag{12.4.8}$$

for all n and $(x, a) \in \mathbb{K}$, and, in addition,

$$\rho_n = \langle (1, 0), (\rho_n, u_n) \rangle \uparrow \sup(P^*). \tag{12.4.9}$$

The following theorem shows that the existence of a suitable maximizing sequence for (P*) implies, in particular, that the *strong duality* condition for (P) holds [see (12.2.39)].

12.4.2 Theorem. [**Solvability of (P*), strong duality and the ACOE.**] *Suppose that Assumptions 11.4.1 and 12.3.1 are satisfied, and, furthermore, there exists a maximizing sequence (ρ_n, u_n) for (P*) with $\{u_n\}$ bounded in the w_0-norm, that is,*

$$\|u_n\|_{w_0} \le k \quad \forall n, \tag{12.4.10}$$

for some constant k. Then:

(a) *The dual problem (P*) is solvable.*

(b) *The strong duality condition holds, that is, $\max(P^*) = \min(P)$.*

(c) *If μ^* is an optimal solution for the primal program (P), then the ACOE (12.3.61) holds $\widehat{\mu}^*$-a.e., where $\widehat{\mu}^*$ is the marginal of μ^* on X; in fact, there is a function h^* in $\mathbb{B}_{w_0}(X)$ and a deterministic stationary policy f_*^∞ such that*

$$\begin{aligned}\rho^* + h^*(x) &= \min_{A(x)} \left[c(x,a) + \int_X h^*(y)Q(dy|x,a) \right] \\ &= c(x, f_*) + \int_X h^*(y)Q(dy|x, f_*)\end{aligned} \tag{12.4.11}$$

for $\widehat{\mu}^*$-almost all $x \in X$.

Proof. (a) By (12.3.62) and Theorem 12.3.4 we have

$$\sup(P^*) = \rho^* = \min(P) \tag{12.4.12}$$

and, moreover, we can write (12.4.9) as

$$\rho_n \uparrow \rho^*. \tag{12.4.13}$$

Now define the function

$$h^*(x) := \limsup_{n \to \infty} u_n(x),$$

which belongs to $\mathbb{B}_{w_0}(X)$, by (12.4.10). Therefore [by (12.4.13) and Fatou's Lemma 8.3.7(b)] taking $\limsup_n$ in (12.4.8) we obtain

$$\rho^* + h^*(x) \le c(x,a) + \int_X h^*(y)Q(dy|x,a) \quad \forall(x,a) \in \mathbb{K}.$$

This yields that (ρ^*, h^*) is feasible for (P*) [see (12.3.22)], which together with the first equality in (12.4.12) shows that (ρ^*, h^*) is in fact *optimal for* (P*).

(b) This part follows from (a) and (12.4.12).

(c) Let us first note that if μ is feasible for (P) and (ρ, u) is feasible for (P*), then

$$\langle L\mu, (\rho, u)\rangle = \langle (1,0), (\rho, u)\rangle = \rho,$$

or, equivalently,

$$\langle \mu, L^*(\rho, u)\rangle = \rho, \qquad (12.4.14)$$

where L^* is the adjoint of L, in (12.3.15). Now let μ^* be an optimal solution for (P), and (ρ^*, h^*) an optimal solution for (P*). By part (b) we have

$$\langle \mu^*, c\rangle = \rho^*,$$

whereas (12.4.14) gives

$$\langle \mu^*, L^*(\rho^*, h^*)\rangle = \rho^*.$$

Thus, subtracting the last two equalities we get

$$\langle \mu^*, c - L^*(\rho^*, h^*)\rangle = 0,$$

i.e.,

$$\int [c(x,a) - L^*(\rho^*, h^*)(x,a)]\mu^*(d(x,a)) = 0. \qquad (12.4.15)$$

By Lemma 9.4.4 we may disintegrate μ^* as $\mu^*(d(x,a)) = \varphi(da|x)\widehat{\mu}^*(dx)$ for some stochastic kernel $\varphi \in \Phi$, and then [using (12.3.15)] we can rewrite (12.4.15) as

$$\int_X \left[c(x, \varphi) - \rho^* - h^*(x) + \int_X h^*(y)Q(dy|x, \varphi)\right] \widehat{\mu}^*(dx) = 0.$$

Therefore, as the integrand is *nonnegative* [by (12.3.22)], we get that for $\widehat{\mu}^*$-a.e. (almost all) x in X

$$\begin{aligned}
\rho^* + h^*(x) &= c(x, \varphi) + \int_X h^*(y)Q(dy|x, \varphi)\\
&= \int_A \left[c(x,a) + \int_X h^*(y)Q(dy|x,a)\right]\varphi(da|x),
\end{aligned}$$

and so

$$\rho^* + h^*(x) \geq c(x, f_*) + \int_X h^*(y)Q(dy|x, f_*) \quad \widehat{\mu}^* - a.a. \ x \in X \quad (12.4.16)$$

for some decision function $f_* \in \mathbb{F}$ whose existence is guaranteed by Lemma 9.4.7. Finally, as (12.3.22) implies

$$\rho^* + h^*(x) \leq \min_{A(x)} \left[c(x,a) + \int_X h^*(y)Q(dy|x,a)\right] \quad \text{for all } (x,a) \in \mathbb{K},$$

we get that, by (12.4.16), for $\widehat{\mu}^*$-a.a. $x \in X$

$$
\begin{aligned}
\rho^* + h^*(x) &\geq c(x, f_*) + \int_X h^*(y)Q(dy|x, f_*) \\
&\geq \min_{A(x)} \left[c(x, a) + \int_X h^*(y)Q(dy|x, a) \right] \\
&\geq \rho^* + h^*(x),
\end{aligned}
$$

and (12.4.11) follows. $\square$

12.4.3 Remark. Theorem 12.4.1 and 12.4.2 remain valid if Assumption 12.3.1 holds, but Assumption 11.4.1 is replaced by the conditions (a), (b), and (c) in Remark 12.3.11. $\square$

Notes on §12.4

1. The results in this sections are from Hernández-Lerma and González-Hernández [1], in which similar results for *multichain* AC-problems are also obtained.

2. Minimizing sequences and policy iteration. Let f^∞ be a deterministic stationary policy for which the transition law $Q_f(\cdot|x) \equiv Q(\cdot|x, f)$ admits an i.p.m. μ_f, that is,

$$
\mu_f(B) = \int_X Q(B|x, f)\mu_f(dx) \quad \forall B \in \mathcal{B}(X). \tag{12.4.17}
$$

Now, for every $x \in X$, let $\delta_{f(x)}(\cdot)$ be the Dirac measure at $f(x)$, and let μ^f be the p.m. on $X \times A$, concentrated on $\mathbb{K}$, given by

$$
\mu^f(B \times C) := \int_B \delta_{f(x)}(C)\mu_f(dx) = \int_B I_C[f(x)]\mu_f(dx)
$$

for B and C in $\mathcal{B}(X)$ and $\mathcal{B}(A)$, respectively. Then the marginal of μ^f on X is $\widehat{\mu}^f = \mu_f$, and, on the other hand,

$$
\langle \mu^f, c \rangle = \int_{\mathbb{K}} c(x, a)\mu^f(d(x, a)) = \int_X c_f(x)\mu_f(dx), \tag{12.4.18}
$$

where $c_f(x) \equiv c(x, f) \equiv c(x, f(x))$ for all $x \in X$. Finally, let $\{f_n^\infty\}$ be the sequence of deterministic stationary policies defined by (10.5.17), (10.5.18). Thus, under the assumptions of Theorem 10.5.2, the sequence $\{\mu^{f_n}\}$ can be seen as a *minimizing sequence* for (P). In particular, observe that (12.4.17) is the same as the equation after (12.3.21), with $\varphi := f$ and $\widehat{\mu} := \mu_f$.

Similarly, it can be seen that the *value iteration* procedure in §5.6 gives a *maximizing sequence* for (P*).

12.5 Finite LP approximations

We will now show a procedure to approximate the AC-related primal linear program (P) by *finite-dimensional* linear programs. For the sake of continuity in the exposition, in this section we describe the procedure and the proof of its convergence is postponed to §12.6.

We will work in essentially the same setting of the previous sections except that now we shall require the spaces X and $\mathbb{K}$ to be *locally compact separable metric* (LCSM) spaces. Hence throughout the following we suppose:

12.5.1 Assumption. Assumptions 11.4.1 and 12.3.1 are satisfied, and in addition X and $\mathbb{K}$ are LCSM spaces.

A sufficient condition for $\mathbb{K}$ to be LCSM is given in Remark 12.3.6(a). On the other hand, the hypothesis that X and $\mathbb{K}$ are LCSM spaces ensures that $C_0(X)$ and $C_0(\mathbb{K})$ are both *separable* Banach spaces [see Remark 12.2.2(b)]. In particular, $C_0(X)$ contains a *countable* subset $\mathcal{C}(X)$ which is dense in $C_0(X)$. This is a key fact to proceed with the first step of our approximation procedure.

A. Aggregation

Let $\mathcal{P}_w(\mathbb{K})$ be the family of *probability measures* (p.m.'s) in $\mathbb{M}_w(\mathbb{K})_+$, which, in other words, is the family of measures μ that satisfy the constraint $\langle \mu, 1 \rangle = \mu(\mathbb{K}) = 1$ in (12.3.19), (12.3.20). Thus we may rewrite (P) as:

(P)   minimize $\langle \mu, c \rangle$

$$\text{subject to: } L_1\mu = 0, \quad \mu \in \mathcal{P}_w(\mathbb{K}), \qquad (12.5.1)$$

where $L_1\mu$ is the signed measure in $\mathbb{M}_{w_0}(X) \subset \mathbb{M}(X)$ defined by (12.3.13). We also have:

12.5.2 Lemma. *Let $\mathcal{C}(X) \subset C_0(X)$ be a countable dense subset of $C_0(X)$. Then the following are equivalent conditions for μ in $\mathcal{P}_w(\mathbb{K})$:*

(a) $L_1\mu = 0$.

(b) $\langle L_1\mu, u \rangle = 0 \quad \forall u \in C_0(X)$.

(c) $\langle L_1\mu, u \rangle = 0 \quad \forall u \in \mathcal{C}(X)$.

Proof. The equivalence of (a) and (b) is due to the fact that $(\mathbb{M}(X), C_0(X))$ is a dual pair—in fact, $\mathbb{M}(X)$ is the topological dual of $C_0(X)$ [Remark 12.2.2(b)]. Finally, the implication (b) $\Rightarrow$ (c) is obvious, whereas the converse follows from the denseness of $\mathcal{C}(X)$ in $C_0(X)$. $\square$

By Lemma 12.5.2, we may further rewrite (P) in the equivalent form:

(P)   minimize $\langle \mu, c \rangle$

subject to: $\langle L_1 \mu, u \rangle = 0$ $\forall u \in C(X); \; \mu \in \mathcal{P}_w(\mathbb{K})$.   (12.5.2)

Observe that (12.5.2) defines an *aggregation* (of constraints) of (P); see Definition 12.2.13(a). In other words, the constraint $L_1 \mu = 0$ in (12.5.1) is "aggregated" into *countably* many constraints $\langle L_1 \mu, u \rangle = 0$ with u in $C(X)$. We will next reaggregate (12.5.2) into *finitely* many constraints as follows.

Let $\{\mathcal{C}_k\}$ be an increasing sequence of *finite* sets $\mathcal{C}_k \uparrow C(X)$. For each k, consider the aggregation

$\mathbb{P}(\mathcal{C}_k)$ minimize $\langle \mu, c \rangle$

subject to: $\langle L_1 \mu, u \rangle = 0$ $\forall u \in \mathcal{C}_k; \; \mu \in \mathcal{P}_w(\mathbb{K})$.   (12.5.3)

This linear program has indeed a *finite number of constraints*, namely, the cardinality $|\mathcal{C}_k|$ of $\mathcal{C}_k$. We also have our first approximation result:

12.5.3 Theorem. *Suppose that Assumption 12.5.1 is satisfied. Then*

(a) $\mathbb{P}(\mathcal{C}_k)$ *is solvable for each* $k = 1, 2, \ldots$; *in fact, the aggregation* $\mathbb{P}(W)$ *is solvable for any subset* W *of* $C_0(X)$.

(b) *For each* $k = 1, 2, \ldots$, *let* μ_k *be an optimal solution for* $\mathbb{P}(\mathcal{C}_k)$, *i.e.,*

$$\langle \mu_k, c \rangle = \min \mathbb{P}(\mathcal{C}_k).$$

Then

$$\langle \mu_k, c \rangle \uparrow \min(\mathrm{P}) = \rho_{\min},   (12.5.4)$$

where the equality is due to Theorem 12.3.3(a). Furthermore, there is a subsequence $\{\mu_m\}$ *of* $\{\mu_k\}$ *that converges in the weak topology* $\sigma(\mathrm{M}(\mathbb{K}), C_b(\mathbb{K}))$ *to an optimal solution* μ^* *for* (P), *i.e.,*

$$\langle \mu_m, v \rangle \to \langle \mu^*, v \rangle   \forall v \in C_b(\mathbb{K});   (12.5.5)$$

in fact, any weak-$\sigma(\mathrm{M}(\mathbb{K}), C_b(\mathbb{K}))$ *accumulation point of* $\{\mu_k\}$ *is an optimal solution for* (P).

Proof. See §12.6.

B. Aggregation-relaxation

The *equality* constraint $\langle L_1 \mu, u \rangle = 0$ in (12.5.3) will now be "relaxed" to inequalities of the form $|\langle L_1 \mu, u \rangle| \le \varepsilon$ with $\varepsilon > 0$.

Let $\mathcal{C}_k \uparrow C(X)$ be as in (12.5.3), and let $\{\varepsilon_k\}$ be a sequence of numbers $\varepsilon_k \downarrow 0$. For each $k = 1, 2, \ldots$, consider the linear program

$\mathbb{P}(\mathcal{C}_k, \varepsilon_k)$ minimize $\langle \mu, c \rangle$

subject to: $|\langle L_1\mu, u \rangle| \leq \varepsilon_k$ $\forall u \in \mathcal{C}_k$; $\mu \in \mathcal{P}_w(\mathbb{K})$.   (12.5.6)

12.5.4 Remark. If $\varepsilon > 0$ and $I \subset C_0(X)$ is a *finite* subset of $C_0(X)$, then [by (12.2.10)] the set

$$N(I, \varepsilon) := \{\nu \in \mathbb{M}(X) | \, |\langle \nu, u \rangle| \leq \varepsilon \ \forall u \in I\}$$

defines a (closed) weak—actually weak*—neighborhood of the "origin" (that is, the null measure) in $\mathbb{M}(X)$. In particular, if we take ε and I as ε_k and $\mathcal{C}_k$, respectively, then the constraint (12.5.6) states that $L_1\mu$ is in the weak* neighborhood $N(\mathcal{C}_k, \varepsilon_k)$, i.e.,

$$L_1\mu \in N(\mathcal{C}_k, \varepsilon_k).   (12.5.7)$$

This provides a natural interpretation of $\mathbb{P}(\mathcal{C}_k, \varepsilon_k)$ as an approximation of the original program (P) in the weak* topology $\sigma(\mathbb{M}(X), C_0(X))$. □

The following result states that Theorem 12.5.3 remains basically unchanged when $\mathbb{P}(\mathcal{C}_k)$ is replaced by $\mathbb{P}(\mathcal{C}_k, \varepsilon_k)$.

12.5.5 Theorem. *Suppose that Assumption 12.5.1 is satisfied. Then*

(a) $\mathbb{P}(\mathcal{C}_k, \varepsilon_k)$ *is solvable for each* $k = 1, 2, \ldots$.

(b) *If* μ_k *is an optimal solution for* $\mathbb{P}(\mathcal{C}_k, \varepsilon_k)$, *i.e.,*

$$\langle \mu_k, c \rangle = \min \mathbb{P}(\mathcal{C}_k, \varepsilon_k) \ \text{for} \ k = 1, 2, \ldots,$$

then $\{\mu_k\}$ *satisfies the same conclusion of Theorem 12.5.3(b); in particular,*

$$\langle \mu_k, c \rangle \uparrow \min(\text{P}) = \rho_{\min}.   (12.5.8)$$

Proof. See §12.6.

C. Aggregation-relaxation-inner approximations

The programs $\mathbb{P}(\mathcal{C}_k)$ and $\mathbb{P}(\mathcal{C}_k, \varepsilon_k)$ have a *finite number of constraints* and give "nice" approximation results—Theorems 12.5.3 and 12.5.5. However, they are still not good enough for our present purpose because the "decision variable" μ lies in the *infinite-dimensional* space $\mathbb{M}_w(\mathbb{K}) \subset \mathbb{M}(\mathbb{K})$. (For the latter spaces to be finite-dimensional we would need the state and action sets, X and A, to both be finite sets.) Now to obtain *finite-dimensional* approximations of (P) we will combine $\mathbb{P}(\mathcal{C}_k, \varepsilon_k)$ with a suitable sequence of *inner approximations* [see Definition 12.2.13(b)]. These are based on the following well-known result (for a proof see, for instance, Billingsley [1, p. 237, Theorem 4] or Parthasarathy [1, p. 44, Theorem 6.3]). We shall use the notation introduced in Remark 12.2.1(b).

12.5.6 Proposition. [Existence of a weakly dense set in $\mathcal{P}(S)$.] *Let S be a separable metric space and, $D \subset S$ a countable dense subset of S. Then the family of p.m.'s whose supports are finite subsets of D is dense in $\mathcal{P}(S)$ in the weak topology $\sigma(\mathbb{M}(S), C_b(S))$.*

We will now apply Proposition 12.5.6 to the space $S := \mathbb{K}$. Let $D \subset \mathbb{K}$ be a *countable* dense subset of $\mathbb{K}$, and let $\{D_n\}$ be an increasing sequence of *finite* sets $D_n \uparrow D$. For each $n = 1, 2, \ldots$, let $\Delta_n := \mathcal{P}(D_n)$ be the family of p.m.'s on D_n; that is, *an element of Δ_n is a convex combination of the Dirac measures concentrated at points of D_n.* Then, as $D_n \uparrow D$, the sets Δ_n for an *increasing sequence* (of sets of p.m.'s) whose limit

$$\Delta := \bigcup_{n=1}^{\infty} \Delta_n \tag{12.5.9}$$

is dense in $\mathcal{P}(\mathbb{K})$ in the weak topology $\sigma(\mathbb{M}(\mathbb{K}), C_b(\mathbb{K}))$; that is, for each p.m. μ in $\mathcal{P}(\mathbb{K})$, there is a sequence $\{\nu_k\}$ in Δ such that

$$\langle \nu_k, v \rangle \to \langle \mu, v \rangle \quad \forall v \in C_b(\mathbb{K}). \tag{12.5.10}$$

Let us now consider a linear program as $\mathbb{P}(\mathcal{C}_k, \varepsilon_k)$ except that the p.m.'s μ in (12.5.6) are replaced by p.m.'s in $\Delta_n \cap \mathcal{P}_w(\mathbb{K})$. That is, instead of $\mathbb{P}(\mathcal{C}_k, \varepsilon_k)$ consider the *finite* program

$\mathbb{P}(\mathcal{C}_k, \varepsilon_k, \Delta_n)$:   minimize $\langle \mu, c \rangle$

subject to: $|\langle L_1\mu, u \rangle| \le \varepsilon_k \quad \forall u \in \mathcal{C}_k, \ \mu \in \Delta_n \cap \mathcal{P}_w(\mathbb{K}). \tag{12.5.11}$

This is indeed a *finite* linear program because it has a finite number $|\mathcal{C}_k|$ of constraints, and a finite number $|D_n|$ of "decision variables", namely, the coefficients of a measure in $\Delta_n \cap \mathcal{P}_w(\mathbb{K})$.

The corresponding approximation result is as follows.

12.5.7 Theorem. [Finite approximations for (P).] *If Assumption 12.5.1 is satisfied then:*

(a) *For each $k = 1, 2, \ldots$, there exists $n(k)$ such that, for all $n \ge n(k)$, the finite linear program $\mathbb{P}(\mathcal{C}_k, \varepsilon_k, \Delta_n)$ is solvable and*

$$\min \mathbb{P}(\mathcal{C}_k, \varepsilon_k) \le \min \mathbb{P}(\mathcal{C}_k, \varepsilon_k, \Delta_n). \tag{12.5.12}$$

(b) *Suppose that, in addition, the cost-per-stage function $c(x, a)$ is continuous. Then for each $k = 1, 2, \ldots$ there exists $n^*(k)$ such that*

$$\min \mathbb{P}(\mathcal{C}_k, \varepsilon_k, \Delta_n) \le \min(P) + \varepsilon_k \quad \forall n \ge n^*(k); \tag{12.5.13}$$

hence [by (12.5.12) and (12.5.8)]

$$\min \mathbb{P}(\mathcal{C}_k, \varepsilon_k, \Delta_n) \to \min(P) = \rho_{\min} \quad as \ k \to \infty, \tag{12.5.14}$$

where of course the limit is taken over values of $n \geq n^*(k)$. Moreover, if μ_{kn} [for $k \geq 1$, $n \geq n^*(k)$] is an optimal solution for $\mathbb{P}(C_k, \varepsilon_k, \Delta_n)$, then every weak accumulation point of $\{\mu_{kn}\}$ is an optimal solution for (P).

Proof. See §12.6.

Notes on §12.5

The notes for this section are given at the end of §12.6.

12.6   Proof of Theorems 12.5.3, 12.5.5, 12.5.7

In this section we prove Theorems 12.5.3, 12.5.5 and 12.5.7.

Proof of Theorem 12.5.3. (a) Let W be an arbitrary subset of $C_0(X)$, and in (12.5.3) replace C_k by W. This yields the following aggregation of (P).

$\mathbb{P}(W)$:   minimize $\langle \mu, c \rangle$

$$\text{subject to: } \langle L_1 \mu, u \rangle = 0 \quad \forall u \in W, \ \mu \in \mathcal{P}_w(\mathbb{K}). \tag{12.6.1}$$

We wish to show that $\mathbb{P}(W)$ is *solvable*. First note that as (P) is consistent [Theorem 12.3.3(a)], so is $\mathbb{P}(W)$. More explicitly, there exists a p.m. μ that satisfies (12.5.1), which, by Lemma 12.5.2(b), yields that μ satisfies (12.6.1). Moreover, as (12.5.1) implies (12.6.1), we have

$$0 \leq \inf \mathbb{P}(W) \leq \min(\mathrm{P}), \tag{12.6.2}$$

where the first inequality holds because $c \geq 0$. Now let $\{\mu_n\}$ be a minimizing sequence for $\mathbb{P}(W)$; that is, each μ_n satisfies (12.6.1) and

$$\langle \mu_n, c \rangle \downarrow \inf \mathbb{P}(W). \tag{12.6.3}$$

Thus, by (12.6.2), there exist constants M and N such that

$$\langle \mu_n, c \rangle \leq M \quad \forall n \geq N.$$

Therefore, applying Theorem 12.2.15 and then Prohorov's Theorem 12.2.6, there is a subsequence $\{\mu_m\}$ of $\{\mu_n\}$ and a p.m. μ on $\mathbb{K}$ such that

$$\langle \mu_m, v \rangle \to \langle \mu, v \rangle \quad \forall v \in C_b(\mathbb{K}). \tag{12.6.4}$$

This implies [as in (12.3.37)]

$$\liminf_{m \to \infty} \langle \mu_m, c \rangle \geq \langle \mu, c \rangle, \tag{12.6.5}$$

and so, by (12.6.3),

$$\langle \mu, c \rangle \leq \inf \mathbb{P}(W).$$

It follows that to prove that $\mathbb{P}(W)$ is solvable, it suffices to show that μ *is*
feasible for $\mathbb{P}(W)$, that is,

> (i) $\langle L_1 \mu, u \rangle = 0 \quad \forall u \in W$, and (ii) μ is in $\mathcal{P}_w(\mathbb{K})$, (12.6.6)

which would yield the reverse inequality $\langle \mu, c \rangle \geq \inf \mathbb{P}(W)$. To prove (12.6.6)
observe that the condition (ii) is obvious because

$$\langle \mu, w \rangle = 1 + \langle \mu, c \rangle < \infty$$

by (12.6.5) and the definition of w in (12.3.5). Concerning (i), first note that,
by Assumption 11.4.1(d), L_1^* maps $C_b(X)$ into $C_b(\mathbb{K})$ and, consequently,
(12.6.4) implies that

$$\langle L_1 \mu_m, u \rangle \rightarrow \langle L_1 \mu, u \rangle \quad \forall u \in C_b(X) \tag{12.6.7}$$

because

$$\langle L_1 \mu_m, u \rangle = \langle \mu_m, L_1^* u \rangle \rightarrow \langle \mu, L_1^* u \rangle = \langle L_1 \mu, u \rangle. \tag{12.6.8}$$

In particular, (12.6.7) holds for all u in $W \subset C_0(X) \subset C_b(X)$, which implies
(ii) since $\langle L_1 \mu_m, u \rangle = 0$ for all $u \in W$.

Thus, as $\mathbb{P}(W)$ is solvable for any subset W of $C_0(X)$, it follows that
$\mathbb{P}(\mathcal{C}_k)$ is solvable for each k.

(b) For each k, let μ_k be an optimal solution for $\mathbb{P}(\mathcal{C}_k)$, that is, μ_k satisfies
(12.5.3) and

$$\langle \mu_k, c \rangle = \min \mathbb{P}(\mathcal{C}_k) \leq \min(\mathrm{P}),$$

where the second inequality follows from (12.6.2). Furthermore, as the se-
quence $\mathcal{C}_k$ is increasing, so is the sequence of values $\langle \mu_k, c \rangle$. Therefore, there
is a number ρ such that

$$\langle \mu_k, c \rangle \uparrow \rho, \quad \text{and} \quad \rho \leq \min(\mathrm{P}). \tag{12.6.9}$$

Thus, as $\langle \mu_k, c \rangle \leq \rho$ for all k, the same arguments used to obtain (12.6.4)
and (12.6.5) yield a subsequence $\{\mu_m\}$ of $\{\mu_k\}$ and a p.m. μ on $\mathbb{K}$ such
that

$$\langle \mu_m, v \rangle \rightarrow \langle \mu, v \rangle \quad \forall v \in C_b(\mathbb{K}), \tag{12.6.10}$$

and

$$\liminf_{m \to \infty} \langle \mu_m, c \rangle \geq \langle \mu, c \rangle.$$

In fact, as $\langle \mu_m, c \rangle$ satisfies (12.6.9), we have

$$\langle \mu, c \rangle \leq \lim_{m \to \infty} \langle \mu_m, c \rangle = \rho \leq \min(\mathrm{P}),$$

so that

$$\langle \mu, c \rangle \leq \min(P). \tag{12.6.11}$$

Therefore, if we can show that μ *is a feasible solution for* (P), then we shall have that $\langle \mu, c \rangle \geq \min(P)$, which combined with (12.6.11) will give that μ *is optimal for* (P), that is, $\langle \mu, c \rangle = \min(P)$. Now to prove that μ is feasible for (F) we need to check that the p.m. μ satisfies (12.5.1); equivalently, by Lemma 12.5.2(c), we need to check

(i) $\langle L_1\mu, u \rangle = 0$ $\forall u \in C(X)$, (ii) μ is in $\mathcal{P}_w(\mathbb{K})$. (12.6.12)

The condition (ii) follows from (for instance) (12.6.11) and the definition of w in (12.3.5), that is, $\langle \mu, w \rangle = 1 + \langle \mu, c \rangle < \infty$. To prove (i) recall first that $C(X)$ is the limit of the increasing sequence $\{C_k\}$, i.e.,

$$C(X) = \bigcup_{k=1}^{\infty} C_k.$$

Therefore, if u is in $C(X)$, then there exist N such that u is in C_k for all $k \geq N$, and so the subsequence $\{\mu_m\}$ in (12.6.10) satisfies

$$\langle L_1\mu_m, u \rangle = 0 \quad \forall m \geq N$$

because μ_m is feasible for $\mathbb{P}(C_m)$. Finally [arguing as in (12.6.7), (12.6.8)], from (12.6.10) we obtain

$$\langle L_1\mu, u \rangle = \lim_m \langle L_1\mu_m, u \rangle = 0,$$

which implies (12.6.12)(ii) since u as an arbitrary function in $C(X)$. This completes the proof that μ is feasible for (P), which, as was already mentioned, yields that μ is an optimal solution for (P). $\square$

Proof of Theorem 12.5.5. Many of the arguments in this proof are very similar to those in the proof of Theorem 12.5.3, and, consequently, in several places we only sketch the main facts.

(a) If a measure μ satisfies (12.5.3), then it obviously satisfies (12.5.6). this fact and Theorem 12.5.3(a) yield

$$0 \leq \inf \mathbb{P}(C_k, \varepsilon_k) \leq \min \mathbb{P}(C_k) \leq \min(P) \quad \forall k = 1, 2, \ldots. \tag{12.6.13}$$

Now fix k, and let $\{\mu_n\}$ be a minimizing sequence for $\mathbb{P}(C_k, \varepsilon_k)$. Then, as in (12.5.3)–(12.6.5), there exists a subsequence $\{\mu_m\}$ of $\{\mu_n\}$ and a p.m. μ on $\mathbb{K}$ such that

$$\langle \mu_m, v \rangle \to \langle \mu, v \rangle \quad \forall v \in C_b(\mathbb{K}), \tag{12.6.14}$$

and

$$\langle \mu, c \rangle \leq \inf \mathbb{P}(C_k, \varepsilon_k). \tag{12.6.15}$$

Thus, to complete the proof of part (a) it suffices to show that μ is feasible for $\mathbb{P}(\mathcal{C}_k, \varepsilon_k)$ since this would yield the reverse inequality in (12.6.15). On the other hand, since (12.6.15) implies $\langle \mu, w \rangle = 1 + \langle \mu, c \rangle < \infty$, to show that μ satisfies (12.5.6) it only remains to prove that

$$|\langle L_1 \mu, u \rangle| \le \varepsilon_k \quad \forall u \in \mathcal{C}_k. \tag{12.6.16}$$

To prove this, observe that

$$
\begin{aligned}
|\langle L_1 \mu, u \rangle| &\le |\langle L_1 \mu, u \rangle - \langle L_1 \mu_m, u \rangle| + |\langle L_1 \mu_m, u \rangle| \\
&\le |\langle L_1 \mu, u \rangle - \langle L_1 \mu_m, u \rangle| + \varepsilon_k \quad \forall u \in \mathcal{C}_k,
\end{aligned}
\tag{12.6.17}
$$

because each μ_m satisfies (12.5.6). Thus letting $m \to \infty$ in the latter inequality we obtain (12.6.16) since [as in (12.6.7), (12.6.8)] the weak convergence (12.6.14) implies

$$\langle L_1 \mu_m, u \rangle \to \langle L_1 \mu, u \rangle \quad \forall u \in C_b(X).$$

(b) For every $k = 1, 2, \ldots$, let μ_k be an optimal solution for $\mathbb{P}(\mathcal{C}_k, \varepsilon_k)$. Clearly the sequence of optimal values $\langle \mu_k, c \rangle = \min \mathbb{P}(\mathcal{C}_k, \varepsilon_k)$ is nondecreasing, which combined with (12.6.13) implies [as in (12.6.9)–(12.6.11)] the existence of a number $\rho \le \min(P)$, a subsequence $\{\mu_m\}$ of $\{\mu_k\}$, and a p.m. μ on $\mathbb{K}$ such that $\mu_m \to \mu$ weakly [as in (12.6.10)] and

$$\langle \mu, c \rangle \le \lim_{m \to \infty} \langle \mu_m, c \rangle = \rho \le \min(P).$$

Hence to prove that μ is optimal for (P) it suffices to show that μ *is feasible for* (P), which would yield $\langle \mu, c \rangle \ge \min(P)$. In fact, as it is evident that μ is in $\mathcal{P}_w(\mathbb{K})$ [i.e., $\langle \mu, w \rangle = 1 + \langle \mu, c \rangle < \infty$], to show that μ satisfies (12.5.1) it only remains to prove that $L_1 \mu = 0$, or equivalently [by Lemma 12.5.2(c)], that

$$\langle L_1 \mu, u \rangle = 0 \quad \forall u \in C(X).$$

This, however, can be proved almost exactly as (12.6.12)(i) [simply replace the equality "$\langle L_1 \mu_m, u \rangle = 0 \; \forall m \ge N$" by the inequality "$|\langle L_1 \mu_m, u \rangle| \le \varepsilon_m \; \forall m \ge N$"], and so we shall omit the details. $\square$

Proof of Theorem 12.5.7. (a) *Fix an arbitrary $k \ge 1$.* Observe that if $\mathbb{P}(\mathcal{C}_k, \varepsilon_k, \Delta_n)$ is solvable for some n, then the inequality (12.5.12) trivially follows from (12.5.11) and (12.5.6), because $\Delta_n \cap \mathcal{P}_w(\mathbb{K}) \subset \mathcal{P}_w(\mathbb{K})$. Thus to prove part (a) we shall concentrate in the first statement, solvability of $\mathbb{P}(\mathcal{C}_k, \varepsilon_k, \Delta_n)$ for all $n \ge n(k)$.

Let μ^* be an optimal solution for (P); that is, μ^* satisfies

$$\langle \mu^*, c \rangle = \min(P) = \rho_{\min} \tag{12.6.18}$$

and [by (12.5.1) and Lemma 12.5.2(c)]

$$\langle L_1 \mu^*, u \rangle = 0 \quad \forall u \in C(X), \text{ and } \mu^* \text{ is in } \mathcal{P}_w(\mathbb{K}). \tag{12.6.19}$$

Now, as the family Δ in (12.5.9) is weakly dense in $\mathcal{P}(\mathbb{K})$ [in the weak topology $\sigma(\mathrm{M}(\mathbb{K}), C_b(\mathbb{K}))$] and

$$\Delta \cap \mathcal{P}_w(\mathbb{K}) \subset \mathcal{P}_w(\mathbb{K}) \subset \mathcal{P}(\mathbb{K}),$$

we see that $\Delta \cap \mathcal{P}_w(\mathbb{K})$ is weakly dense in $\mathcal{P}_w(\mathbb{K})$. Hence, there is a sequence $\{\mu_j\}$ in $\Delta \cap \mathcal{P}_w(\mathbb{K})$ such that

$$\langle \mu_j, v \rangle \to \langle \mu^*, v \rangle \quad \forall v \in C_b(\mathbb{K}).$$

This implies [as in (12.6.7), (12.6.8)]

$$\langle L_1 \mu_j, u \rangle \to \langle L_1 \mu^*, u \rangle \quad \forall u \in C_b(X);$$

in particular, by (12.6.19),

$$\langle L_1 \mu_j, u \rangle \to \langle L_1 \mu^*, u \rangle = 0 \quad \forall u \in \mathcal{C}_k \tag{12.6.20}$$

because $\mathcal{C}_k \subset \mathcal{C}(X) \subset C_0(X) \subset C_b(X)$. Therefore, as $\mathcal{C}_k$ is a *finite* set, letting ε_k be as in (12.5.6), we see from (12.6.20) that there exists $j(k)$ such that

$$|\langle L_1 \mu_j, u \rangle| \leq \varepsilon_k \quad \forall u \in \mathcal{C}_k \text{ and } j \geq j(k). \tag{12.6.21}$$

That is, for all $j \geq j(k)$, the measure $\mu_j \in \Delta \cap \mathcal{P}_w(\mathbb{K})$ is feasible for $\mathbb{P}(\mathcal{C}_k, \varepsilon_k)$. On the other hand, as Δ is the limit of the increasing sequence Δ_n [see (12.5.9)], there exists $n(k)$ such that $\mu_{j(k)}$ is in

$$\Delta_{n(k)} \cap \mathcal{P}_w(\mathbb{K}) \subset \Delta_n \cap \mathcal{P}_w(\mathbb{K}) \quad \forall n \geq n(k).$$

It follows from (12.6.21) and (12.5.11), that $\mathbb{P}(\mathcal{C}_k, \varepsilon_k, \Delta_n)$ is *consistent* for all $n \geq n(k)$, which in turn implies that $\mathbb{P}(\mathcal{C}_k, \varepsilon_k, \Delta_n)$ *is solvable for all* $n \geq n(k)$ because it is a *finite* linear program with a finite value (≥ 0)—see any reference on elementary (finite-dimensional) LP. This completes the proof of part (a).

(b) As in the proof of part (a), fix an arbitrary $k \geq 1$, and let μ^* be an optimal solution for (P)—see (12.6.18), (12.6.19). Moreover, for each $j = 1, 2, \ldots$, consider the set

$$E_j := \{(x, a) \in \mathbb{K} | c(x, a) \leq j\}.$$

As $c \geq 0$ is l.s.c. and strictly unbounded [by Assumption 11.4.1(b), (c)], E_j is a closed set contained in some compact set; hence, E_j itself is compact. Now let μ_j be the p.m. on $\mathbb{K}$ defined by

$$\mu_j(B) := \mu^*(B \cap E_j)/\mu^*(E_j) \quad \text{for} \quad B \in \mathcal{B}(\mathbb{K}),$$

which of course is well defined [that is, $\mu^*(E_j) > 0$] for all j sufficiently large. Furthermore, as

$$E_j \uparrow \mathbb{K} \quad \text{and} \quad \mu^*(E_j) \uparrow \mu^*(\mathbb{K}) = 1, \tag{12.6.22}$$

μ_j converges weakly to μ^* in the weak topology $\sigma(\mathrm{M}(\mathbb{K}), C_b(\mathbb{K}))$, i.e.,

$$\langle \mu_j, v \rangle = \mu^*(E_j)^{-1} \int_{E_j} v \, d\mu^* \to \langle \mu^*, v \rangle \quad \forall v \in C_b(\mathbb{K}). \qquad (12.6.23)$$

Thus [as in (12.6.7), (12.6.8)]

$$\langle L_1 \mu_j, u \rangle \to \langle L_1 \mu^*, u \rangle \quad \forall u \in C_b(X).$$

This implies [as in (12.6.20)]

$$\langle L_1 \mu_j, u \rangle \to 0 \quad \forall u \in C_k.$$

Therefore, as C_k is a *finite* set, there exists $j_1(k)$ such that

$$|\langle L_1 \mu_j, u \rangle| \leq \varepsilon_k/2 \quad \forall u \in C_k \text{ and } j \geq j_1(k). \qquad (12.6.24)$$

On the other hand, by (12.6.22) and the definition of μ_j,

$$\langle \mu_j, c \rangle \to \langle \mu^*, c \rangle.$$

Hence, there exists $j_2(k)$ such that

$$\langle \mu_j, c \rangle \leq \langle \mu^*, c \rangle + \varepsilon_k/2 \quad \forall j \geq j_2(k). \qquad (12.6.25)$$

Now fix $j_0 \geq \max\{j_1(k), j_2(k)\}$, and let $\mathcal{P}(E_{j_0})$ be the family of p.m.'s on $\mathbb{K}$, concentrated on E_{j_0}. Then μ_{j_0} is a p.m. in $\mathcal{P}(E_{j_0})$ and satisfies (12.6.24) and (12.6.25). We now wish to approximate μ_{j_0} by a suitable sequence $\{\nu_n\}$ in $\Delta \cap \mathcal{P}(E_{j_0})$.

Let $D \subset \mathbb{K}$ be the countable dense subset in the definition of $\mathbb{P}(C_k, \varepsilon_k, \Delta_n)$. Then $D \cap E_{j_0}$ is dense in E_{j_0}, and so, by Proposition 12.5.6,

$$\Delta \cap \mathcal{P}(E_{j_0}) = \mathcal{P}(D \cap E_{j_0})$$

is dense in $\mathcal{P}(E_{j_0})$ in the weak topology $\sigma(\mathrm{M}(\mathbb{K}), C_b(\mathbb{K}))$. Therefore, there exists a sequence $\{\nu_n\}$ in $\Delta \cap \mathcal{P}(E_{j_0})$ such that

$$\langle \nu_n, v \rangle \to \langle \mu_{j_0}, v \rangle \quad \forall v \in C_b(\mathbb{K}), \qquad (12.6.26)$$

and [as in (12.6.7), (12.6.8)]

$$\langle L_1 \nu_n, u \rangle \to \langle L_1 \mu_{j_0}, u \rangle \quad \forall u \in C_b(X).$$

In particular, as $C_k \subset C_b(X)$ is a *finite* set, there exists $n_1(k)$ such that

$$|\langle L_1 \nu_n, u \rangle - \langle L_1 \mu_{j_0}, u \rangle| \leq \varepsilon_k/2 \quad \forall u \in C_k \text{ and } n \geq n_1(k).$$

Combining this fact with (12.6.24), taking $j = j_0$, we obtain

$$|\langle L_1 \nu_n, u \rangle| \leq \varepsilon_k \quad \forall u \in C_k \text{ and } n \geq n_1(k). \qquad (12.6.27)$$

On the other hand, since the restriction of c to E_{j_0} is a continuous bounded function, (12.6.26) yields

$$\langle \nu_n, c \rangle \to \langle \mu_{j_0}, c \rangle.$$

From this fact and (12.6.25), with $j = j_0$, there exists $n_2(k)$ such that

$$\langle \nu_n, c \rangle \leq \langle \mu^*, c \rangle + \varepsilon_k \quad \forall n \geq n_2(k), \tag{12.6.28}$$

which in particular implies that ν_n is in $\mathcal{P}_w(\mathbb{K})$, that is,

$$\langle \nu_n, w \rangle = 1 + \langle \nu_n, c \rangle < \infty. \tag{12.6.29}$$

Finally, to verify that ν_n satisfies (12.5.11), note that since each ν_n is in the set $\Delta \cap \mathcal{P}(E_{j_0}) \subset \Delta$, (12.6.29) and (12.5.9) imply the existence of n_3 such that

$$\nu_n \in \Delta_{n_3} \cap \mathcal{P}_w(\mathbb{K}) \subset \Delta_n \cap \mathcal{P}_w(\mathbb{K}) \quad \forall n \geq n_3. \tag{12.6.30}$$

To summarize, define $n^*(k) := \max\{n_1(k), n_2(k), n_3\}$. Then (12.6.30) and (12.6.27) give that ν_n is a feasible solution for $\mathbb{P}(\mathcal{C}_k, \varepsilon_k, \Delta_n)$ for all $n \geq n^*(k)$, which together with (12.6.28) and part (a) yields (12.5.13) and (12.5.14).

It only remains to prove the last statement in (b). To do this, for each $k \geq 1$ fix $n \geq n^*(k)$ and let $\mu_k := \mu_{kn}$ be an optimal solution for $\mathbb{P}(\mathcal{C}_k, \varepsilon_k, \Delta_n)$, that is, μ_k satisfies (12.5.11) and

$$\langle \mu_k, c \rangle = \min \mathbb{P}(\mathcal{C}_k, \varepsilon_k, \Delta_n).$$

Then, by (12.5.14) and the argument used to obtain (12.6.4) and (12.6.5), there exists a subsequence $\{\mu_m\}$ of $\{\mu_k\}$ and a p.m. μ on $\mathbb{K}$ such that

$$\langle \mu_m, v \rangle \to \langle \mu, v \rangle \quad \forall v \in C_b(\mathbb{K}) \tag{12.6.31}$$

and

$$\min(\mathrm{P}) = \liminf_{m \to \infty} \langle \mu_m, c \rangle \geq \langle \mu, c \rangle, \tag{12.6.32}$$

which in particular gives that μ is in $\mathcal{P}_w(\mathbb{K})$. Thus, to conclude that μ is an optimal solution for (P), it only remains to check that μ satisfies $L_1 \mu = 0$ in (12.5.1), or, equivalently [by Lemma 12.5.2(c)] that

$$\langle L_1 \mu, u \rangle = 0 \quad \forall u \in \mathcal{C}(X). \tag{12.6.33}$$

This would yield that μ is feasible for (P), so that $\langle \mu, c \rangle \geq \min(\mathrm{P})$, which together with (12.6.32) would show that μ is optimal for (P). To prove (12.6.33), first note that (12.6.31) implies [as in (12.6.7), (12.6.8)]

$$\langle L_1 \mu, u \rangle = \lim_{m \to \infty} \langle L_1 \mu_m, u \rangle \quad \forall u \in C_b(X) \supset \mathcal{C}(X). \tag{12.6.34}$$

Now fix an arbitrary function u in $C(X)$ and note that, as $C_k \uparrow C(X)$, there exists N such that u is in C_k for all $k \geq N$. Therefore, by (12.5.11),

$$|\langle L_1 \mu_m, u \rangle| \leq \varepsilon_m \to 0 \text{ as } m \to \infty,$$

which combined with (12.6.34) gives $\langle L_1 \mu, u \rangle = 0$. Thus, as $u \in C(X)$ was arbitrary, (12.6.33) follows. □

Notes on §12.5 and §12.6

1. Sections 12.5 and 12.6 are based on Hernández-Lerma and Lasserre [16]. A similar approach, combining aggregations, relaxations and inner approximations, can be used to approximate general (not necessarily MCP-related) infinite linear programs, as in Hernández-Lerma and Lasserre [15]. These two papers provide many related references.

2. The approximation schemes in §12.5 are somewhat similar in spirit to schemes proposed by Vershik [1] and Vershik and Temel't [1]; but with a basic difference. Namely, we use *weak* and *weak** topologies (see Remark 12.5.4 and Lemma 12.5.6), whereas Vershik and Temel't use stronger—for instance, normed—topologies. This is a key fact because we only need "reasonable" things, for example (12.6.4) and (12.6.7), whereas their context would require convergence in the *total variation norm*, which is obviously too restrictive. For instance, for uncountable metric space, the density result in Proposition 12.5.6—with finitely supported measures—is, in general, virtually impossible to get in the total variation norm.

Finally, it is worth noting that the approach in §12.5 can be used to approximately compute an i.p.m. for a noncontrolled Markov chain on a LCSM space whose transition kernel satisfies the (weak) Feller condition in Assumption 11.4.1(d). The idea would be to introduce an "artificial" MCP with a *singleton* control set A and with a continuous "cost" function that satisfies the hypothesis of Theorem 12.5.7.

References

Alden, J. M. and Smith, R. L.
[1] Rolling horizon procedures in nonhomogeneous Markov decision processes. *Oper. Res.* **40** (1992), 183–194.

Altman, E.
[1] *Constrained Markov Decision Processes*. Chapman & Hall/CRC, Boca Raton, FL, to appear.

Anderson, E. J. and Nash, P.
[1] *Linear Programming in Infinite-Dimensional Spaces*. Wiley, Chichester, U.K., 1987.

Ash, R. B.
[1] *Real Analysis and Probability*. Academic Press, New York, 1972.

Balder, E.
[1] *Lectures on Young Measures*, Cahiers de mathématiques de la décision. CEREMADE, Université Paris IX-Dauphine, Paris, 1995.

Baykal-Gürsay, M. and Ross, K. W.
[1] Variability sensitive Markov decision processes. *Math. Oper. Res.* **17** (1992), 558–571.

Bensoussan, A.
[1] Stochastic control in discrete time and applications to the theory of production. *Math. Programm. Study* **18** (1982), 43–60.

Bertsekas, D.P.
[1] *Dynamic Programming: Deterministic and Stochastic Models.* Prentice-Hall, Englewood Cliffs, N. J., 1987.

Bertsekas, D. P. and Shreve, S. E.
[1] *Stochastic Optimal Control: The Discrete Time Case.* Academic Press, New York, 1978.

Bes, C. and Lasserre, J. B.
[1] An on-line procedure in discounted infinite-horizon stochastic optimal control. *J. Optim. Theory Appl.* **50** (1986), 61–67.

Bes, C. and Sethi, S. P.
[1] Concepts of forecast and decision horizons: applications to dynamic stochastic optimization problems. *Math. Oper. Res.* **13** (1988), 295–310.

Bhattacharya, R. N. and Majumdar, M.
[1] Controlled semi-Markov models—the discounted case. *J. Statist. Plann. and Inference* **21** (1989), 365–381.

Billingsley, P.
[1] *Convergence of Probability Measures.* Wiley, New York, 1968.

Blackwell, D.
[1] Memoryless strategies in finite-stage dynamic programming. *Ann. Math. Statist.* **35** (1964), 863–865.

Bourbaki, N.
[1] *Intégration*, Chap. IX. Hermann, Paris, 1969.

Brezis, H.
[1] *Analyse Fonctionnelle: Théorie et Applications*, 4e tirage. Masson, Paris, 1993.

Brown B. W.
[1] On the iterative method of dynamic programming on a finite space discrete time Markov process. *Ann. Math. Statist.* **33** (1965), 719–726.

Cavazos-Cadena, R.
[1] Finite-state approximations to denumerable discounted Markov decision processes. *Appl. Math. Optim.* **14** (1986), 1–26.

Cavazos-Cadena, R. and Montes-de-Oca, R.
[1] Optimal stationary policies in controlled Markov chains with the expected total-reward criterion. Preprint, Departamento de Matemáticas, UAM-Iztapalapa, México, 1997.

Chen, C.-T.
[1] *Linear System Theory and Design.* Holt, Rinehart and Winston, New York, 1984.

Craven, B. D. and Koliha, J. J.
[1] Generalizations of Farkas' theorem. *SIAM J. Math. Anal. Appl.* **8** (1977), 983–997.

Dekker, R., Hordijk, A. and Spieksma, F.
[1] On the relation between recurrence and ergodicity properties in denumerable Markov chains. *Math. Oper. Res.* **19** (1994), 539–559.

Derman, C.
[1] *Finite State Markovian Decision Processes*. Academic Press, New York, 1970.

Derman, C. and Strauch, R. E.
[1] A note on memoryless rules for controlling sequential control processes. *Ann. Math. Statist.* **37** (1966), 276–278.

Doob, J. L.
[1] *Measure Theory*. Springer-Verlag, New York, 1994.

Duflo, M.
[1] *Méthodes Récursives Aléatoires*. Masson, Paris, 1990.

Dutta, P. K.
[1] What do discounted optima converge to? A theory of discount rate asymptotics in economic models. *J. Economic Theory* **55** (1991), 64–94.

Dynkin, E. B. and Yushkevich, A. A.
[1] *Controlled Markov Processes*. Springer-Verlag, New York, 1979.

Easley, D. and Spulber, D.F.
[1] Stochastic equilibrium and optimality with rolling plans. *International Econ. Rev.* **22** (1981), 79–103.

Fernández-Gaucherand, E., Ghosh, M. K., and Marcus, S. I.
[1] Controlled Markov processes on the infinite planning horizon: weighted and overtaking cost criteria. *ZOR: Math. Methods in Oper. Res.* **39** (1994), 131–155.

Filar, J. A., Kallenberg, L. C. M. and Huey-Miin, L.
[1] Variance-penalized Markov decision processes. *Math. Oper. Res.* **14** (1989), 147–161.

Flynn, J.
[1] On optimality criteria for dynamic programs with long finite horizons. *J. Math. Anal. Appl.* **76** (1980), 202–208.

Gale, D.
[1] Cn optimal development in a multi-sector economy. *Rev. of Economic Studies* **34** (1965), 1–19.

Glynn, P. W.
[1] *Simulation output analysis for general state space Markov chains*. Ph. D. Dissertation, Dept. of Operations Research, Stanford University, 1989.
[2] Some topics in regenerative steady-state simulation. *Acta Appl. Math.* **34** (1994), 225–236.

Glynn, P. W. and Meyn, S. P.
[1] A Lyapunov bound for solutions of the Poisson equation. *Ann. Prob.* **24** (1996), 916-931.

González-Hernández, J. and Hernández-Lerma, O.
[1] Envelopes of sets of measures, tightness, and Markov control processes. *Appl. Math. Optim.*, to appear.

Gordienko, E. and Hernández-Lerma, O.
[1] Average cost Markov control processes with weighted norms: value iteration. *Appl. Math.* (Warsaw) **23** (1995), 219–237.
[2] Average cost Markov control processes with weighted norms: existence of canonical policies. *Appl. Math.* (Warsaw) **23** (1995), 199–218.

Gordienko, E., Montes-de-Oca, R. and Minjárez-Sosa, A.
[1] Average cost optimization in Markov control processes with unbounded costs: ergodicity and finite horizon approximation. Preprint, Departamento de Matemáticas, UAM-Iztapalapa, México, 1995.

Hall, P. and Heyde, C. C.
[1] *Martingale Limit Theory and Its Applications*. Academic Press, New York, 1980.

Haviv M. and Puterman, M. L.
[1] Bias optimality in controlled queueing systems. *J. Appl. Prob.* **35** (1998), 136–150.

Hernández-Lerma, O.
[1] *Adaptive Markov Control Processes*. Springer-Verlag, New York, 1989.

Hernández-Lerma, O., Carrasco, G. and Pérez-Hernández, R.
[1] Markov control processes with the expected total-cost criterion: optimality, stability, and transient models. Reporte Interno, Depto. de Matemáticas, CINVESTAV-IPN, 1998. (Submitted.)

Hernández-Lerma, O. and González-Hernández, J.

[1] Infinite linear programming and multichain Markov control processes in uncountable spaces. *SIAM J. Control Optim.* **36** (1998), 313–335.

Hernández-Lerma, O. and Lasserre, J. B.

[1] *Discrete-Time Markov Control Processes: Basic Optimality Criteria.* Springer-Verlag, New York, 1996.

[2] Invariant probabilities for Feller-Markov chains. *J. Appl. Math. and Stoch. Anal.* **8** (1995), 341–345.

[3] Existence of bounded invariant probability densities for Markov chains. *Stat. Prob. Lett.* **28** (1996), 359–366.

[4] Cone-constrained linear equations in Banach spaces. *J. Convex Anal.* **4** (1997), 149–164.

[5] Existence and uniqueness of fixed points for Markov operators and Markov processes. *Proc. London Math. Soc.* **76** (1998), 711–736.

[6] On the probabilistic multichain Poisson equation. LAAS Report N° 97155, LAAS-CNRS, Toulouse, 1997.

[7] Existence of solutions to the Poisson equation in L_p spaces. *Proc. IEEE Conference on Decision and Control* (CDC), Kobe, Japan, 1996, vol. 4, pp. 4190–4195.

[8] Error bounds for rolling horizon policies in discrete-time Markov control processes. *IEEE Trans. Autom. Control* **35** (1990), 1118–1124.

[9] Value iteration and rolling plans for Markov control processes with unbounded rewards. *J. Math. Anal. Appl.* **177** (1993), 38–55.

[10] A forecast horizon and a stopping rule for general Markov decision processes. *J. Math. Anal. Appl.* **132** (1988), 388–400.

[11] Policy iteration for average cost Markov control processes on Borel spaces. *Acta Appl. Math.* **47** (1997), 125–154.

[12] New criteria for positive Harris recurrence of Markov chains. LAAS Report N° 97431, LAAS-CNRS, Toulouse, 1997.

[13] Ergodic theorems and ergodic decomposition for Markov chains. *Acta Appl. Math.* **54** (1998), 99–199.

[14] Linear programming and average optimality for Markov control processes in Borel spaces—unbounded costs. *SIAM J. Control Optim.* **32** (1994), 480–500.

[15] Approximation schemes for infinite linear programs. *SIAM J. Optim.* **8** (1998), 973–988.

[16] Linear programming approximations for Markov control processes in metric spaces. *Acta Appl. Math.* **51** (1998), 123–139.

Hernández-Lerma, O., Montes-de-Oca, R. and Cavazos-Cadena, R.

[1] Recurrence conditions for Markov decision processes with Borel state space: a survey. *Ann. Oper. Res.* **28** (1991), 29–46.

Hernández-Lerma, O. and Vega-Amaya, O.

[1] Infinite-horizon Markov control processes with undiscounted cost criteria: from average to overtaking optimality. *Appl. Math.* (Warsaw) **25** (1998), 153–178.

Hernández-Lerma, O., Vega-Amaya, O. and Carrasco, G.

[1] Sample-path optimality and variance-minimization of average cost Markov control processes. *SIAM J. Control Optim.*, to appear.

Hinderer, K.

[1] *Foundations of Non-Stationary Dynamic Programming with Discrete-Time Parameter.* Lecture Notes in Oper. Res. and Math. Syst. **33**, Springer-Verlag, Berlin, 1970.

Hinderer, K. and Hübner, G.

[1] An improvement of J. F. Shapiro's turnpike theorem for the horizon of finite stage discrete dynamic programs. In *Trans. 7th Prague Conf. on Information Theory, Statist. Dec. Func. and Random Proc.*, Vol. A, 1974 (Academia, Prague, 1977), pp. 245–255.

Hopp, W.

[1] Identifying forecast horizons in non-homogeneous Markov decision processes. *Oper. Res.* **37** (1989), 339–344.

Hordijk, A.

[1] *Dynamic Programming and Markov Potential Theory*, 2nd ed. Mathematical Centre Tracts No. **51**, Mathematisch Centrum, Amsterdam, 1977.

Hordijk, A. and Lasserre, J. B.

[1] Linear programming formulation of MDPs in countable state space: the multichain case. *Z. Oper. Res.* **40** (1994), 91–108.

Hordijk, A. and Spieksma, F.

[1] A new formula for the deviation matrix. In *Probability, Statistics and Optimization* (F. P. Kelly, ed.), Wiley, New York, 1994, pp. 497–507.

Johansen, L.

[1] *Lectures on Macroeconomic Planning.* North-Holland, Amsterdam, 1977.

Kallenberg, L. C. M.

[1] *Linear Programming and Finite Markovian Control Problems.* Mathematical Centre Tracts No. **148**, Mathematisch Centrum, Amsterdam, 1983.

Kartashov, N. V.

[1] Criteria for uniform ergodicity and strong stability of Markov chains with a common phase space. *Theory Probab. and Math. Statist.* **30** (1985), 71–89.

[2] Inequalities in theorems of ergodicity and stability for Markov chains with common phase space. I. *Theory Probab. Appl.* **30** (1985), 247–259.

[3] Inequalities in theorems of ergodicity and stability for Markov chain with common phase space. II. *Theory Probab. Appl.* **30** (1985), 507–515.

[4] Strongly stable Markov chains. *J. Soviet Math.* **34** (1986), 1493–1498.

[5] *Strong Stable Markov Chains*. VSP, Utrecht, The Netherlands, 1996.

Kleinman, D.

[1] An easy way to stabilize a linear control system. *IEEE Trans. Autom. Control* **15** (1970), p. 692.

Kurano, M.

[1] Markov decision processes with a minimum-variance criterion. *J. Math. Anal. Appl.* **123** (1987), 572–583.

Kurano, M. and Kawai, M.

[1] Existence of optimal stationary policies in discounted decision processes: approaches by occupation measures. *Computers Math. Appl.* **27** (1994), 95–101.

Kwon, W. H., Bruckstein, A. M. and Kailath, T.

[1] Stabilizing state feedback design via the moving horizon method. *Internat. J. Control* **37** (1983), 631–643.

Lasota, A. and Mackey, M. C.

[1] *Chaos, Fractals, and Noise: Stochastic Aspects of Dynamics*, 2nd ed. Springer-Verlag, New York, 1994.

Lasserre, J. B.

[1] Existence and uniqueness of an invariant probability measure for a class of Feller-Markov chains. *J. Theoret. Prob.* **9** (1996), 595–612.

[2] Invariant probabilities for Markov chains on a metric space. *Stat. Prob. Lett.*, to appear.

[3] Sample-path average optimality for Markov control processes. *IEEE Trans. Autom. Control*, to appear.

Lippman, S. A.

[1] On dynamic programming with unbounded rewards. *Manage. Sci.* **21** (1975), 1225–1233.

Luenberger, D. G.

[1] *Optimization by Vector Space Methods*. Wiley, New York, 1969.

Makcwski, A. M. and Shwartz, A.
[1] On the Poisson equation for countable Markov chains. Tech. Rpt., Dept. of Electrical Engineering, University of Maryland, 1994.

Mandl, P.
[1] On the variance in controlled Markov chains. *Kybernetika* **7** (1971), 1–12.
[2] A connection between controlled Markov chains and martingales. *Kybernetika* **9** (1973), 237–241.

Mandl, P. and Lausmanová, M.
[1] Two extensions of asymptotic methods in controlled Markov chains. *Ann. Oper. Res.* **28** (1991), 67–80.

McKenzie, L. W.
[1] Turnpike theory. *Econometrica* **44** (1976), 841–865.

Metivier, M. and Priouret, P.
[1] Théorèmes de convergence presque sure pour une classe d'algorithmes stochastique à pas decroissant. *Prob. Th. Rel. Fields* **74** (1987), 403–428.

Meyn S. P.
[1] The policy iteration algorithm for average reward Markov decision processes with general state space. *IEEE Trans. Autom. Control* **42** (1997), 1663–1679.

Meyn, S. P. and Tweedie, R. L.
[1] *Markov Chains and Stochastic Stability.* Springer-Verlag, London, 1993.
[2] Computable bounds for geometric convergence rates of Markov chains. *Ann. Appl. Prob.* **4** (1994), 981–1011.

Mokkadem, A.
[1] Sur un modèle autorégressif nonlinéaire. Ergodicité et ergodicité géométrique. *J. Time Series Anal.* **8** (1987), 195–205.

Neveu, J.
[1] *Mathematical Foundations of the Calculus of Probability.* Holden-Day, San Francisco, 1965.

Nowak, A. S.
[1] Stationary overtaking optimal strategies in Markov decision processes with general state space. Preprint, Institute of Mathematics, Technical University of Wroclaw, Poland, 1992.

Nowak, A. S. and Vega-Amaya, O.
[1] A counterexample on overtaking optimality. Preprint, Institute of Mathematics, Technical University of Wroclaw, Poland; Departamento de Matemáticas, Universidad de Sonora, México, 1997.

Nummelin, E.

[1] *General Irreducible Markov Chains and Non-Negative Operators*. Cambridge University Press, Cambridge, 1984.

[2] On the Poisson equation in the potential theory of a single kernel. *Math. Scand.* **68** (1991), 59–82.

Orey, S.

[1] *Limit Theorems for Markov Chain Transition Probabilities*. Van Nostrand Reinhold, London, 1971.

Parthasarathy, K. R.

[1] *Probability Measures on Metric Spaces*. Academic Press, New York, 1967.

Piunovski, A. B.

[1] General Markov models with the infinite horizon. *Problems of Control and Infor. Theory* **18** (1989), 169–182.

Pliska, S. R.

[1] On the transient case for Markov decision chains with general state spaces. In: Puterman [2], pp. 335–349.

Puterman, M. L.

[1] *Markov Decision Processes*. Wiley, New York, 1994.

[2] (Editor) *Dynamic Programming and Its Applications*. Academic Press, New York, 1979.

Quelle, G.

[1] Dynamic programming of expectation and variance. *J. Math. Anal. Appl.* **55** (1976), 239–252.

Ramsey, F. P.

[1] A mathematical theory of savings. *Economic J.* **38** (1928), 543–559.

Rempala, R.

[1] Forecast horizon in a dynamic family of one-dimensional control problems. *Diss. Math.* **315** (1991).

Revuz, D.

[1] *Markov Chains*, revised ed. North-Holland, Amsterdam, 1984.

Rieder, U.

[1] Measurable selection theorems for optimization problems. *Manuscripta Math.* **24** (1978), 115–131.

[2] On optimal policies and martingales in dynamic programming. *J. Appl. Prob.* **13** (1976), 507–518.

[3] On Howard's policy improvement method. *Math. Operationsforsch. Statist., Ser. Optimization* **8** (1977), 227–236.

Robertson, A. P. and Robertson, W.

[1] *Topological Vector Spaces*. Cambridge University Press, Cambridge, UK, 1964.

Ross, S. M.

[1] *Applied Probability Models with Optimization Applications*. Holden-Day, San Francisco, 1970.

Royden, H. L.

[1] *Real Analysis*, 2nd ed. Macmillan, New York, 1968.

Rudin, W.

[1] *Real and Complex Analysis*, 3rd ed. McGraw-Hill, New York, 1986.

Schäl, M.

[1] Conditions for optimality and for the limit of n-stage optimal policies to be optimal. *Z. Wahrs. Verw. Geb.* **32** (1975), 179–196.

Schweitzer, P. J.

[1] On undiscounted Markovian decision processes with compact action spaces. *RAIRO Rech. Opér./Oper. Res.* **19** (1985), 71–86.

Shapiro, J. M.

[1] Turnpike planning horizons for a Markovian decision model. *Manage. Sci.* **14** (1968), 292–300.

Spieksma, F. and Tweedie, R. L.

[1] Strengthening ergodicity to geometric ergodicity for Markov chains. *Stoch. Models* **10** (1994), 45–75.

Strauch, R.

[1] Negative dynamic programming. *Ann. Math. Statist.* **37** (1966), 871–890.

Syski, R.

[1] Ergodic potential. *Stoch. Proc. Appl.* **7** (1978), 311–336.

Tijms, H. C.

[1] *Stochastic Models: An Algorithmic Approach*. Wiley, New York, 1994.

Tijms, H. C. and Wessels, J.

[1] (Editors) *Markov Decision Theory*. Mathematical Centre Tracts No. 93, Mathematisch Centrum, Amsterdam, 1977.

Tweedie, R. L.

[1] Sufficient conditions for ergodicity and geometric ergodicity of Markov chains on a general state space. *Stoch. Proc. Appl.* **3** (1975), 385–403.

[2] The existence of moments for stationary Markov chains. *J. Appl. Probab.* **20** (1983), 191–196.

van Hee, K. M. Hordijk, A., and van der Wal, J.
[1] Successive approximations for convergent dynamic programming. In: Tijms and Wessels [1], pp. 183–211.

van Nunen, J. A. E. E. and Wessels, J.
[1] A note on dynamic programming with unbounded rewards. *Manage. Sci.* **24** (1978), 576–580.
[2] Markov decision processes with unbounded rewards. In: Tijms and Wessels [1], pp. 1–24.

Vega-Amaya, O.
[1] Overtaking optimality for a class of production-inventory systems. Preprint, Departamento de Matemáticas, Universidad de Sonora, México, 1996.
[2] *Markov Control Processes In Borel Spaces: Undiscounted Criteria.* Doctoral Thesis, UAM-Iztapalapa, México, 1998. (In Spanish.)
[3] Sample path average optimality of Markov control processes with strictly unbounded cost. *Appl. Math.* (Warsaw), to appear.

Veinott, A. F. (Jr.)
[1] Discrete dynamic programming with sensitive discount optimality criteria. *Ann. Math. Statist.* **40** (1969), 1635–1660.
[2] On finding optimal policies in discrete dynamic programming with no discounting. *Ann. Math. Statist.* **37** (1965), 1284–1294.

Vershik, A. M.
[1] Some remarks on the infinite-dimensional problems of linear programming. *Russian Math. Surveys* **29** (1970), 117–124.

Vershik, A. M. and Temel't, V.
[1] Some questions concerning the approximation of the optimal value of infnite-dimensional problems in linear programming. *Siberian Math. J.* **9** (1968), 591–601.

von Weizsäcker, C. C.
[1] Existence of optimal programs of accumulation for an infinite horizon. *Rev. of Economic Studies* **32** (1965), 85–104.

Wakuta, K.
[1] Arbitrary state semi-Markov decision processes with unbounded rewards. Optimization **18** (1987), 447–454.

Wessels, J.
[1] Markov programming by successive approximations with respect to weighted supremum norms. *J. Math. Anal. Appl.* **58** (1977), 326–335.

Yosida, K.
[1] *Functional Analysis*, 6th ed. Springer-Verlag, Berlin, 1980.

Yushkevich, A. A.
[1] Blackwell optimality in Borelian continuous-in-action Markov decision processes. *SIAM J. Control Optim.* **35** (1997), 2157–2182.

Abbreviations

a.a.	almost all
a.e.	almost everywhere
a.s.	almost surely
i.i.d.	independent and identically distributed
l.s.c.	lower semicontinuous
u.s.c.	upper semicontinuous
p.m.	probability measure
i.p.m.	invariant probability measure
AC	average cost
ACOE	Average Cost Optimality Equation
ACOI	Average Cost Optimality Inequality
ADO	asymptotic discount optimality
DC	discounted cost
DCOE	discounted cost optimality equation
DP	dynamic programming
ETC	expected total cost
IFS	iterated function system
LCSM	locally compact separable metric
LP	linear programming
LLN	Law of Large Numbers
MCM	Markov control model
MCP	Markov control process
O.O.	overtaking optimality
P.E.	Poisson equation

PI	policy iteration
PIA	policy iteration algorithm
RH	rolling horizon
VI	value iteration

Glossary of notation

□	end of proof or example or remark
:=	equality by definition
I_B	indicator function of a set B, defined as

$$I_B(x) := \begin{cases} 1 & \text{if } x \in B, \\ 0 & \text{otherwise.} \end{cases}$$

$r^+ := \max(r, 0), \ r^- := -\min(r, 0)$

Chapter 7

Section 7.1

X	Borel (state) space
$\mathcal{B}(X)$	Borel σ-algebra

Section 7.2

w	weight function
$\|u\|$	sup norm of a function u
$\|u\|_w$	w-norm of a function u
$\mathbb{B}(X)$	Banach space of bounded measurable functions on X
$\mathbb{B}_w(X)$	Banach space of measurable functions on X with finite w-norm

$\|\mu\|_{TV}$	total variation norm of a measure μ
$\|\mu\|_w$	w-norm of a measure μ
$\mathbb{M}(X)$	Banach space of signed measures μ on X with $\|\mu\|_{TV} < \infty$
$\mathbb{M}_w(X)$	Banach space of signed measures μ on X with $\|\mu\|_w < \infty$
$Q(B\|x)$	signed kernel on X

$$Qu(x) := \int_X u(y)Q(dy|x)$$

$$\mu Q(B) := \int_X Q(B|x)\mu(dx)$$

$\|Q\|_w$	w-norm of a signed kernel Q

δ_x — Dirac measure at $x \in X$

QR — composition of the signed kernels Q and R

$Q^n := QQ^{n-1}$

$Q^0(E|x) := \delta_x(B)$

Section 7.3

$P(B|x)$ — transition probability function of a Markov chain

$P^t(B|x)$ — t-step transition probability

τ_B — hitting time of the set B

η_B — occupation time of the set B

$L(x, B) := P_x(\tau_B < \infty)$

$U(x, B) := E_x(\eta_B)$

$\nu \ll \mu$ — the measure ν is absolutely continuous with respect to the measure μ

λ — maximal irreducibility measure

$\mathcal{B}(X)^-$ — family of sets $B \in \mathcal{B}(X)$ with $\lambda(B) > 0$

$\int v d\mu \equiv \mu(v) \equiv \langle \mu, v \rangle$

$C_b(X)$ — Banach space of continuous bounded functions on X

Chapter 8

Section 8.2

$\mathcal{M} := [X, A, \{A(x)|x \in X\}, Q, c]$ — Markov control model

$\mathbb{K}$ — set of feasible state-action pairs

$\mathbb{F}$ — set of decision functions (or selectors)

Φ — set of randomized decision functions

H_t — family of admissible histories up to time t

π — control policy

φ^∞ — randomized stationary policy, $\varphi \in \Phi$

f^∞ — deterministic stationary policy, $f \in F$

Π — set of all control policies

Π_{RM} — set of randomized Markov policies

Π_{RS} — set of randomized stationary policies φ^∞

Π_D — set of deterministic policies

Π_{DM} — set of deterministic Markov policies

Π_{DS} — set of deterministic stationary policies f^∞

$c(x, \varphi) \equiv c_\varphi(x) := \int_A c(x, a)\varphi(da|x)$ for $\varphi \in \Phi$

$c(x, f) \equiv c_f(x) := c(x, f(x))$ for $f \in F$

$Q(\cdot|x, \varphi) \equiv Q_\varphi(\cdot|x) := \int_A Q(\cdot|x, a)\varphi$ $(da|x)$ for $\varphi \in \Phi$

$Q(\cdot|x, f) \equiv Q_f(\cdot|x) := Q(\cdot|x, f(x))$ for $f \in F$

$(\Omega, \mathcal{F})$ — canonical measurable space

P^π_ν — p.m. on $(\Omega, \mathcal{F})$ determined by the policy π and the initial distribution ν

$P^\pi_x := P^\pi_\nu$ if $\nu = \delta_x$

E^π_ν — expectation with respect to P^π_ν

$E^\pi_x := E^\pi_\nu$ if $\nu = \delta_x$

Section 8.3

$V(\pi, x)$ — α-discounted cost ($0 < \alpha < 1$) when using the policy π, given the initial state x

$V^*(x)$ — α-discount value function

$V_n(\pi, x)$ — α-discounted n-stage expected cost

$v_n(x)$ — α-value iteration (α-VI) function, $n = 1, 2, \ldots$

$\|Q\|_w$ — w-norm of the transition law Q

T_α DP operator

Section 8.4

$\mathbb{F}_n$ set of α-VI decision function, $n = 1, 2, \ldots$

$\mathbb{F}_*$ set of α-discount optimal decision functions

$D(x, a)$ α-discount discrepancy function

$A_*(x)$ α-discount optimal control actions in the state x

$A_n(x)$ α-VI optimal control actions in the state x; $n = 1, 2, \ldots$

$D_n(x, a)$ α-VI discrepancy function, $n = 1, 2, \ldots$

Section 8.5

$\mathbb{L}(X)$ family of l.s.c. functions on X

$\mathbb{L}_w(X) := \mathbb{L}(X) \cap \mathbb{B}_w(X)$

$C(X)$ family of continuous functions on X

$C_w(X) := C(X) \cap \mathbb{B}_w(X)$

$C_b(X)$ family of continuous bounded functions on X

Chapter 9

Section 9.1

$V_1(\pi, x)$ expected total cost (ETC) when using the policy π, given the initial state x

$V_1^*(x)$ ETC value function

Section 9.2

$\overline{R}$ extended real numbers

$r^+ := \max(r, 0)$

$r^- := \max(-r, 0) = -\min(r, 0)$

Section 9.3

$J_n(\pi, x)$ n-stage ETC when using the policy π, given the initial state x

$J_n^*(x)$ n-stage optimal ETC

$V_\alpha(\pi, x) \equiv V(\pi, x)$ α-discounted cost, $0 < \alpha < 1$

$V_\alpha^*(x) \equiv V^*(x)$ α-discount value function

$V_1^{(+)}(\pi, x) := E_x^\pi \left(\sum_{t=0}^\infty c_t^+ \right)$

$V_1^{(-)}(\pi, x) := E_x^\pi \left(\sum_{t=0}^\infty c_t^- \right)$

$V_1^n(\pi, x)$ ETC from time n onwards when using the policy π, given the initial state $x_0 = x$

Section 9.4.

$V_1(\pi, \nu)$ ETC when using the policy π, given the initial distribution ν

μ_ν^π ETC-expected occupation measure on $X \times A$ when using the policy π, given the initial distribution ν

$\widehat{\mu}_\nu^\pi$ marginal of μ_ν^π on X

$\mu_{\nu,t}^\pi$ distribution of (x_t, a_t) when using the policy π, given the initial distribution ν, $t = 0, 1, \ldots$

$\widehat{\mu}_{\nu,t}^\pi$ marginal of $\mu_{\nu,t}^\pi$ on X

$\pi^{(1)}$ 1-shift policy determined by π

Section 9.5

$T := T_1$ dynamic programming operator (when $\alpha = 1$)

$\mathcal{U}$ see Definition 9.5.1

$D_1(x, a)$ ETC-discrepancy function

$\pi^{(n)}$ n-shift policy determined by π ($n = 0, 1, \ldots$)

M_n^* see (9.5.12)

Section 9.6

$Q_\varphi(\cdot|x) \equiv Q(\cdot|x, \varphi)$
$\qquad := \int_A Q(\cdot|x, a)\varphi(da|x)$
$Q_t(\cdot|x) := Q_{\varphi_t}(\cdot|x)$
$Q_\pi^t := Q_0 Q_1 \cdots Q_{t-1}$ for $t = 1, 2, \ldots$
$\|Q_\varphi\|_v$ w-norm of Q_φ

Chapter 10

Section 10.1

$OC(\pi, x)$ opportunity cost of policy π, given the initial state x

$D(\pi, x)$ Dutta's criterion

$J(\pi, x)$ expected average cost (AC) when using the policy π, given the initial state x

$J^*(x)$ optimal expected AC

Section 10.2

$Q_f^t(\cdot|x) \equiv Q^t(\cdot|x, f)$ t-step transition probability, $t = 0, 1, \ldots$

$\|Q_f\|_u$ w-norm of Q_f

μ_f i.p.m. of Q_f

$L_1(\mu) = L_1(X, \mathcal{B}(X), \mu)$

$\mu_f(u) = \int_X u \, d\mu_f$

$c_f(x) \equiv c(x, f) := c(x, f(x))$

$J(f) := \mu_f(c_f)$

h_f bias function of $f \in \mathbb{F}$

Section 10.3

$\mathbb{F}_{AC}$ family of AC-optimal decision functions

$\mathbb{F}_{ca}$ family of canonical decision functions

$\mathbb{F}_{bias}$ family of bias-optimal decision functions

(ρ^*, h^*) solution of the ACOE

$J_n(\pi, x, h)$ n-stage ETC with terminal cost function h

$J_n^*(x, h)$ value function for $J_n(\pi, x, h)$

λ irreducibility measure

$\widehat{h}(x)$ optimal bias function

$A^*(x)$ AC-canonical control actions at state x

Section 10.4

$u_\alpha(\cdot) := V_\alpha^*(\cdot) - V_\alpha^*(z)$

$\rho(\alpha) := (1 - \alpha)V_\alpha^*(z)$

Section 10.7

$\mathcal{M}_{bias}$ see (10.7.3)

Chapter 11

Section 11.1

$J_n^0(\pi, \nu)$ n-stage sample-path cost when using the policy π, given the initial distribution ν

$J^0(\pi, \nu)$ long-run sample-path AC

$J^I(\pi, \nu)$ limit-infimum expected AC

$\text{Var}\,(\pi, \nu)$ limiting average variance when using the policy π, given the initial distribution ν

$\rho_{\min}$ see (11.1.12)

$\widehat{\rho}$ see (11.1.15)

ρ^* see (11.1.23)

$\mathcal{P}(X)$ set of probability measures on X

$\mathcal{P}_\delta(X) := \{\delta_x | x \in X\}$

Section 11.2

$P^{(n)}(\cdot|x)$ expected average occupation measure

$\sigma^2(c, x)$ limiting average variance for a Markov chain

σ_c^2 see (11.2.11)

$\psi(x)$ see (11.2.12)

Y_t see (11.2.17)

$$M_n := \sum_{t=1}^{n} Y_t$$

$\mathcal{F}_t := \sigma\{x_0, \ldots, x_t\}$

Section 11.3

$\mathcal{P}_w(X)$	see (11.3.2)
(ρ^*, h_*)	solution of the ACOE
$\underline{J}^0(\pi, x)$	lim-inf sample path AC
$\psi(x, a)$	see (11.3.5)
$\psi_f(x) := \psi(x, f(x))$	
(σ_*^2, V_*)	see (11.3.17)
$v(x) := w(x)^{1/2}$	
$\mathcal{F}_t(\pi, x)$	see (11.3.23)
$Y_t(\pi, x)$	see (11.3.24)
$M_n(\pi, x)$	see (11.3.25)
$\widehat{D}(x, a)$	AC-discrepancy function
$\Psi_f(X)$	see (11.3.35)
$\sigma^2(f) := \mathrm{Var}\,(f^\infty, \cdot)$	
$\mathcal{M}\mathrm{var}$	MCM for the variance minimization problem

Chapter 12

Section 12.2

$(\mathcal{X}, \mathcal{Y})$	dual pair of vector spaces $\mathcal{X}, \mathcal{Y}$
$\sigma(\mathcal{X}, \mathcal{Y})$	weak topology on $\mathcal{X}$
$\sigma(\mathcal{Y}^*, \mathcal{Y})$	weak* topology on $\mathcal{Y}^*$
$C_0(S)$	Banach space of continuous functions vanishing at infinity
$\langle \mu, u \rangle := \int u d\mu$	
G^*	adjoint of linear map G
w_0	weight function on X
w	weight function on $X \times Y$
$\mathbb{M}_w(X)_+$	positive cone in $\mathbb{M}_w(X)$
$\mathbb{B}_w(X)_+$	positive cone in $\mathbb{B}_w(X)$
K^*	dual cone of the positive cone K
$\widehat{\mu}$	marginal on X of the measure μ on $X \times Y$

$\mathbb{P}$	primal linear program
$\inf \mathbb{P}$	value of $\mathbb{P}$
$\min \mathbb{P}$	optimal value of $\mathbb{P}$
$\mathbb{P}^*$	dual of $\mathbb{P}$
$\sup \mathbb{P}^*$	value of $\mathbb{P}^*$
$\max \mathbb{P}^*$	optimal value of $\mathbb{P}^*$

Section 12.3

$w(x, a)$	weight function on $\mathbb{K}$; see (12.3.5)
$w_0(x)$	weight function on X; see (12.3.5)
$L_0\mu$	see (12.3.12)
$L_1\mu$	see (12.3.13)
$L\mu$	see (12.3.14)
$L^*(\rho, \mu)$	see (12.3.15)
(P)	AC-related primal linear program
(P*)	dual of (P)
$\mathcal{L}$	perturbation of L

Section 12.5

$\mathcal{C}(X)$	a countable dense subset of $C_0(X)$
$\{\mathcal{C}_k\}$	increasing sequence of finite sets $\mathcal{C}_k \uparrow \mathcal{C}(X)$
$\mathbb{P}(\mathcal{C}_k)$	aggregation of constraints of (P)
$\{\varepsilon_k\}$	sequence of numbers $\varepsilon_k \downarrow 0$
$\mathbb{P}(\mathcal{C}_k, \varepsilon_k)$	aggregation-relaxation of (P)
D	a countable dense subset of $\mathbb{K}$
$\{D_n\}$	increasing sequence of finite sets $D_n \uparrow D$
$\Delta_n := \mathcal{P}(D_n)$	
$\Delta := \cup_{n=1}^{\infty} \Delta_n$	
$\mathbb{P}(\mathcal{C}_k, \varepsilon_k, \Delta_n)$	aggregation-relaxation-inner approximation of (P)

Section 12.6

$E_j := \{(x, a) \in \mathbb{K} | c(x, a) \le j\}$

Index

The numbers in this index refer to sections

Applications of Mathematics

(continued from page ii)